VON NEUMANN, MORGENSTERN, AND THE CREATION OF GAME THEORY

Drawing on a wealth of new archival material, including personal correspondence and diaries, Robert Leonard tells the fascinating story of the creation of game theory by Hungarian Jewish mathematician John von Neumann and Austrian economist Oskar Morgenstern. Game theory first emerged amidst discussions of the psychology and mathematics of chess in Germany and fin-de-siècle Austro-Hungary. In the 1930s, on the cusp of anti-Semitism, and political upheaval, it was developed by von Neumann into an ambitious theory of social organization. It was shaped still further by its use in combat analysis in World War II and during the Cold War. Interweaving accounts of the period's economics, science, and mathematics, and drawing on the private lives of von Neumann and Morgenstern, Robert Leonard provides a detailed reconstruction of a complex historical drama.

Robert Leonard writes about the history of twentieth-century economics and the social sciences in scientific and cultural contexts. His work has appeared in a range of journals that address economics and the history of science, including the *Economic Journal, History of Political Economy*, and *Isis*. His 1995 article in the *Journal of Economic Literature*, from which the present book grew, won the Best Article Award of the History of Economics Society. Leonard is Professor of Economics at the Université du Québec à Montréal.

Historical Perspectives on Modern Economics

General Editor: Craufurd D. Goodwin, *Duke University*

This series contains original works that challenge and enlighten historians of economics. For the profession as a whole, it promotes better understanding of the origin and content of modern economics.

Other books in the series:

William J. Barber, *Designs within Disorder: Franklin D. Roosevelt, the Economists, and the Shaping of American Economic Policy, 1933–1945*

William J. Barber, *From New Era to New Deal: Herbert Hoover, the Economists, and American Economic Policy, 1921–1933*

Filipo Cesarano, *Monetary Theory and Bretton Woods: The Construction of an International Monetary Order*

Timothy Davis, *Ricardo's Macroeconomics: Money, Trade Cycles, and Growth*

Jerry Evensky, *Adam Smith's Moral Philosophy: A Historical and Contemporary Perspective on Markets, Law, Ethics, and Culture*

M. June Flanders, *International Monetary Economics, 1870–1960: Between the Classical and the New Classical*

J. Daniel Hammond, *Theory and Measurement: Causality Issues in Milton Friedman's Monetary Economics*

Samuel Hollander, *The Economics of Karl Marx: Analysis and Application*

Lars Jonung (ed.), *The Stockholm School of Economics Revisited*

Kyun Kim, *Equilibrium Business Cycle Theory in Historical Perspective*

Gerald M. Koot, *English Historical Economics, 1870–1926: The Rise of Economic History and Mercantilism*

David Laidler, *Fabricating the Keynesian Revolution: Studies of the Inter-War Literature on Money, the Cycle, and Unemployment*

Odd Langholm, *The Legacy of Scholasticism in Economic Thought: Antecedents of Choice and Power*

Harro Maas, *William Stanley Jevons and the Making of Modern Economics*

Philip Mirowski, *More Heat Than Light: Economics as Social Physics, Physics as Nature's Economics*

Philip Mirowski (ed.), *Nature Images in Economic Thought: "Markets Read in Tooth and Claw"*

D. E. Moggridge, *Harry Johnson: A Life in Economics*

Mary S. Morgan, *The History of Econometric Ideas*

Takashi Negishi, *Economic Theories in a Non-Walrasian Tradition*

Heath Pearson, *Origins of Law and Economics: The Economists' New Science of Law, 1830–1930*

Malcolm Rutherford, *Institutions in Economics: The Old and the New Institutionalism*

Esther-Mirjam Sent, *The Evolving Rationality of Rational Expectations: An Assessment of Thomas Sargent's Achievements*

Yuichi Shionoya, *Schumpeter and the Idea of Social Science*

Juan Gabriel Valdes, *Pinochet's Economists: The Chicago School of Economics in Chile*

Karen I. Vaughn, *Austrian Economics in America: The Migration of a Tradition*

E. Roy Weintraub, *Stabilizing Dynamics: Constructing Economic Knowledge*

Von Neumann, Morgenstern, and the Creation of Game Theory

From Chess to Social Science, 1900–1960

ROBERT LEONARD

Université du Québec à Montréal

CAMBRIDGE
UNIVERSITY PRESS

CAMBRIDGE UNIVERSITY PRESS
Cambridge, New York, Melbourne, Madrid, Cape Town,
Singapore, São Paulo, Delhi, Mexico City

Cambridge University Press
32 Avenue of the Americas, New York, NY 10013-2473, USA

www.cambridge.org
Information on this title: www.cambridge.org/9781107609266

First published 2010
Reprinted 2010, 2011
First paperback edition 2012

A catalog record for this publication is available from the British Library.

Library of Congress Cataloging in Publication Data

Leonard, Robert, 1962 –
Von Neumann, Morgenstern, and the creation of game theory : from chess
to social science, 1900–1960 / Robert Leonard.
p.   cm. – (Historical perspectives on modern economics)
Includes bibliographical references and index.
ISBN 978-0-521-56266-9 (hardback)
1. Game theory – History. 2. Von Neumann, John, 1903–1957. 3. Morgenstern,
Oskar, 1902–1977. I. Title. II. Series.
HB144.L46   2010
519.3–dc22   2009034837

ISBN 978-0-521-56266-9 Hardback
ISBN 978-1-107-60926-6 Paperback

Contents

Acknowledgments

If scholarly work of this kind is essentially a solitary activity, it is nonetheless dependent upon the existence of an extended and variegated web of support. For the financial aid that permitted indispensable archival research, translation, and research assistance, I am grateful to the Université du Québec à Montréal (UQAM), the provincial *Fonds québécois de la recherche sur la société et la culture* (FQRSC), and the Social Science and Humanities Research Council of Canada. For access to various collections of papers, I thank the Rare Book, Manuscript, and Special Collections Library of Duke University (Oskar Morgenstern and Karl Menger); the Library of Congress, Washington, D.C. (John von Neumann and Oswald Veblen); the library of the American Philosophical Association, Philadelphia (Stan Ulam); and the Rockefeller Archive Center (Austrian Institute for Business Cycle Research). For translation from German and Hungarian, I thank Cornelia Brandt-Gaudry and Michael Szirti. For further documents and photographs, and for their kind hospitality, I thank Harold Kuhn, the late Dorothy Morgenstern-Thomas, Rosemary Gilmore, Laszló Filep, and Marina von Neumann Whitman.

During the preparation of this book, I have been sustained by the intellectual and moral support of many colleagues. Rather than search for superlatives, let me simply thank Craufurd Goodwin of Duke University and Scott Parris of Cambridge University Press. In the face of my prevarication and reticence to "rush" into print, they have shown extraordinary patience and generosity. Like Craufurd, Roy Weintraub went from being a dissertation advisor to a professional colleague and friend. This book owes a great deal to him – even more, I suspect, than he realizes. Beyond that, many others have been important at various stages. In no particular order, and with no claim to comprehensiveness, I think of Bruce Caldwell, Mary Morgan, José Luís Cardoso, Philippe Fontaine, Albert Jolink, Kevin Hoover, Ted

Porter, Neil De Marchi, Harro Maas, John Davis, Philippe Le Gall, Amy Dahan, Margaret Schabas, Avi Cohen, Esther-Mirjam Sent, Phil Mirowski, Loïc Charles, Steve Medema, Roger Backhouse, Judy Klein, Thomas Uebel, Hansjörg Klausinger, Karl Sigmund, Friederich Stadler, Elisabeth Nemeth, Gilles Dostaler, Robert Nadeau, Maurice Lagueux, Benoît Pépin, Yves Gingras, Philippe Mongin, Francois Gardes, Keith Jakee, Mike Harrison, Antoin Murphy, and Harvey Shoolman.

Various parts of this book have been presented at seminars at Duke, Michigan State, New York, Harvard, York, and Laval universities; the University of Toronto; Trinity College Dublin; University College Cork; École normale supérieure à Cachan; PHARE in Paris; the University of Turin; the University of California at Los Angeles; Institut National de la Statistique et des Études Économiques; the Centre Koyré CRHST in Paris; and the Joseph Schumpeter Society of Vienna, as well as at various academic conferences. To participants at all of these events, I express my thanks, as I do to my colleagues at UQAM, and my graduate students, Jerry Lawless, Paolo Longato, Catia Corriveau, Jérome Thiollier, Fabrice Thierry, Jean-Guillaume Forand, Patrick Lebonnois, Marc-André Bacon, Martin Gauthier, Magdalena Planeta, and Yan-Olivier Charest.

Moving within the confines of the family circle, I thank, all in Dublin, my parents, Patrick and Joan Leonard, and sisters Noelle and Miriam for their great constancy. Finally, I thank my wife, Valérie Cauchemez, and our daughters, Niamh and Aoife, for stoically facilitating the writing of this book and, at the same time, making it all worthwhile. To them, this book is dedicated.

VON NEUMANN, MORGENSTERN, AND THE CREATION OF GAME THEORY

Introduction

Game theory, it may reasonably be claimed, has proved to be one of the more significant scientific contributions of the twentieth century. Albeit haltingly and unevenly, and in a manner quite unforeseeable in 1944 when the *Theory of Games and Economic Behavior* was published, it has affected not only economics and political science but also evolutionary biology, ethics, and philosophy proper. Within economics, particular areas such as microeconomic theory, industrial organisation, international trade, and experimental economics have all been reshaped under the theory's influence. Although game theory initially came from outside as a critical contribution, it has now been completely embraced by the economics discipline, as indicated by the awarding of the Nobel Memorial Prize in Economics to John Nash, John Harsányi, and Reinhard Selten in 1994, and to Robert Aumann and Thomas Schelling in 2005.

Various aspects of this development have received the attention of historians of economics and others. In 1992, under the editorship of Roy Weintraub, an exploratory set of essays titled *Towards a History of Game Theory* featured both historical accounts and reminiscences. Building upon their contribution to that volume, a 1996 book by Robert Dimand and MaryAnn Dimand provided a historical survey of the various game-theoretic contributions in the first half of the century. In his 2003 monograph on the evolution of economic rationality, Nicola Giocoli devotes considerable attention to game theory, particularly as it affected the neoclassical conception of the economic agent. A similar theme, treated differently, is central to Philip Mirowski's *Machine Dreams* of 2002, which casts the history of game theory as part of the rise of "cyborg" thinking, linked in an essential manner to von Neumann's work on computing and automata.

With the appearance in 1998 of Sylvia Nasar's biography of John Nash, *A Beautiful Mind*, and its subsequent adaptation as a Hollywood film, the

1

audience for the history of game theory has grown to include the general public. Tom Siegfried's (2006) *A Beautiful Math*, seeks to explain to a broad audience the success of Nash's game theory across the scientific spectrum, and a 2007 BBC documentary, *The Trap*, attempts to show how contemporary social and political life have been shaped by the adoption of game-theoretic conceptions of rationality in the policy sphere. The film traces a direct line from early game theory to World War II, RAND, McNamara's Pentagon and, ultimately, Margaret Thatcher's reform of the British welfare state.[1]

Although each of these accounts has its strengths, none of them, I think it fair to say, does justice to the rich historical process by which von Neumann and Morgenstern were led to the creation of *The Theory of Games and Economic Behavior*. Few of them treat the authors as flesh-and-blood figures, and none of them considers the cultural and political context of *fin-de-siècle* and interwar Europe, without which, in my opinion, the making of game theory cannot be understood. In some, von Neumann is given short shrift, and treated as a necessary stepping-stone towards John Nash; in others, Morgenstern is omitted entirely and von Neumann's game theory is forced to herald a future of machines and computation.

It is a commonplace that every historical account must strike a balance between the fact that the narrator is "omniscient" and the fact that, at any given historical moment, the future was neither fully determined nor known. The present account is an attempt to tell the story of von Neumann, Morgenstern, and the creation of game theory in a way that exploits authorial omniscience only minimally. Every attempt is made to restore the historical specificity of the subject, to hold the future in abeyance, and to allow our characters to develop over time in response to changing circumstances. The story carries the reader back to the world of Budapest and Vienna in the first half of the twentieth century, with its chess cafés, debates over the nature and purpose of economics, and intense concern with politics. It then moves to the very different world of Princeton in the 1930s and the postwar United States. When we treat von Neumann of the 1920s, it is not in the awareness that he would later come to the United States or work on the atomic bomb or the computer, but by deliberately suspending our knowledge of those developments. When we consider Morgenstern in interwar Vienna, we treat him not as future co-author of a book at wartime

---

[1] See Weintraub (ed.) (1992); Dimand and Dimand (1996); Giocoli (2003); Nasar (1998); Mirowski (2002); and Siegfried (2006).

Princeton, but as a dissenting Austrian economist in a particular cultural environment.

In a 1992 essay, published in the aforementioned volume edited by Weintraub, I portrayed von Neumann's 1928 paper on games as an isolated contribution, bearing little relation to the work or interests of his contemporaries. I have since changed that opinion considerably, having been led by a passing remark by mathematician Ernest Zermelo, to see the paper as connected to the rich discussion of the psychology and mathematics of chess in the first decades of the twentieth century. Our story, therefore, begins at the chessboard. Chapter 1 discusses the cultural importance of *Schach* in Central Europe and the emerging interest in the psychology of the game, and the idea, perhaps best illustrated in the writings of German chess champion and mathematician Emanuel Lasker, that chess might provide some insights into economic and social interaction. Chess emerges as a fruitful wellspring of mathematical, psychological, and sociological reflection, and the source of an emerging discourse of "equilibrium and struggle".

In the chapters dealing with von Neumann's Hungarian background, we devote our attention to two intertwined groups: the country's assimilated Jewish community and its mathematicians. The relevance of the latter is obvious: for a country of small size and limited state of development, Hungary produced a remarkable mathematical culture, and the prodigious von Neumann remained proud of his origins to the very end. As for the theme of Hungarian Jewry, the subject is broached here for both its cultural relevance at the time and its importance, especially later, in von Neumann's life. Chapter 4 considers his journey in the 1920s from Budapest to Göttingen, where he made his initial foray into the mathematics of games. As an attempt to provide a mathematical treatment of an unusal field, the paper bears the imprint of Göttingen's Hilbert, an influence that would remain important in von Neumann's work in the area.

Switching to Vienna in Chapters 5 and 6, we consider the early career of the very different Oskar Morgenstern. He began as a conventional student of the nonmathematical Austrian School of economics, and the influences to which he was exposed speak to the richness of debate in the period: the now-forgotten epistemological critique of Hans Mayer; the Romantic Universalist fantasies of Othmar Spann; and Ludwig von Mises' admixture of theoretical critique and political didacticism. In time, Morgenstern broke ranks with most of these economist mentors, stimulated, in part, by Viennese mathematician Karl Menger, who also happened to be – and the irony was not lost on him – the son of the founder of the Austrian School.

If gaining access to the papers of Morgenstern was easy, it was a different matter for those of Karl Menger. Indeed, it took months of diplomatic negotiation before I found myself one Friday evening in Chicago, in a dusty storage room at the Illinois Institute of Technology, with nothing before me but the weekend and twenty boxes of Menger papers, virtually undisturbed since his death in 1985. That exploration not only revealed ribbon-bound treasures relating to the life of his economist father, but allowed me to approach Menger from a different angle. It became clear that there were subtle connections between Menger's disparate spheres of activity, including the debates on the foundations of mathematics, his formal theory of ethics, and his political involvements in Vienna of the 1930s (see Leonard 1998). As we show in Chapter 7, politics was an important factor throughout, whether in shaping Menger's dissociation from L. E. J. Brouwer or in provoking his 1934 book, *Morality, Decision and Social Organization*. Morgenstern's view in the latter of the glimmer of a mathematical solution to what he regarded as weaknesses in the orthodox treatment of the rational economic agent showed, further, that only two short steps separated Viennese politics and the debate on economic theory.

This complex intertwining of economic, mathematical, and political themes is pursued in Chapter 8, where we explore the alliance that formed between Morgenstern and Menger. It owed much to a common search for "purity" and was not unconnected to local power struggles in Viennese economic research. Rejecting what they perceived as the infiltration of economics by politics, epitomized in the work of von Mises to the right and Neurath to the left, they felt that the use of mathematics, to the extent that it demanded clear thinking and logical demonstration, could make economic analysis more "value-free".

Menger's account of how his construction of this "geometry" of society was rooted in the political tumult of 1933–34 also sparked my interest in what we may call individual creativity: in this case, that of the creative mathematician in a particular context, moving in psychological time, one might say, from the "blank page" to the finished construction or proof. By what mysterious path, particularly when writing on such "reflexive" matters as the fields of rationality or social interaction, did the mathematician proceed from initial steps to a result deemed worthy of the name? The knowledge that the page was, of course, never quite blank did nothing to reduce the power of this image for me. Nor was I deterred by the realization that the inner creative process could never be completely recovered: to catch glimpses of it would be sufficient.

It was in the light of this experience with Menger that I renewed my probing of von Neumann. If, as a result of the examination of the world of chess, his initial interest in games now made sense, there still remained the puzzle of why he returned to game theory at the close of the 1930s, after a hiatus of more than a decade. The conventional wisdom among historians of economics seemed to rest upon two ideas: that he was always interested in the subject, as evidenced by the appearance of the minimax technique in his 1937 paper on equilibrium economic growth, and that it was Morgenstern who brought him back to the subject when they met in the late 1930s. A valuable paper by Danish historian of mathematics, Tinne Kjeldsen (2002), put paid to the first idea, by deconstructing the connections between the von Neumann papers of 1928 and 1937: the first did *not* involve a fixed-point theorem, and the second, which did, was observed by von Neumann to bear only an accidental relationship to the first. The technique-driven history was wrong, Kjeldsen showed. The idea that von Neumann was stimulated by Morgenstern's theoretical puzzles, presented at Fine Hall afternoon teas, while certainly true, somehow didn't seem adequate in explaining the mathematician's Herculean 600-page effort, in 1940–1941, in the middle of the Second World War.

To get beyond this impasse, I read and reread *The Theory of Games and Economic Behavior*, and at the same time dug more deeply into von Neumann's life in the 1930s. The exploration of new correspondence – some furnished by Marina von Neumann Whitman, more newly translated – brought a much-needed human dimension to his activities, and further reading on Hungarian social and political history shed light on his concerns as a Jewish expatriate at the time. It gradually became clear that the reawakening of von Neumann's interest in games owed something to his experience of the political tumult of the late 1930s, a truly dramatic period during which social questions drew much of his attention. Our examination of this interlude in Chapter 9 allows us to understand von Neumann's return to game theory, as well as certain emphases in his new mathematics of coalitions, such as the idea of multiple social orders and the stabilizing role played by conventions concerning social discrimination. This phase of von Neumann's creative life now made sense from both personal and historical points of view, and it was clear that game theory was shaped only partly by the discussion of economic theory. The collaboration with Morgenstern and the effect of this experience on the latter are discussed in Chapters 10 and 11.

The "emotional shock" that affected von Neumann during this time was the Nazi destruction of the world he had known, that socially heterogeneous

Mitteleuropean culture of mathematics and science. If his 1940 development of a new mathematics of society was in part a reaction to this – internal and symbolic in nature – so too was his simultaneous decision to "go to war", discussed in Chapter 12. Although von Neumann's wartime involvements gave expression to an "apocalyptic" attitude often evoked in connection with the Hungarian expatriates, they also had the more mundane effect of seeing a small part of game theory being used in operations research, on problems of strategic bombing and submarine search, worlds away from the abstract mathematics of stable sets and discriminatory social orders.

It was in this instrumental capacity that game theory received the postwar approval of a scientific community extending beyond its original authors. Following the war, it became a defining element in the worldview of the RAND Corporation. Adopted initially as an analytical tool to help in problems of strategic bombing or fighter pursuit, as RAND evolved, game theory became an integral element in a complex web of social–scientific activities at that institution, many of them reflective of the prevailing Cold War culture. This is discussed in our final chapter, 13.

This book is the result of a long and absorbing period of "detective work", an investigation that, to the understandable chagrin of my editor and many colleagues, has been every bit as enjoyable as the publication of results. A hint revealed in an article footnote or archived letter would open up a new vista, giving rise to months of new reading. The book was thus shaped by the results, often serendipitous, of research and reading, with new forays into the archives being conducted as late as a year before submission of the final manuscript. It took a while for me to realize that not all the vistas, or all the findings, could necessarily be featured in the same book.

In the story that follows, I have deliberately tried to bring the reader as close as possible to the "thinking" of von Neumann, Morgenstern, and others, all the while portraying the broader intellectual and social contexts in which they lived and worked. This attempt to relate biography, scientific creativity, and the evolution of external events has been one of the more rewarding aspects of writing this book. I hope that will be evident to the reader as he or she takes the time to read it.

PART ONE

STRUGGLE AND EQUILIBRIUM: FROM LASKER TO VON NEUMANN

ONE

"The Strangest States of Mind"

Chess, Psychology, and Emanuel Lasker's *Kampf*

Introduction

I was in one of those moods where danger is attractive. Hence I plunged from the start into a combination the outcome of which was exceedingly doubtful. For the gain of a pawn I risked to retard the development and to accelerate that of the opponent. Mr. Speijer wisely sacrificed also the exchange, and opened a concentrated fire upon my King; but once he missed the best continuation, and therefore lost quickly. Games of this character, where every move counts for much, are best suited to entertain spectators, and they are of great value for the ripening of the "position judgment". He who relies solely upon tactics that he can wholly comprehend is liable, in the course of time, to weaken his imagination. And he is at a disadvantage against an opponent who tries to win through bold venture, yet does not step beyond the finely drawn boundary of what is sound.

Emanuel Lasker (1908), quoted in Hilbert (2001), p. 5

Thus wrote world chess champion Emanuel Lasker in his *Evening Post* chess column in late December 1908, following his victorious third and final game in an Amsterdam match against the Dutch champion, Abraham Speijer. Having come to the city from Vienna, where he had been playing exhibition games the previous week, Lasker played Speijer in a pavillion in an Amsterdam park, watched by an audience of 150. The German beat the Dutchman in the first of three games but, to the delight of the Dutch audience, was held to a draw in the second. In the third game, shunning textbook play and avoiding safe continuations, Lasker won in twenty-seven moves.

Lasker's remark that "he who relies solely upon tactics that he can wholly comprehend is liable, in the course of time, to weaken his imagination" gets to the heart of what distinguished him as a chessplayer. This German mathematician was a dominant figure in the world of chess in the late nineteenth and early twentieth centuries, during which he had an extraordinarily long reign as world champion. Regarded by some as the player who put psychology at the centre of the game, he was known for his unconventional play, being inattentive to game openings but given to heroic struggles in the mid-game. His chess writings, insofar as they tended to promote the conventional, textbook strategies set out by his compatriot Wilhelm Steinitz, from whom he had taken the world title in 1894, did not espouse the psychologically daring play for which he himself became known. However, those writings were very broad, revealing Lasker to be not only a player but also a humanist philosopher, keen to draw upon the lessons of chess in order to facilitate social understanding and progress. Lasker's life and writings provide a window onto the world of chess in the period during which he competed. They also are enlightening regarding why several of Lasker's fellow mathematicians, both German and Hungarian, began to take an interest in the structure of chess and parlour games.

The Perfect Strategist

In the first decades of the twentieth century, chess enjoyed great visibility in many parts of Continental Europe. The game was important in England, France, Germany, and Russia, and particularly so in the countries of the Austro-Hungarian Empire. Amongst the Jewish communities in those countries, it commanded special interest. From London to Moscow, the grand masters enjoyed great visibility and prestige, and the game was played in the chess cafés of the capitals, such as the Marienbrücke in Vienna and the famous Café de la Régence in Paris. Against a background of high tournament drama, chessmasters such as Lasker and Siegbert Tarrasch wrote manuals on strategy, and the influence of the game was felt in many dimensions of scientific and literary culture. Thus psychologists investigated the thought processes required in chess, and mathematicians wondered whether so human an activity could be made amenable to formal treatment. Others speculated philosophically about the relationship of chess to life in general, and the game was a source of inspiration for several writers, including Vladimir Nabokov, author of *The Defense* in 1929, and Viennese exile Stefan Zweig, whose *Schachnovelle* was the last thing he wrote before his suicide in Brazil in 1942.[1]

---

[1] See Stefan Zweig, *The Royal Game and Other Stories* (New York: Harmony Books, 1981), orig. *Schachnovelle*, written in late 1941-early 1942. Translated as *The Royal Game* (New York: Viking Press, 1944).

Looming large over the world of chess at this time was the figure of Emanuel Lasker (1868–1941), who held the title of world champion for an unprecedented twenty-four years, from 1897 to 1921. Son of a cantor and nephew of a rabbi, Lasker came from a German Jewish family of modest means. He trained as a mathematician – his mentors included David Hilbert and Max Noether – and he completed a doctorate in mathematics at Erlangen in 1902 with his dissertation on the theory of vector spaces. He shared these algebraic interests with several others, including Gyula König of Budapest and, at Göttingen, his teacher's daughter, Emmy Noether, who later developed Lasker's algebraic work further.[2]

Lasker's short path to chess supremacy began when he interrupted his mathematical studies to play the game for money. Although he long tried to obtain an academic post as mathematician, he was unable to do so as a Jew and was condemned, so to speak, to live by chess. Admired by Albert Einstein, who knew him in Berlin, Lasker was regarded as the player who introduced psychological considerations into the game. In this, he stood in particular contrast to previous world champion, Wilhelm Steinitz, and German champion, Siegbert Tarrasch, both of whom advocated a highly logical approach to chess and the idea that, for every position, there existed a theoretically optimal move, independent of the character of one's opponent.[3] Lasker took the game one stage further: like most players of the time, he had completely assimilated the analysis and prescriptions of Steinitz and he used that knowledge to seek to destabilize his adversary by playing moves that were unexpected and even, on the face of it, inferior. The effect was to provoke confusion and induce errors in his opponent's play, leading some to say that Lasker played, not the game at hand, but the man in front of him.

Of Lasker, British chess champion and commentator Gerald Abrahams writes: "If he had a style ... it is revealed in a desire for an unbalanced game; a different type of imbalance from that sought by Alekhine, and possibly a greater strain on playing power ... In the battles he fought he was conscious of the truth that there need not be 'a best move'. The consequence is that Lasker played a type of chess that is difficult to describe. His vision was very great ... Consequently he was always dissatisfied ... and frequently sought

---

[2]   On Lasker's work in mathematics, see Neumann (2000) and the essay by Lang in Sieg and Dreyer (eds.) (2001), *Emanuel Lasker: Schach, Philosophie und Wissenschaft*, Berlin: Philo.
[3]   On Lasker, see Hannak (1959). This book tends to be criticised for its hagiographic treatment of its subject. Recently, Lasker has been the focus of renewed attention, with the formation of a *Lasker Gesellschaft* in Germany. See Sieg and Dreyer (eds.) (2001).

to unbalance the game because of the possibilities that he saw – the battle after the skirmish, the course of the war beyond the battle".[4]

This opposition between the logical and psychological approaches to the game became a feature of chess discussions of the period and, as we shall see later, it was against this background that chess-playing mathematicians such as Ernest Zermelo, who knew Lasker, became interested in a formal, deliberately nonpsychological, analysis of the game.

Mathematician by training, chessplayer by necessity, and humanist philosopher by inclination, Lasker became a prominent figure in the culture of Weimar Berlin. He was married to Martha Cohn, a writer of popular novels published under the pseudonym L. Marco. With his brother, Berthold, a medical doctor who, for a short while, was married to the well-known poet Else Lasker-Schuler, Lasker wrote a play, "Vom Menschen die Geschichte" (On the History of Mankind), which dealt with the eternal theme of how to live an ethically good life. He was a prolific author. In his writings about chess and other topics, he showed himself keen to draw connections between the game and other realms of activity and experience. In 1907, the year before the Amsterdam meeting, for example, he published *Kampf*, an eighty-page pamphlet in which he he extended the metaphor of the "game" to the analysis of the social realm, developing an embryonic "science of struggle".

"What is struggle and victory?" asks Lasker. "Do they obey laws which reason is able to capture and establish?"[5] Analysing the place of struggle in various fields, Lasker makes a speculative yet systematic attempt to apply the metaphor of the "game" to the analysis of social and economic life. Significantly, economic ideas are the thread that binds the work together, with the book's short chapters bearing titles such as "Strategy", "The Work Principle", "The Economy Principle", and "Equilibrium and Dominance".[6]

Lasker called his science of struggle Machology, after "Machee", the classical Greek term for a fight. The notion of struggle is understood broadly, encompassing any form of struggle against resistance, thus being applicable to the efforts not only of sentient beings but also plants, nations, races, and even languages. Any struggle involves several centres of activity, termed

---

[4] Abrahams (1974), p. 146.
[5] Lasker (2001), p. 11.
[6] Lasker also wrote about this in his 1919 work, "Die Philosophie des Unvollendbaren", which, in the 1960s, was recognized by German cyberneticist and successful chessplayer, Georg Klaus, as prefiguring game theory. See Klaus (1965). I am grateful to Sybilla Nikolow for bringing this essay to my attention.

strata. Thus, in a war, the strata are soldiers, the canon, the fleet, and so on. Each stratem, in turn, is made up of "jonts". In a marine conflict, for example, a battleship is composed of captain, sailors, equipment, and so on.

Lasker felt the most important objection against any attempt to construct a science of struggle was the infinite number of events in a struggle, and the uncertainty surrounding them. Chess demonstrates the numerous ways in which a struggle can develop from a certain position, according to Lasker. One would think it impossible to establish a law amidst such multiplicity. But even in the game of chess at an ordinary level, he says, it is evident that the choice of moves is very much limited by their need to be useful. Amongst chessmasters, there is an even greater reduction of possibilities and as the game progresses, the number of possibilities diminishes more and more.

Throughout *Kampf*, Lasker makes many references to the economic realm and to value, and gives central place to the figure of *homo economicus*. Just as the best merchant is one whose buying and selling is conducted to his best monetary advantage, so too can one speak of a best way to engage in a struggle. For example, a military general striving to achieve a certain objective will do so in such a way as to keep his losses in military value as small as possible. In the domain of struggle, says Lasker, the parallel to money in commercial life may be called energy. *Macheeiden*, or perfect strategists, will be infinitely economical with the energy of the struggle at their disposal.[7] "Let's look at the struggle of a merchant", he writes:

The social utility of his activity or of his goods, the work they save society, represent his strata capable of attack. The announcement [i.e., advertisement], in any form, constitutes an army of very flexible strata, which serve for attack and which are sent forward to points of great enemy pressure. His money and credit are his flexibility. Book-keeping is the wall. The enemy is his [competitive] better. Other enemy strata are tasks to be done, such as taking orders, distribution and receiving cash.

The machee-ide solves them following the principle of economy which will be treated later on, and the field of struggle which is determined by the consuming and buying society, its legislation, and the purchasing power of money.[8]

---

[7] Lasker says that even atoms to the extent that they obey Gauss's principle of least resistance and other minimal principles, are probably *Macheeiden*. Similarly, "instinct" is a *Macheeide* because plants and animals instinctively behave economically when faced with attacks. A species makes an effort only when facing some resistance: when the resistance disappears, the organ used to combat it will be directed by nature towards some other purpose. Likewise, the energy consumption for a change in lifestyle is, and always has been, infinitely economical and the principle of development of all life is necessarily deduced from that.

[8] Lasker, *Kampf*, pp. 27–28.

Lasker then goes on to describe what he calls the work principle (*Principe der Arbeit*) or the value principle (*Werthprincip*). When strata are involved in a struggle, aiming to achieve an objective, they do "struggle work". Lasker says this is no more difficult to measure than the work in mechanical, thermal, electric, or other forms of power. In a war, for example, the work or value of a run of bullets is its expected number of hits. In chess, it is the capture of pieces, the domination of fields of escape of the king, and so on. Lasker continues:

> The genius of the strategist lies simply in his ability to accomplish as much work as possible with his army of strata . . . The capacity of a group of strata to perform will be called its "value".
> The perfect strategist will obtain more work from a group of strata the higher its value. One can prove this exactly. Suppose that at the beginning of a struggle the strategist has the choice of two groups of strata, A and B, which he can include in his army. Because A and B can only be of an advantage to the strategist insofar as they help him with the resolution of his tasks, so the strategist will choose without doubt the group of strata with the greater performance capacity. The strategist cannot be wrong because he can calculate in advance (*im Voraus*) the optimal path (*eumachische Bahn*) of the upcoming struggle. This is why he will really take great advantage of the group of strata with the highest capacity to perform.[9]

Lasker continues with a verbal account of how, in order to derive the greatest utility from an army, one must give a small task to strata of little worth, because then the duty to perform a task diminishes the flexibility of a stratem and therefore any other work one could get out of it. The strata with the highest utility will be most likely to be attacked by the enemy, so they have to be placed and protected in a way such that it would cost the enemy enormous efforts to force them to retreat or to eliminate them. Any manoeuvre costs an effort and has to be compensated by an increase of utility.

According to Lasker, the "principle of economy" applies to all areas of creation, artistic and scientific. Good work in any of these areas is that which achieves the most with the means at one's disposal:

> The perfect strategist is by nature infinitely economical with the energy at his disposition. One who longs to come as close as possible to being a perfect strategist will therefore critically examine all actions or manoeuvres, even if a way to obtain an advantage with little effort is clearly available . . . The perfect strategist therefore is completely free of panic-like emotions of fear. He is always objective. This statement is so obvious that it sounds banal. It is nonetheless very rarely taken into account.[10]

---

[9] *Ibid*, pp. 30–31.
[10] *Ibid*, pp. 35–36.

The principle of economy applies not only to a struggle between opponents, but also to the realms of art and science:

A man full of the creative impulse struggles with an idea which imperatively demands to be artistically expressed or be examined by him with scientific rigour. The artist masters the technical means of his art – words, colours, sounds, building materials – and he wants to create a work, which puts the feelings in a certain motion. The scientist sees a puzzle and wants to make it understandable. The field of struggle is the emotional or spiritual life of society. If they – the artist like the scientist – embody the idea in all its dimensions, but do so with the most economical means, which is to say "eumachisch", then they create a work of art or make a scientific advance. Every lack of economy is felt to be ugly. Every unmotivated or superfluous effort is ugly. And every absolutely economical creation is, in terms of beauty (or scientifically), of lasting importance.[11]

Lasker then explores notions of "equilibrium" and "dominance" arising from the principles just discussed. If army A is sent to contain another army, B, then an equilibrium is reached when A, acting as a perfect strategist and obeying the principle of economy, devotes just the right amount of resources to the task. Too much implies resources are being wasted; too little effort means the task will not be accomplished. Between these two extremes can be found the number of jonts that will allow him to accomplish the task in a economic manner. Any disturbance of such an equilibrium, he says, will lead to strategic dominance of one player over another.

In a passage worth quoting at length, given our particular interests, Lasker introduces probability into the picture:

I beg your pardon, dear reader. Maybe I should have explained from the beginning how I see the definition of chance, luck and misfortune. It is true, even a perfect strategist can be the victim of bad luck, despite all his precautions. That which can happen has, from time to time, to become a reality. But even a random event has to obey the laws of probability . . .

Therefore, *in a struggle where chance plays a role, the perfect strategist A will consider all these random events and their probabilities in order to find a solution for the given task, so that he contains B, in the sense that the danger of losing is as big for B as for A. And when either of these undertakes a manoeuvre the advantage of which represents, according to probability, a particular value, then he will suffer, for all of the reasons mentioned above, a loss whose expected value, according to probability, equals the expected advantage. In other words, when the struggle between A and B, being a strategic equilibrium, is repeated often, then the conquered values of A will be as great as those of B, as long as B behaves in a perfectly strategic way.* Otherwise, they will be greater. If A, on the other hand, has predominance, then, under these conditions,

---

[11] *Ibid*, pp. 36–38.

the values that A has won will always exceed those of B, no matter if B maneouvres strategically or not.[12]

Not only does Lasker thus embody the cultural importance of chess at this time, but he is a central figure in the emergence of a discourse that combined elements of games, mathematics, and social interaction. Though it contained little trace of the complex psychological jousting for which Lasker's own chess play was known, emphasizing instead deliberation, rationality, and economy, his exploratory science of struggle reached out to draw connections between *kampf* at the chessboard and in society. For Lasker, the games relevant to social interaction were structures in which *Homo ludens* became *Homo economicus*, with two players confronting each other, each armed with a set of strategic resources and motivated by considerations of strategic perfection, probability, and equilibrium.

In 1908, the year after the publication of *Kampf*, Lasker beat Tarrasch in the World Championship in Germany, bringing an end to the great rivalry between them and unambiguously placing the mathematician at the international pinnacle of chess. In 1909, in St. Petersburg, he took first place against the great Polish player Akiba Rubinstein.[13]

Mathematics and the Endgame

In this context, we can easily understand to what Ernest Zermelo was referring when, in 1912, he introduced a mathematical paper on chess saying that he wanted to consider the game without any reference to psychology and perfect play.[14] A keen chessplayer, Zermelo (1871–1953) was of the same generation of German mathematics students as Lasker, and Hilbert taught both.[15] At Göttingen, where Zermelo worked from 1897 until 1910, there

---

[12] *Ibid*, p. 46 (emphasis added).

[13] Rubinstein was one of the more enigmatic figures in the chess world of the turn of the century. He suffered from a nervous psychological disorder termed anthrophobia, or fear of men and society, source of lifelong trouble for a player of such a public game. As his condition worsened over the years, he complained about his tournament concentration being disrupted by a fly – which no one else in the room could see. He became reticent about receiving visitors to his home and his wife would welcome guests with a greeting, no doubt designed to put them at ease: "Do not stay long, for if you do he will leave by way of the window". Rubinstein was a partial inspiration for Luzhin, the protagonist in Nabokov's *The Defense*, readers of which will remember the brilliant final passage in which Luzhin, rendered suicidal, chooses precisely defenestration – the "icy air" gushing into his mouth. See Nabokov (1964).

[14] Zermelo (1913) in Schwalbe and Walker (2001), p. 133.

[15] On Zermelo, see Segal (2003), pp. 467–69.

was considerable interest in chess. Indeed, that university was also home to the oldest surviving handwritten document on the game, the *Göttingen Manuscript*, a Latin treatise on chess problems and openings, written by Portugese player, Lucena, in the late fifteenth century.

Zermelo studied mathematics, physics, and philosophy at Berlin, Halle, and Freiburg and his teachers included Frobenius, Max Planck, Lothar Schmidt, and Edmund Husserl. After an 1894 dissertation on the calculus of variations at the University of Berlin and a further two years there as Planck's assistant, Zermelo went to Göttingen, where he completed his *Habilitation* and was appointed *Dozent* in 1899. As of 1902, he began publishing on set theory, which was becoming an important field at Göttingen, with Russell's paradoxes appearing in 1903. Zermelo's 1904 proof that every set can be well-ordered was celebrated, earning him a professorial appointment at Göttingen a year later. It was also controversial work, relying, as it did, on the axiom of choice, which was contested by intuitionist mathematicians. In 1908, Zermelo produced an axiomatics of set theory, which, when improved by Fraenkel in the early 1920s, became a widely accepted system. In 1910, he took a chair at Zurich.

In 1912, Zermelo presented his "On an Application of Set Theory to the Theory of the Game of Chess" to the International Congress of Mathematicians at Cambridge.[16]

The following considerations are independent of the special rules of the game of Chess and are valid in principle just as well for all similar games of reason, in which two opponents play against each other with the exclusion of chance events; for the sake of determinateness they shall be exemplified by Chess as the best known of all games of this kind. Also they do not deal with any method of practical play, but only with the answer to the question: can the value of an arbitrary position, which could possibly occur during the play of a game, as well as the best possible move for one of the playing parties be determined or at least defined in a mathematically objective manner, without having to make reference to more subjective-psychological notions such as the "perfect player" and similar ideas?[17]

Assuming that only a finite number of positions are possible (in the sense that the number of squares and the number of pieces are both limited), and without assuming any stopping rules (thereby implicitly allowing for infinite sequences of moves), Zermelo asks two questions. First: what does

---

[16] See Zermelo (1913). See also Schwalbe and Walker (2001), which provides an English translation of the Zermelo paper, with an extensive introduction, and clarifies the subsequent contributions by Dénes König (1927) and Laszló Kalmár (1928/29) in relation to Zermelo's.

[17] Zermelo (1913) in Schwalbe and Walker (2001), p. 133.

it mean for a player to be in a "winning position", and is it possible to define this mathematically? Second: if a player is in a winning position, is it possible to determine the number of moves necessary to ensure the win?

Zermelo shows a player is in a winning position if and only if a particular set is non-empty – namely the set containing all the sequences of moves that guarantee a player can win independently of how the opponent plays. Were this set empty, then the best the player could hope for would be a draw. Thus, Zermelo defines a different set, containing all sequences of moves that would allow the player to postpone his loss indefinitely, thereby implying a draw. If this set is empty, then this is the same as implying that the opponent can force a win. As for the second question, Zermelo answers it, employing a proof by contradiction and showing that the number of moves in which a player in a winning position is able to force a win can never exceed the number of positions in the game. Were White, say, able to win in a number of moves greater than the number of positions, then at least one of the "winning positions" would have had to appear twice, in which case White could have adopted those winning moves when the winning position appeared the first time round, rather than wait until the second.

Obviously, Zermelo's analysis of chess was purely formal, an attempt to say something minimal about the game, without any consideration of the tactical and psychological features that made the game interesting to play. It was neither intended to be, nor, one suspects, was it of any value to the chessplayer. As Zermelo noted, closing the paper, the question of whether the game's starting position could guarantee a win for one of the players remained open, and answering it would imply that chess would lose its gamelike character. Yet, the paper was regarded as brilliant by several Hungarian mathematicians immersed in set theory, and they sought to improve it after World War I was over. By that time, the psychological dimensions of chess were commanding greater interest than ever.[18]

The "Strangest States of Mind"

The 1920s opened with Lasker handing over the title to Capablanca in 1921, though he continued to be a dominant figure internationally. The decade was also marked by the appearance of the Hypermodern Movement, at

---

[18] As for Zermelo himself, during the war, with no sign of his lung ailment improving, he resigned his chair at Zurich, left academia, and moved to Germany's Black Forest, where he taught private classes for a decade. In 1926, he was given an honorary position at Freiburg. Later, his 1935 refusal to give the Hitler salute provoked a controversy at that university, causing him to withdraw from all teaching activity.

the instigation of Richard Réti, a Hungarian trained in mathematics and physics at the University of Vienna, along with compatriot Gyula Breyer, Svelly Tartakower, and Akiba Rubinstein. Responding to declarations that the possibilities of chess had been exhausted, they broke with the Classical style, personified by Tarrasch, and created a new approach to the game, introducing ideas so radical that, in the eyes of many players, they bordered on the irrational. The strategic essentials of hypermodernism were laid out by Réti in his 1923 *Modern Ideas in Chess*[19] and by Nimzovich in his 1925 *My System*: (1) attacking the centre squares from far away with knights and bishops instead of occupying them with pawns and pieces in the usual manner, (2) blockading isolated pawns with knights, and (3) deliberate overprotection.[20] Some hypermodernists saw their approach as the manifestation in chess of the French Surrealist spirit of Marcel Duchamp, himself another *Schach* fanatic.

The middle of the decade also saw the appearance of *Lasker's Manual of Chess*.[21] Like no other chess book of the period, it stood in particular contrast to Tarrasch's more conventional *The Game of Chess*, published not long afterwards.[22] Although certainly a manual, Lasker's book continued his reflections on the interrelations amongst chess, mathematics, economics, and social life. For example, the proposition that "The Plus of a Rook suffices to win the game" (i.e., having a one-Rook advantage), becomes an opportunity for Lasker to explore the importance of the *ceteris paribus* condition and, with that, the relationship between chess and mathematics. His discussion of many propositions concerning the strength of different pieces and the ability to force a win in various situations is imbued with mathematical language: "This demonstration is mathematical";[23] "The question is one of pure mathematics . . . ".[24] His discussion of the "Exchange-Value of

---

[19] See Richard Réti (1923a). See also his "Do 'New Ideas' Stand Up in Practice?" which was published in Russian in the *Chessplayer's Calendar* and then in the October 1987 issue of the *Chess Bulletin*, and recently translated by R. Tekel and M. Shibut in the Sept./Oct. 1993 issue of the *Virginia Chess Newsletter* (1923b). Réti burst onto the international scene toward the end of the war, sharing first prize at Budapest in 1918, coming first at Rotterdam and Amsterdam in 1919 and Vienna in 1920, and winning an important international tournament in Gothenburg in the same year. In 1924, he caused a storm in New York by becoming the first person to beat Capablanca in ten years, shaking the latter sufficiently to cause him to lose against Lasker later in the tournament. Réti died in Prague in 1929, shortly before his masterpiece, *Masters of the Chessboard*, went to press.

[20] See Nimzovich (1974, orig. 1925).

[21] See Lasker (1976).

[22] See Tarrasch (1940).

[23] Lasker (1976), p. 15.

[24] *Ibid*, p. 22.

the Pieces" combines economics and the *ceteris paribus* condition, and, as in *Kampf*, his discussion of the aesthetic effect in chess is based on notions of economy.

In a closing chapter, "Final Reflections on Education in Chess", Lasker continues his speculations on the embryonic science of struggle, applicable to the realm of social interaction: "It is easy to mould the theory of Steinitz into mathematical symbols, by expressing a kind of Chess, the rules and regulations of which are expressed by mathematical symbols... In such a game, the question whether thorough analysis would confirm the theory of Steinitz or not, presumably could be quickly solved, because the power of modern mathematics is exceedingly great. The instant that this solution is worked out, humanity stands before the gate of an immense new science which prophetic philosophers have called the mathematics or the physics of contest".[25] Such knowledge, he says, could transform political life. It could even lead to outlawing war, because it would provide an alternative way to settle disputes. The mathematics of chess would not eliminate contests of life, which are necessary to the functioning of society, but it would couch such problems "in precise terms and point to a solution":

The science of contest will progress irresistibly, as soon as its first modest success has been scored.

It is desirable that institutes to further these ends should be erected. Such institutes would have to work upon a mass of material already extant: theory of mathematical games, of organisation, of the conduct of business, of dispute, or negotiation: they would have to breed teachers capable of elevating the multitude from its terrible dilettantism in matters of contest; they would have to produce books on instruction...

Such an institute should be founded by every people who want to make themselves fit for a sturdier future and at the same time to aid the progress and the happiness of all humankind.[26]

Throughout the book, psychological considerations loom repeatedly. Lasker says a combination is born in the player's mind, surviving amongst many jostling thoughts, true and false, sound and unsound, and achieving victory over its rivals when it is transformed into a movement on the board: "Does a Chess-master really cogitate as just outlined? Presumably so, but with detours and repetitions. However, it matters not by what process he conceives an idea; the important point to understand is that an idea takes

[25] *Ibid*, p. 340.
[26] *Ibid*, pp. 340–41.

hold of the master and obsesses him. The master, in the grasp of an idea, sees that idea suggested and almost embodied on the board".[27]

The cogitations of the chessmaster's mind were very topical in 1925, when Lasker published his *Manual*. For example, that year saw the appearance of a Russian silent film, *Shakmatnaya goryachka*, or "Chess Fever", in which the mental stability of the protagonist is threatened when he tries to play the game against himself.[28] The film has cameo appearances of Réti and others, who were participating that year in a big international tournament in Moscow, at which Lasker came second after the Russian Bogoljubow. Also, the occasion of that tournament was used by researchers at Moscow's Psychotechnics Institute for an important experimental study of the game. Taking a group of the participating chessmasters, psychologists Djakow, Petrowski, and Rudik subjected them to tests in an attempt to determine what exactly it was that mentally distinguished the good player from the common mortal.[29]

At that point, the only comparable study was French psychologist Alfred Binet's 1894 *Psychologie des Grands Calculateurs et Joueurs d'Échecs*, which dealt with blind chess.[30] Many observers in London and Paris had been astounded in 1859 by American player Paul Morphy, the youthful Mozart of the game, who, on top of his remarkable performance in normal chess, played simultaneous blind games. Blind chess involved playing without being able to see the board, against an opponent who could do so. The blind player's knowledge of the board was based solely on his mental retention of moves played, with the opponent's moves being called out. Simultaneous

---

[27] *Ibid*, p. 114.

[28] Again, a similar theme is exploited in Nabokov's *The Defense*, in which Luzhin, denied the possibility of playing by his respectable family-in-law, resorts to playing in his head. Soon, he sees his own life as one large chess game, in which every social encounter is interpreted in terms of a move or counter-move. His descent into madness follows.

[29] See Djakow, Petrowksi, and Rudik (1927).

[30] See Binet (1894). Binet (1857–1911) initially was a student of Jean Martin Charcot (1825–1893) at the Hôpital Salpêtrière, Paris, where he worked on hypnosis. Breaking with Charcot in 1890, he moved to the Sorbonne's Laboratory of Physiological Psychology, directed by Henri Beaunis, where he began working on the study of cognitive processes. In 1895, with Beaunis, he founded the first French psychology journal, *L'Année psycho-logique*, and then became director of the psychology laboratory. Unreceptive towards the laboratory methods employed in German psychological work, Binet preferred the use of questionnaires and interviews over that of laboratory instruments and measurement apparatus. In 1905, he opened a laboratory for the study of children and pedagogy. His *L'Étude expérimentale de l'intelligence* (Paris: Schleicher Frères et Cie, 1903) was a person-ality study of his two daughters. Appointed to a ministerial commission to study retarded school children, with Theodore Simon he developed the Binet-Simon intelligence test for which he remains best known. See Wolf (1973) and Pollack (ed.) (1995).

Figure 1.1. Position of pieces in Sittenfeld game. *Credit:* From Alfred Binet (1894), p. 300.

blind play involved playing against several opponents, visiting each table for one move at a time only. The blind player thus had to retain in his memory the evolving states of several boards simultaneously. Such performances, which lasted for hours, demanded extraordinary stamina on the part of the blind player. Following Morphy in 1859, the record number of simultaneous blind games had been pushed steadily upwards.

Curious as to how they did it, Binet conducted a study by means of a questionnaire distributed to a small number of players, including Tarrasch in Germany and Blackburne in England. The study focused on the importance of *visual* representation in the player's mind. From the responses of ten players concerning the way they thought and reasoned during blind play, Binet concluded three factors were important: experience (*érudition*), imagination (*imagination*), and memory (*mémoire*). Imagination involved the capacity to visualize a position, the ability to see the chessboard clearly in one's mind, something that was emphasised by all but three of the players. Memory, too, involved visual representation, and this is where Binet saw the psychological originality of his study. For example, when asked to reproduce what he saw in his mind when he played, one player, Sittenfeld, drew a picture. Here is the actual position on the board, followed by Sittenfeld's rendition of his mental image of same (Figures 1.1. and 1.2).

Apart from the conclusions drawn from the study, Binet's book contains many interesting reflections and asides on the game itself, as seen by this psychologist – no chessplayer himself – towards the close of the nineteenth century. Insisting on the analogy between chess and mathematics, Binet notes that many well-known historical figures, including Voltaire,

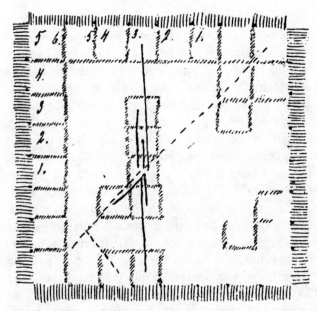

Figure 1.2. Visual image as drawn by Sittenfeld. *Credit:* From Alfred Binet (1894), p. 301.

Rousseau, and J. S. Mill, had been chessplayers, but that women tended not to excel in the game, as it required physical vigour and a taste for combat. Also, chessplayers tended to be rather vain: those remaining passive in victory and defeat, such as Morphy, were exceptional. Binet also notes that, up to the end of the eighteenth century, chess tended to be dominated by Latin players: the Italians, Spanish, and Portugese. These were then bypassed by the Germans, Slavs, Anglo-Saxons, and, especially, Jews. Showing a table classifying prominent players by country, religion, and race, Binet noted that of sixty-two players, eighteen were either Polish or Hungarian Jews. Furthermore, almost all the strong Jewish players were professional, "which shows clearly the seriousness of the race".[31]

In their study in postrevolutionary Moscow, the Russian psychologists referred to Binet's work, but more important to them by 1925 was the appearance in the interim of psychoanalysis and attempts to apply it to chess. By the mid-1920s, psychoanalysis had fallen afoul of the Soviet authorities, and Djakow et al were clearly concerned to retrieve the game

[31] Binet (1894) p. 222.

from the clutches of the psychoanalysts and show why chess, notwithstanding its individualistic and combative elements, could serve important social and educational functions in the new Russia.[32]

The psychoanalytical study Djakow et al had in mind probably was Alexander Guerbstman's (1925) *Psichoanaliz sacmatnoj igri* [The Psychoanalysis of Chess. An Interpretative Essay], which, unfortunately, remains untranslated.[33] However, the essence of the psychoanalytical approach to chess can be distilled from several other contributions of the period, including the landmark paper on the subject by Freud's British disciple, another keen player, Ernest Jones.[34]

In "The Problem of Paul Morphy: A Contribution to the Psycho-Analysis of Chess", which he read before the British Psycho-Analytical Society in November 1930, Jones put it clearly: "the unconscious motive actuating the players is not the mere love of pugnacity characteristic of all competitive games, but the grimmer one of father-murder".[35] The appeal of chess lay in its ability to gratify hostile Oedipal impulses. To checkmate the King was to render him immobile and sterile, the symbolic expression of the player's desire to overcome the father in an acceptable way, aided by the mother (Queen). The mathematical quality of the game, said Jones, the "exquisite purity and exactness of the right moves", the "unrelenting pressure" and then "merciless *dénouement*", all combined to give the game a particular anal-sadistic nature: "The sense of overwhelming mastery on the one side matches that of unescapable helplessness on the other".[36]

Jones recalled some of Paul Morphy's qualities: his ability to play impassively from morning till midnight for several days running with no signs of fatigue. On Morphy's famous European trip, when he played eight opponents blindfold at the Café de la Régence, it took seven hours before the first of them was beaten, and another three before the match ended, throughout all of which the American neither ate nor drank. At seven the next morning, Morphy promptly called his secretary and dictated to him every move

---

[32] On the history of psychoanalysis in the Soviet Union, see Etkind (1997) and Martin A. Miller (1998). Whereas neither of these books discusses chess *per se*, they do confirm the Bolshevik state's rejection of psychoanalytic methods by 1925. See, for example, Etkind *op cit*, p. 179ff.

[33] See Jacques Berchtold (ed.) (1998), especially the editor's introductory essay.

[34] See Ernest Jones (1931). See also the essay on the Hamlet figure in chess by another disciple of Freud, the Swiss pastor, Oskar Pfister (1931), which appeared in the *Psychoanalytische Bewegung* alongside the German translation of Jones' paper "Das Problem Paul Morphy". See also Coriat (1941), and Fleming and Strong (1943).

[35] Jones (1931) p. 168.

[36] *Ibid*, p. 170.

in all of the games, discussing the possible consequences of hundreds of hypothetical variations. Where Binet had found evidence of remarkable visual memory, Jones saw "a very exceptional level of sublimation, for a psychological situation of such a degree of freedom can only mean that there is no risk of its stimulating any unconscious conflict or guilt".[37]

Noting that Morphy's stellar success in chess had begun just a year after the shock of his father's sudden death, Jones surmised "that his brilliant effort of sublimation was, like Shakespeare's *Hamlet* and Freud's *Traumdeutung*, a reaction to this critical event".[38] Jones pursues a psychoanalytical reading of Morphy's European performance, including his vain three-month effort to lure to a challenge the British champion, Staunton, who had become, for Morphy, the "supreme father *imago*". That Staunton took to criticizing Morphy in the press as a monetary adventurer, all the while refusing to play him, accentuated the frustration felt by the young American. To Jones, Morphy's case was illustrative of the connection between genius and mental instability. The artistic conscience was characterized by rigour, sincerity, and purity, but the psychical integrity of the artist was vulnerable to any of these being disturbed. Jones felt Morphy's chess-playing ability reflected his capacity for sublimation of parricidal and homosexual impulses, all of which served a defensive function for him. When Staunton persistently refused to accept Morphy's challenge, this sublimation broke down, the defensive function failed, and Morphy could no longer use his talent as a means of guarding against overwhelming *id* impulses. Stripped bare, the player collapsed.

The Russian psychologists, Djakow et al, rejected the attempt to explain chess in Freudian terms as forced and one-sided, emphasizing instead the social dimensions of the game. Frequently invoking Lasker, they said chess was a struggle. It provided gratification through the activity of playing, not merely in the resulting victory. The game was an "expression of social life, its specific spirit the expression of social desires in the specific form of social activity".[39] It satisfied the desire for friendly company, for public display of strength. If there were biological roots to the game, they were to be found, not in the Oedipal interpretation, but in the "much larger biosocial foundations of life, in which struggle is a fundamental law. To reduce every struggle to a struggle for a woman would mean adopting an extremely one-sided analytical approach... The essential figure, the King,

---

[37] *Ibid*, p. 173.
[38] *Ibid*, p. 180.
[39] Djakow et al (1927), p. 16.

has an all-too-clear historical origin to permit any attempt whatsoever to explain it using the sexual desires of individual psychology".[40]

They felt the energies and emotions flowing through chess stemmed not from individualistic desires but from "much deeper and more general instincts of great social significance, such as the instinct of activity, of creativity, to display one's power and superiority, social acceptance, and the instinct of struggle or of competition as a basis of personal and social life".[41] Unlike gymnastics and physical exercise, chess was a synthesis of functions, capturing complete episodes of life itself, an activity in which the personality could dissolve yet which offered diverse satisfactions. Chess left no room for chance; success in it depended on intensive solitary work. The rhythm of the game gave rise to "a rich alternation of the strangest states of mind".[42]

In a section titled "Game theories", Djakow et al describe how chess provides pleasure by facilitating the flight from daily effort and work. They are also keen, in Moscow in the mid-1920s, not to insist too much on competition between individuals. "We cannot but share the view of Dr. Em. Lasker that chess is a struggle and that every human being feels the desire to fight – in sports, at the card table, while playing boardgames.... But in our opinion the moment of struggle is only one aspect of this phenomenon. Besides that, or perhaps even prior to that, is the moment of solitude, of isolation, of plunging into an entirely different world, which is filled with the purely intellectual struggle".[43]

The game was characterized by psychological tension. It required creativity, action, real impulses of will. This was what distinguished it from, for example, enjoying a work of art. In an age when the individual was increasingly subordinate to machines and technology, the authors felt, games allowed for the relief of monotony and the rupture of routine. Unlike engaging with inert material, playing chess meant encountering a flesh-and-blood adversary and adapting to his movements. "Here, rarely is something foreseeable for sure. Chess is the highest level of such struggle of 'foresight.' At every moment two ideas, two intentions, collide with each other. Therefore, no advance plan exists; the plan and its realisation emerge during the game. One would need enormous insight (*übersicht*) to foresee just a little bit... One needs to invent something new all the time".[44] The game thus

---

[40] *Ibid*, p. 17.
[41] *Ibid*, p. 18.
[42] *Ibid*, p. 9.
[43] *Ibid*, p. 9.
[44] *Ibid*, p. 14.

Figure 1.3. Endgame. *Credit:* From Djakow et al (1927), p. 29.

produced a heightening of psychic tone (*Tonus der Psyche*) and elevated self-feeling (*Selbstgefühl*).

Contrary to the case in the Binet study, the identities of the international chessmasters participating in the Moscow experiments remain unknown. They were subjected to a range of tests intended to examine the functions of memory, attention, higher intellectual processes, imaginative power, and intellectual character. All tests were conceived with a view to determining the psychological make-up of a good player.[45] In one of the memory tests, for example, the player was shown, for a minute, a chessboard with coloured counters. He then had two minutes to reconstruct the configuration, winning points for correct positioning and correct colours. In another, he was briefly shown the above endgame (in which White can force a win in three moves) and asked afterwards to reconstitute it as exactly as possible.

The attention tests required examining various shapes and then reproducing them in the correct order. Intellectual processes were gauged with reference to combinatorics and intellectual function. For example, to a board with two "queens", the player had to add five more in such a way that no queen could take any other. Or the player was shown ten numerical equations and had to say, in a limited time interval, whether they were right or wrong. When it came to measuring the player's powers of imagination and psychological type, the Russians employed Rohrschach tests. Players were shown ten images of random ink blots, and their responses were

[45] It is unclear whether or not a control group was used. The results in the study indicate that there were twelve subjects. De Groot (1965) says that the study failed to consider the performance of non-chessplayers (p. 10).

interpreted as providing information about power of imagination, will-power, and other attributes.

The results obtained were varied. In terms of general memory of num-bers, the players were no better than average people. In retaining geometrical forms, they were inferior. Regarding attention, players were average in their ability to reproduce, in the proper order, the shapes shown on cards, but they were superior when it came to the range and dynamics of attention. For example, when shown eight boards with three pieces in sequence, they showed a strong ability to reconstruct the movement of the pieces through-out the sequence. The thought processes of the players were not particularly fast, but the purely logical side of their reasoning was more developed than in non-chessplayers.

Asking what attributes were required to become a chessmaster, the Rus-sians come up with a portrait, the "psychogram". It included much: phys-ical strength, strong nerves, self-control, a perceptive type of psyche, a high level of intellectual development, concrete thinking ability (which was not the same as logical or mathematical thinking), objective thinking, a very strong chess memory (which was not the same thing as psychological memory), disciplined will, emotive and affective discipline, and awareness of one's strengths. Intellectual development, objective thinking, and chess memory could be encouraged, the authors said; the rest were innate.

Djakow et al close their monograph by quoting Réti, the Hungarian who insisted that it was not a player's ability to think ten or twenty moves ahead that was important:

Chessplayers, who ask me from time to time, how many moves in advance I cal-culate, are astonished to hear me answer: usually none. A bit of mathematics will show us that it is impossible and even useless to foresee an exact sequence of moves. When you try to calculate in advance 3 moves of White and Black, the number of variations mounts already to $3^6 = 729$; to calculate this is thus practically impossible. . . . Every chessplayer – the weakest like the strongest – has consciously or unconsciously well-defined principles through which he is guided in the choice of his moves.[46]

If Binet, in Paris in 1894, had applauded his subjects' refined and dignified use of geometric visual memory, in Moscow in the mid-1920s, Djakow and colleagues saw in chess an opportunity to shape the body politic. The game of chess was striking proof of the possibility of unlimited development of single sides of the human psyche when one had sufficient drive and interest. The "dialectics of chess" showed that it provided an objective measure of

---

[46] Djakow et al (1927), p. 59.

our own reason; the game *"deprives us of the possibility and the right of appealing to something higher with even more authority. It destroys in the case of defeat our last hope of self-justification.* Such is the tragedy of chess".[47] The characteristics associated with good chessplay are good for society: *"From its essence as well as the history of its evolution, chess merits without doubt becoming a game of the people".*[48]

Conclusion

In the first decades of the twentieth century, chess held an important place in the cultural life and imagination of *Mitteleuropa*. The chessboard was a locus for rich speculation concerning the psychological and psychoanalytical features of effective chessplay and the parallels between the game and social life in general. The former is captured in various contributions to literature and film and, especially, in the psychological investigations of Binet in France and Djakow in Moscow, and the psychoanalytical contributions of Ernest Jones and other Freudians. The latter is embodied in the writings of German Jewish mathematician and chess champion, Lasker, whose "science of struggle" was nothing less than a speculative attempt to consider the actions of individuals in society in terms of strategic economy.

It was in the midst of this that mathematician Zermelo turned to the formal consideration of chess and, deliberately casting subjective psychological considerations aside, analyzed the endgame in set-theoretic terms. As we shall see, this work was again taken up in the 1920s by Hungarian mathematicians interested in set theory, and close to Göttingen, including Denés König, Laszló Kalmár, and John von Neumann.

[47] *Ibid*, p. 60 (emphasis in original).
[48] *Ibid*, p. 61 (emphasis in original).

TWO

"Deeply Rooted, Yet Alien"

Hungarian Jews and Mathematicians

Of König, Kalmár, and von Neumann it may be said that they came from a world, now lost, composed of two social sets and their intersection: Hungary's mathematicians and its assimilated Jewry. Not all the former were of Jewish origin, of course, but, as in the chess world, a significant number of them were. If, over the next few pages, we examine this milieu, it is not only because it allows us to view the world in which the young von Neumann emerged, but also because it provides critical background against which we can understand later developments. Von Neumann's engagement with social science in the 1930s was bound up with the dissolution of the community described here.

Hungarian Jewry

A characteristic emphasised in many histories of the Jews of Hungary is the degree to which, beginning in the mid-ninteenth century, they achieved integration into Hungarian society.[1] A Jewish community had been present in Hungary since the tenth century, its numbers growing at the end of the eleventh with the arrival of refugees escaping pogroms. The first Jewish law in the history of Hungary was passed when King Béla put the community under his protection, with taxes being paid to the court. The second half of the nineteenth century saw a large wave of refugees from further pogroms in Russia and the eastern part of the Monarchy, so that, between 1840 and 1890, the Jewish proportion of the Hungarian population rose from 2 percent to almost 5 percent. The emancipation of Hungary's Jews began in 1849, with the law passed that year forming the basis for a more substantial

[1] On the history of the Hungarian Jews, see Braham (1981), Katzburg (1981), Patai (1996) and Frojimovic et al (1999).

law in 1867. This was the year of the *Ausgleich*, or Compromise, when the Hapsburg Monarchy, in the face of nationalist pressure, granted greater autonomy to Hungary, marking the beginning of a flourishing period for the country. Law XVII of that year, on the "emancipation of the inhabitants of the Israelite faith of the country", allowed Jews to hold various commercial licenses, practice certain professions, and enter parts of the public service. Thus Hungary's efflorescence was accompanied by the assimilation of many Jews into the economic and cultural life of the country.

In December 1868, in Pest, the First National Israelite Congress created an organization of Jewish congregations. Because there was disagreement over observance of the Jewish codex (*Sculchan Aruch*), a split occurred, with the liberal majority forming the Neolog community and the traditional group the Orthodox one. In 1895, the Hungarian parliament accepted the Jewish religion as "bevett vallás" – that is, recognized by the laws of the country, with the congregations and associated cultural institutions receiving support from the state and municipalities. From the 1870s onwards, assimilation was greatest amongst the less religiously strict Neolog Jews, amongst whom it became quite common, for example, to educate children at non-Jewish schools, to change one's surname in favour of a more Hungarian-sounding one, and even go so far as to choose Christian baptism. By the late nineteenth century, quite a few prominent Jewish businessmen and professionals were awarded titles of nobility for their services to the Austro-Hungarian Empire. Many adopted the mores and aristocratic lifestyle of the Hungarian nobility and intermarried with their families. From emancipation until the dissolution of Austro-Hungary with World War I, this liberal project of assimilation saw the emergence of a tacit social alliance between the assimilated Jews, who represented the country's economic, financial, and industrial interests, and the Magyars, or indigenous Hungarians, whose semi-feudal aristocracy tended to be dominated by landowners, army officers, and higher civil servants.

Miksa (Max) Neumann, father of János (John), was one such assimilated Jew.[2] He was relatively unknown until the mid-1890s, when he began

---

[2] Max Neumann (1870–1923) had arrived in Budapest at age 10 from Pecs in the Southwest. Trained as a lawyer, he married Margit Kann, daughter of Jacob Kann. The Kanns were a wealthy family whose fortune had been made selling agricultural equipment and hardware to Hungary's large farms. The Kann-Heller firm was located on the ground floor of 62 Vaczi Boulevard (later renamed Bajcsy-Zsilinszky St.), with the rest of the building being divided into apartments. The Hellers occupied the second floor and the Kanns the remaining two floors. The top-floor, eighteen-room apartment went to Max and Margit, to whom John von Neumann was born. See Macrae (1992), pp. 37–46.

working at the *Jelzáloghitel* (or Mortgage) Bank. He did well there, in part because Kalman Szell, head of the bank and Neumann's personal protector, became prime minister of Hungary in 1899. In 1913, Neumann acquired a title of nobility from the emperor Franz Joseph, becoming von Margittai Neumann. In time, as historian William McCagg puts it, Max von Neumann became "as redolent of new wealth as the new baron Henrik Ohrenstein or as József Lukács, the philosopher's father, his colleagues in the banking community".[3] As Halmos (1973, p. 382) points out, the full name in Hungarian was Margittai Neumann János, meaning John Neumann of Margitta. In German translation, the genitive form of the place name was dropped, yielding Jansci von Neumann. In America, he was quite happy to answer to "Johnny", but always used the "von". Although John Harsányi, Hungarian émigré and game theorist of the next generation, did not know von Neumann personally, he knew well the society from which he came. According to Harsányi, it was characteristic of von Neumann, "that he always used the title 'von' and was sort of offended if somebody didn't use it. Not only that... but if the sentence starts with von Neumann... then you don't capitalize the 'von', because you don't change the original spelling. And von Neumann insisted even on that... This, of course, was just a very minor human weakness... Of course, this was not uncommon in the Austro-Hungarian monarchy: that Jewish people who were either rich or famous, or both, would try, first of all, to become Christian, and try to acquire aristocratic titles or at least titles of the nobility, and he was obviously influenced by this".[4]

The von Neumann family was part of a merchant and financial community that, though Jewish, saw itself as patriotically Hungarian. Thus John von Neumann was educated, not at Hebrew school, but at the Lutheran Gymnasium, along with other Jews including Jené (Eugene) Wigner, later a physicist, and Vilmos (William) Fellner, who became an economist. Like later physicists Tódor (Theodore) von Kármán and Ede (Edward) Teller at the Minta Gymnasium, these assimilated Jews were conscious of their cultural inheritance, yet felt themselves to be Hungarian through and through.[5] Thus there was observance of rites on special occasions and the shared allusions and language of Central European Jewish humour, but the Orthodox traditions of Talmudic scholarship and devotion were not part of their lives. With successive generations of assimilation, their consciousness of being

---

[3] William O. McCagg Jr. (1972), p. 69.
[4] Harsányi interview with author, Apr. 16, 1992, Berkeley, California.
[5] The Minta had been founded by Mór von Kármán, Tódór's father, and was affiliated with the University of Budapest.

different quietly faded into the background. Thus John von Neumann's brother, Nicholas, recalls their other brother, Michael, questioning the family's ambiguous religious stance, to which Max von Neumann replied that it was simply a matter of tradition. Stan Ulam, a Polish Jew and close friend of von Neumann, recalled that the tradition of Talmudic Judaic scholarship was "quite conspicuously absent from von Neumann's makeup", but he remembered his indulging in Jewish jokes and banter. "The goys have the following theorem...". (1976), p. 97.[6] The first hints that this might no longer be possible emerged at the end of World War I, when John von Neumann was in his teens, and old Hungary was broken up. In 1920, the Treaty of Trianon saw the dismantling of the Austro-Hungarian empire, with Hungary required to sacrifice no less than two-thirds of its lands and, with them, one-third of ethnic Hungarians, to the successor states, Czechoslovakia, Rumania, Yugoslavia, and Austria. The result was a greatly reduced "rump Hungary" and the birth of revisionist ambitions to regain the lost territories. (See Plate 2.)

After Trianon, the position of the Hungarian Jews began to change. With the disappearance of other large ethnic groups with the surrendered regions, the integrated Jews lost part of their political function in the Jewish–Magyar alliance. Jewish numbers were swelled by further immigrants from the east, many of them Orthodox, keen, as they were, to remain within Hungary. Following the loss of the territories, there was also a rise in the proportion of positions held by Jews in the business, legal, and medical professions. The result was a sharpening of focus on Hungarian Jewry, with all the usual contradictions inherent in such scapegoating. Thus, even if most assimilated Jews were opposed to Bela Kun's short-lived Communist insurgency in the summer of 1919 – including the von Neumanns, who left for a holiday home on the Adriatic – the fact that a majority of Kun's revolutionary Commissars were Jewish intellectuals contributed to the popular image of the "Jewish Bolshevik". When Admiral Nikolas Horthy regained power and cracked down on Kun's supporters in the White Terror of 1920, a great many of those killed, tortured, or forced to flee were

---

[6] Another assimilated Jew from the same background as the von Neumanns was banker's son and Communist, György (George) Lukács (1885–1971): "The Leopoldstadt families were completely indifferent to all religious matters. Religion only interested us as a matter of family convention, since it played a certain role at weddings and other ceremonies... we all regarded the Jewish faith with complete indifference" (1983), p. 26. "At the Protestant Gymnasium I attended, children from Leopoldstadt played the role of the aristocracy. So I was regarded as a Leopoldstadt aristocrat, not as a Jew. Hence the problems of the Jews never came to the surface. I always realized that I was a Jew, but it never had a significant influence on my development" (*Ibid*, p. 29).

Jewish. At the same time, the visible Jewish presence in commercial life, coupled with the extravagant display of riches of a few, served to reinforce the popular perception of enormous Jewish wealth. Some Hungarian anti-Semites called the country's capital "Judapest". It was in this context that Horthy's Hungary, in 1920, passed the first piece of anti-Semitic legislation in twentieth-century Europe. Ostensibly designed to control university registrations in general, the key clause of Law 1920: XXV, the *Numerus Clausus*, was one intended to restrict Jewish access to higher education, and thus the professions, to a level corresponding to their proportion of the population.

Thus, whereas their merchant and banking parents had flourished during the Golden Age that followed emancipation and the *Ausgleich*, the more highly educated generation of Jewish youth that matured around the time of World War I were to be less settled. Not only was Hungary already small and limited in terms of opportunities in science and education, but there was also the added discriminatory element. By the time the 1920 law passed, many young Hungarian Jews had, in fact, already begun looking abroad for opportunities. For example, mathematician Gyorgy (George) Pólya completed a doctorate in Budapest before heading off, in 1911, to Vienna and then to Göttingen, Paris, and Zurich. He eventually went to Stanford. Chemist Michael Polanyi, having already studied in Germany, left Budapest in 1919 to return to Karlsruhe. He eventually settled in England. He was helped by Theodore von Kármán, then already professor of aerodynamics in Aachen and a source of guidance to many émigrés. Von Kármán himself later ended up at the California Institute of Technology. Polanyi, in turn, helped several others find jobs, including physicists Leo Szilard and Imre Brody.[7] Although John von Neumann would have been accepted within the quota imposed, he became one of several thousand Hungarian Jewish students who went abroad to study in Austria, Czechoslovakia, Germany, Italy, and Switzerland. In contrast to the desultory university environment they were leaving behind, these students found an enthusiastic welcome abroad, especially amongst the university mandarins in the tolerant climate of Weimar Germany.

The Mathematicians

Amongst the fields in which assimilated Jewry played an important role in late-nineteenth century Europe, mathematics stands out. This was equally

---

[7] See Frank (2001).

true of Hungary, and von Neumann was always particularly proud of this small country's remarkable mathematical culture. Prior to the *Ausgleich*, Hungary's mathematicians were few, the best known of them being the Bolyais, father and son.[8] Parkas Bolyai studied at Göttingen, where he was a fellow student of Carl Friedrich Gauss, making original contributions in Euclidean geometry.[9] As for Janos Bolyai, it was not until after his death in 1860, that his work received international attention, with Felix Klein and Henri Poincaré both drawing upon it. C. B. Halsted at the University of Texas was particularly active in translating and promoting Bolyai's work.

Were one to sketch the Hungarian mathematical "family tree" to which Dénes König, Kalmár, and von Neumann belonged, it might look like the following, running from the Bolyais in the early nineteenth century through to Erdós, "The Man who loved Numbers", in the late twentieth.

Parkas Bolyai (1775–1856)
|
Janos Bolyai (1802–1860)
|
Gyula König (1849–1913)   Jósef Kürschák (1864–1933)
|
Lipót Fejér (1880–1959)   Frigyes Riesz (1880–1956)
Dénes König (1884–1944)
|
Alfred Haar (1885–1933)   Rudolf Ortvay (1885–1945)
Bela Kjerekárto (1898–1946)   Gabor Szego (1895–1985)
|
John von Neumann (1903–1957)   László Kalmár (1905–1976)
Rózsa Péter (1905–1977)
|
Paul Turán (1910–1976)   Paul Erdós (1913–1996)

In the generation after Bolyai, several names stand out, both for their scientific work and their role in the eventual cultivation of a national mathematical culture. Gyula König, father of Dénes, who worked in algebra,

---

[8] On the world of Hungarian mathematics, see Hersh and John-Steiner (1993).
[9] Bolyai wrote an important textbook and taught at the Reformed College in Marosvasarhely, Transylvania. His son Janos also worked on the problem of parallels, based on Euclid's Fifth Postulate, and was one of the independent creators, along with Gauss, of non-Euclidean, "hyperbolic" geometry. See Hersh and John-Steiner, *op cit*.

number theory, geometry, and set theory, was an influential teacher at the Technical University of Budapest, founded in 1874.[10] At the same institution, József Kürschák worked in the fields of geometry, calculus of variations, and linear algebra.

Eighteen ninety-four was a pivotal year in the development of Hungarian mathematics, for this was the year in which Baron Eötvös Loránd (1848–1919), physicist and founder of the Mathematical and Physical Society of Hungary, became Minister of Education. Eötvös epitomized the Magyar liberalism of the late nineteenth century, under which the Hungarian Jews eagerly sought assimilation and became attached to Hungary. He supported the establishment of not only a college for the training of mathematics teachers, but both the Eötvös Competition in mathematics for secondary school students and the *Kozeposkolai Mathematicai Lapok*, or *"KöMaL"* for short, a monthly *Mathematics Journal for Secondary Schools*. The contribution to Hungarian mathematical culture of these two institutions – the competition and the magazine – is universally acknowledged.[11]

Promoted by Gyula König in particular, the Eötvös Competition was an annual examination intended to identify students of ability. The best gymnasium students were groomed months in advance, in preparation for the great day when, in a closed room, they faced a series of written questions of increasing degrees of difficulty. Winning it conferred great prestige on both the student and his teachers. Von Kármán (1881–1963), who, in his day, won the prize, said that the toughest questions demanded true creativity and were intended to signal the potential for a mathematical career. A compilation of the problems was published in 1929 by József Kürschák, and later translated, in 1961, as the *Hungarian Problem Book*.[12] Over the years, in addition to von Kármán, the winners of the Eötvös Prize included Lipót Fejér, Gyula König's mathematician son Dénes, Alfred Haar, Edward Teller, Marcel Riesz, Gabor Szego, Laszl Redei, László Kalmár, and John Harsányi.

---

[10] König completed his doctorate at Heidelberg in 1870, working in the area of elliptic functions. The story is told of his dramatic claim in 1904 to have proved that the continuum hypothesis was false. At the International Congress of Mathematicians, at Heidelberg, where the claim was announced, all other conference sessions were cancelled so that König could be given complete attention. The proof was soon found by Ernst Zermelo to contain an error. For König and other mathematicians cited here, see the Web site in the history of mathematics at the University of St. Andrews, Scotland (www-history-mcs.st-andrews.ac.uk/history/Mathematicians.html). Each of the site's biographical entries is a synthesis of several sources, all cited and, in the present case, many of them in Hungarian.

[11] See, for example, Radó (1932) and Hersh and John-Steiner, *op cit.*

[12] See Kürschák (1963).

The mathematics magazine *KöMaL* was founded in 1894 by Gyór school-teacher Daniel Arány, in order "to give a wealth of examples to students and teachers". Each issue contained general mathematical discussion, a set of problems of varying degrees of difficulty, and the readers' most creative or elegant solutions to the questions of the previous issue. Eagerly awaited in the postbox by many Hungarian students, it brought prestige to those who were successful and, like the Eötvös Prize, contributed to the cultivation of a general interest in mathematics amongst the Hungarian young.

As the previous discussion suggests, schoolteachers of mathematics such as Arány, and also László Rátz and Mikhail Fekete of Budapest, played a critical role in Hungary. Whereas mathematical education was of the first order, university positions were few, with the result that many mathematicians of ability found themselves teaching at secondary level or providing private tutorials to Budapest gymnasium students. There, they noticed and groomed young, talented pupils and, in this small community in which everyone knew everyone else, guided them onwards towards their university colleagues. Von Neumann's mathematics teacher at the Lutheran Gymnasium, Lászlo Rátz, played an important mentoring role, stressed by many commentators on this period. He acted as tutor to von Neumann, who also appears to have received private classes from Gabor Szegö and, later, Michael Fekete.[13]

If the Eötvös Prize and *KöMaL* are the first two factors often cited in discussions of the Hungarian mathematics phenomenon, the third was a person: Lipót Fejér, who was probably the most influential figure in the generation following König and Kürschák.[14] Initially with the surname, Weiss, which he later changed for the Hungarian equivalent, Fejér was born in Pécs, the same town as his friend, Max von Neumann. He distinguished himself in his contributions to *KöMaL* and became known to Lászlo Rátz. Winning the Eötvös Prize in 1897, he studied mathematics and physics at the University of Budapest until 1902. Following a doctoral thesis on Fourier series, he taught in that city for three years, spending some time at Göttingen and Paris. Then, following several years on the university faculty in Koloszvár, in 1911, Fejér won an appointment to a chair in Budapest, where he spent the rest of his career. He is remembered for his work on Fourier series, the theory of general trigonometric series, and the theory of functions of a complex variable, as well as for some contributions to theoretical physics and differential equations.

---

[13] See Lax (1990) and also the discussion with Eugene Wigner in Aspray et al (1989).
[14] On Fejér, see Turán (1949b and 1960).

Budapest graph theorist Paul Turán later credited Fejér with the entire creation of a coherent mathematical school. Another wrote that "a whole culture developed around this man. His lectures were considered the experience of a lifetime, but his influence outside the classroom was even more significant".[15] Hersh and John-Steiner recall some of the legends surrounding Fejér. For example, one concerned Poincaré's 1905 visit to Budapest, to accept the Bolyai Prize. Greeted upon his arrival by various ministers and high-ranking officials, Poincaré looked around and asked: "Where is Fejér?" "Who is Fejér?" the ministers replied. "Fejér", said Poincaré, "is the greatest Hungarian mathematician, one of the world's greatest mathematicians". Within a year, Fejér was appointed to the professorship at Koloszvár.[16] Apparently, Fejér would sit in coffee houses frequented by the mathematicians, such as the Erzsébet café in Buda or the Mignon in Pest, regaling his students with stories about mathematics and mathematicians he had known. A regular dinner guest at the von Neumann household, Fejér enjoyed the friendship of creative people of all sorts, including Endré Ady, the revered Hungarian poet. Fejér had a lasting influence on many younger Hungarian mathematicians, beyond von Neumann, including Pólya, Marcel Riesz, Szegô, Erdós, Turán, László Kalmár, and Rozsa Péter. Although Turán intimates that the events of 1919–23, namely the Kun Revolution and Horthy's White Terror, weighed heavily on Fejér – that he was not quite the same man afterwards – Fejér remained prominent until his death in 1959. He continued to enjoy an international reputation as one of the two recognized leaders of the Hungarian school of analysis, the other being his friend and close collaborator, Frigyes Riesz, who presided over the university at Szeged and was another important mentor for von Neumann.

Born in Györ, Riesz studied at Zurich Polytechnic, Budapest, and Göttingen before completing a doctorate in the Hungarian capital. After several years' school-teaching, in 1911, he took Fejér's post at Koloszvár when the latter left for Budapest. Amongst the contributions for which Riesz is remembered are his representation theorem on the general linear functional on the space of continuous functions; his work on the theory of compact linear operators; his reconstruction of the Lebesgue integral

---

[15] G. L. Alexanderson et al (1987) quoted in Hersh and John-Steiner, *op cit*, p. 18. On a later occasion, in 1911, Fejér's candidacy for the aforementioned chair at Budapest was apparently opposed by faculty anti-Semites who were aware that his name had been Weiss. One asked cynically whether Fejér was related to the distinguished university theologian Father Ignatius Fejér, to which Eötvös, then a faculty member, immediately replied: "Yes. Illegitimate son". The appointment went through without difficulty.

[16] Hersh and John-Steiner, *op cit*.

without measure theory; and the famous Riesz-Fisher theorem, which was a central result in the area of abstract Hilbert space and was essential to proving the equivalence between Schrödinger's wave mechanics and Heisenberg's matrix mechanics.[17] Riesz's brother, Marcel, was also a mathematician of repute. Part of the Hungarian diaspora of the period, he made his career in Stockholm, Sweden.

Also in the Szeged group were Alfred Haar and someone about whom we shall have more to say later, mathematical physicist Rudolf Ortvay. In the mid-1920s, they were joined by topologist Bela Kjerekárto and two new assistants, István Lipka and Laszló Kálmár. Together, the Szeged mathematicians formed the János Bolyai Mathematical Institute[18] and established the *Acta Scientiarum Mathematicarum Szeged*, or *Acta Szeged* for short, which published articles in the international languages and quickly became a mathematics journal of international reputation.

Amongst the younger mathematicians closer in age to von Neumann were Dénes König, László Kalmár, and Rozsa Péter. The first of these studied at Budapest and Göttingen, obtaining his doctorate in 1907, then becoming a teacher at his father's institution, the Budapest Technische Hochschule. His work represented an important stream in Hungarian mathematical research – that of discrete mathematics, which includes graph theory, combinatorics, and number theory.[19] König lectured on graph theory and

---

[17] See Frigyes Riesz, "Obituary"; Edgar R. Lorch (1993). Lorch was one of the visitors who Riesz attracted from abroad in the 1930s. A Columbia PhD, he was finishing a postdoctoral stay at Harvard with Marshall Stone when, on health grounds, no less, he turned down a prospectively gruelling position as assistant to von Neumann, by then at the Institute for Advanced Study in Princeton, and went to Szeged instead: "If John von Neumann was the acknowledged genius of modern mathematics, Frederick Riesz was the dean of functional analysts. He was not well known to the world at large, but the cognoscenti had the highest respect for him" (Lorch, *op cit*, p. 222). Lorch and Riesz collaborated on one paper during that year, on a problem in transformations in Hilbert space to which John von Neumann and Marshall Stone had already contributed. Lorch's article also provides a nice portrayal of Szeged in the 1930s. Prior to the First World War, the most important universities in Hungary were at Budapest and in the Transylvanian town of Kolozsvár, home to the Franz Joséf University. Kolozsvár was the country's second city and home of several administrative offices. After Trianon, however, in 1921, when Transylvania was handed over to Rumania and Kolozsvár renamed Cluj, that university found itself without a Hungarian home. The entire faculty moved temporarily to Budapest before being transferred permanently to Szeged, a provincial garrison town of 120,000 in the south of the country, lying less than ten miles from the triple border with the hostile Yugoslavia and Rumania.

[18] Bolyai Institute, "A Short History of the Bolyai Institute" (no date), available at server .math.u-szeged.hu.

[19] On König, see Gallai (1936).

published a foundational book on it in 1936.[20] Kalmár was a student of Kürschák and Fejér, specializing in the field of logic. After a stay at Göttingen, he took a position at Szeged, initially serving as assistant to both Haar and Riesz.

As for Rózsa Péter, she was one of the very few women mathematicians of the period. Born Rózsa Politzer, she began studying chemistry at Loránd Eötvös University in Budapest, but switched to mathematics after attending lectures by Fejér.[21] Like Kalmár, with whom she was close, she graduated in 1927, specializing in number theory, but, as a Jew and a woman, she was particularly handicapped in obtaining a post as secondary school teacher. Depressed by the discovery that some of her theorems had already been proved by foreign mathematicians, Politzer actually abandoned mathematics, concentrating her energies on poetry and translation. It was Kalmár who encouraged her to return to the fold at the beginning of the 1930s, pointing to Gödel's recent results on incompleteness, which Politzer was then apparently able to reach using different methods. This led her to explore, in their own right, the recursive functions that had served as an important tool in Gödel's work, and she began presenting results in 1932, publishing several papers and eventually joining the editorial board of the *Journal of Symbolic Logic* in 1937. Despite changing her name during this decade from Politzer to the more Hungarian and less Jewish-sounding Péter, she remained without a post for a long time, making a living as a private tutor.

Conclusion

Von Neumann came of age in two intermeshed Hungarian communities, those of assimilated Jewry and the country's mathematicians. By the time he was born, the degree of assimilation was so great that the subject of the distinctiveness of his Jewish identity was not particularly important. It gradually became so, however, a significant landmark being the institutionalization of discrimination through legislation, passed in 1920 when von Neumann was a student. The community with which von Neumann felt an arguably clearer affiliation was that of the Hungarian mathematicians, which, through the creation of educational practices and institutions, was deliberately constructed, beginning in the late nineteenth century, and quickly became a defining feature of the country's intellectual and

---

[20] König (1936). In the area of discrete mathematics, König's successors in the next generation were Paul Turán and Paul Erdós.
[21] On Péter, see Morris and Harkleroad (1990).

scientific culture. The thread of Jewishness and social identity runs strongly through the history of these mathematicians, many of whom were of Jewish background and sought assimilation through the Magyarization of family names.

I have described this background because of its importance to this story up to the 1940s. In the 1920s, because of the existence of limited facilities in Hungary and the growing importance there of identity politics, von Neumann came of age as part of a generation of students-in-exile, studying abroad in Switzerland, Germany, and elsewhere. This, as we shall next see, allowed him to shine in the German mathematical and scientific community, where he continued his early work in mathematics, turned to mathematical physics, and became interested in the mathematics of games. By the late 1930s, identity politics was firmly an issue in both Germany and Hungary, and von Neumann, as we shall see several chapters hence, observed developments from Princeton, from his position of "double-exile", so to speak. It was this stark confrontation with the dissolution of the society that had shaped him that prompted him to return to games in search of a new mathematics of social structure.

THREE

From Budapest to Göttingen

An Apprenticeship in Modern Mathematics

Introduction

Amongst the Hungarian mathematicians, Jansci von Neumann stood out.[1] From a young age, there were stories of strange abilities: dividing two eight-digit numbers in his head at six; proficient in calculus at eight; reading Borel's *Théorie des Fonctions* at twelve. Stories abound about a photographic memory and an ability to apparently recall complete novels and pages of the telephone directory. He also accumulated an encyclopaedic knowledge of history, in time being able to recall the most minute details of the Peloponnesian Wars, the trial of Joan of Arc, and Byzantine history. Many years later in the U.S., when travelling south from Princeton, New Jersey, to Duke University, North Carolina, he astounded his fellow travellers, including mathematicians Albert Tucker and Stan Ulam, with his

---

[1] Existing biographical treatments include articles of appreciation, such as Goldstine and Wigner (1957); Halmos (1973); Morgenstern (1958); Ulam (1958) and (1980), which is a fragment of a draft biography; and books dealing with particular aspects of his scientific work, including Heims (1980), Nagy (1987), Goldstine (1972), Aspray (1990), Dore, Goodwin, and Chakravarty (eds.) (1989), and Poundstone (1992). There also are Vonneumann (1987), by his brother, Nicholas, and Macrae's (1992) book. The latter began in 1980 as a biography to be written jointly by Stan Ulam, von Neumann's Polish mathematician friend and collaborator on the atomic bomb, and Stephen White, Vice-President of the Sloan Foundation. Ulam was to be responsible for the scientific sections; White for the more biographical and social parts. The two soon fell out, however, with Ulam disagreeing over the excessively popular level at which White intended to pitch the book (See Letter, Ulam to Nicholas Vonneumann, May 27, 1980, Ulam–White correspondence, Stan Ulam Papers, American Philosophical Society Library, hereafter SUAP). Macrae, alone, took over the project from White. Finally, as the present manuscript goes to press, there has appeared Israel and Gasca (2009), a much-anticipated book on von Neumann, one of the authors of which also co-wrote the excellent Ingrao and Israel (1991).

recollection of the most precise details of Civil War battles fought at sites along the route.[2]

Although Max von Neumann would have preferred his son to become a well-paid financier rather than a mathematician, he was open to the encouragements of Fejér and Ortvay and finally acquiesced, letting von Neumann pursue his interests and financing his studies abroad. Von Neumann, in return, became the shining, often absent, star of the Fejér circle in Budapest. As a Gymnasium student, he caught the attention of Laszló Rátz and was tutored in university-level mathematics by Mikhail Fekete. By the time he enrolled at the University of Budapest in 1921, he had already written a paper with Fekete and, according to Ulam, was essentially recognized as a mathematician.[3]

At the University, he worked on set theory under the guidance of Fejér, but he did so largely *in absentia*, reappearing there only at the end of each term to take exams. He was part of the Jewish exodus during these years, with 1921–23 being spent at the University of Berlin, where he worked on set theory with Erhard Schmidt and took courses in physics, including statistical mechanics from Einstein. He also attended Zurich's *Technische Hochschule* in 1925, where he studied Hilbert's theory of consistency with Hermann Weyl, and took a parallel degree in chemical engineering. In 1926, he was awarded his doctorate in mathematics at Budapest, *summa cum laude*, and he applied to the Rockefeller-financed International Education Board for a six-month fellowship with David Hilbert at Göttingen, to begin that autumn.[4]

In what follows, we will consider the context in which von Neumann found himself when he arrived at Göttingen in 1926. There was Hilbert's promotion of the axiomatic method as an approach both within particular fields of mathematics and in the mathematical analysis of scientific fields such as quantum mechanics. Independently, there was Hilbert's particular position in the debate on the foundations of mathematics, Formalism, the

---

[2] See Ulam 1958 (Bulletin), p. 6.

[3] According to Ulam, the paper was concerned with a generalization of Fejér's theorem on the location of the roots of a certain kind of polynomial (see Ulam 1958, p. 8).

[4] See letter, von Neumann to International Education Board, Apr. 11, 1926, International Education Board [I.E.B.] Archives, Series 1, Subseries 3, Box 55, Folder 896 John L. Newmann [sic] 1926–1938. The letter mentions Hungarian, German, English, French, and Italian as languages spoken, and it was accompanied by reference letters from Richard Courant, Weyl, and Hilbert, three of Germany's leading mathematicians. On the activities of the I.E.B., see Siegmund-Schultze (1994).

issue that initially drew von Neumann to Hilbert's side. By considering von Neumann's involvement in both the foundations debate and quantum mechanics, we will draw a portrait of him as mathematician. That will allow us to then consider in context his foray into the mathematics of games.

David Hilbert

Over the course of the previous two decades, Hilbert had established Göttingen as one of Germany's two centres of mathematical research and learning, the other being Berlin. He was associated with the axiomatic method in different fields of mathematics and science, an approach first introduced fully in his 1899 book, *Grundlagen der Geometrie*, and steadily promoted and developed by him thereafter.

Through his meticulous exploration of the Hilbert papers at the Göttingen archives, historian of mathematics Leo Corry has provided a very useful archaeology of the axiomatic method as exploited by Hilbert over the first quarter of the century (see Corry 1999; 2000; 2004). Beginning with *Geometrie* and going on to physics and other fields, Hilbert saw the axiomatic approach as providing clarity and rigour. The ideal he had in mind was to establish, for the field under scrutiny, the set of basic postulates that were both necessary and sufficient to generate the laws of the domain. By ensuring that such axioms were independent, there was no redundancy in them. By ensuring their consistency, they could not be a source of contradictory results. If there is one message upon which Corry abundantly insists, it is that, for Hilbert, the underlying inspiration for the choice of axioms in a field was empirical: axioms were to be initially chosen according to their intuitive appeal and plausibility. Hilbert had little interest in axiom systems developed merely for the purposes of examining their logical qualities; he was uninterested in the merely syntactical exploration of postulate systems. The perception that Hilbert viewed mathematics – as distinct from metamathematics – as a syntactical formal 'game', involving the mere manipulation of signs, devoid of contact with the world, bears little relationship to the mathematician's vision, as presented by Corry.

The latter also makes clear Hilbert's scope and ambition. Beyond geometry and a range of areas in physics, including mechanics, thermodynamics, the kinetic theory of gases, relativity, and quantum mechanics, Hilbert saw the axiomatic approach as being applicable to very disparate fields. For example, in his lectures at Göttingen in 1905, he outlined how the axiomatic method might be used in the elaboration of the field of insurance mathematics: for purposes of insurance, a person could be characterized by a function $p(x, y)$, defined for $y > x$, which expressed the probability that a person of age x would reach age y. Such a function, which expressed everything concerning an individual that was relevant to insurance, had to satisfy the following axiom: the probabilities $p(x, y)$, $p(x', y')$ associated with two different individuals are independent for all pairs x, y, x', y' of positive numbers. From there, Hilbert suggested, one could develop a life function, expressing the number of living people of age x, in relationship to the aforementioned probabilities. Similarly, in psychophysics, Hilbert turned to existing work on the psychological theory of colour perception by German astronomer Egon Ritter von Oppolzer, which was based on the Weber-Fechner law relating the magnitude of stimulus and sensation. Hilbert drew out the axioms that he saw as being implicit in Oppolzer's treatment.[5]

Hilbert's influence was very great: "One cannot exaggerate the significance of the influence exerted by Hilbert's thought and personality of all who came out of this institution", writes Corry (2004, p. 4). There emerged a definite Hilbert "style", characterized by a belief in the axiomatic approach; a concern for the empirical basis of any axioms selected; an emphasis, in principle if not always in practice, upon the completeness and consistency of the sets of axioms elaborated for the field under study; and a belief that this approach could be applied to both the physical and nonphysical sciences. The whole was informed by a faith in a pre-established harmony between mathematics and the world, which the mathematician potentially could uncover. Although, this, in itself, was hardly a new idea, what characterized Hilbert's style was the manner in which it sought to balance the empirical and the abstract emphases. The concern for empirical fidelity was constant, but it was supplemented by a distinctly "modern" belief in the autonomous power of abstract mathematics to reveal truths about the world.[6] This alternation between the empirical and abstract emphases was an important part of what von Neumann brought to social science, although not so much in his first intervention in game theory as in the late 1930s at Princeton, when he felt compelled to plumb the mathematics for social insights.

Mathematics in Doubt
The axiomatic style just described promoted a particular way of approaching a field of inquiry, be it an area of mathematics such as geometry; a field

---

[5] On Hilbert's treatment of insurance mathematics and psychophysics, see Corry (2004), pp. 171–72 and pp. 175–78, respectively.
[6] See Ingrao and Israel (1991), pp. 183ff.

in physics such as quantum mechanics, where von Neumann was active; or some less "rigorous" areas such as psychophysics or insurance. It concerned the use of mathematics as both tool of clarification and engine of discovery in the world, and should not be confused, as Roy Weintraub points out, with Hilbert's particular stance in the debate on *metamathematics*, namely Formalism. It was the latter that initially drew von Neumann to Hilbert, and an important part of the story of von Neumann in the period from the mid-1920s to 1931 involves his embracing and then abandoning Hilbert's position in the foundations debate, while remaining faithful throughout, and beyond, to the Hilbertian, axiomatic style in mathematics proper and science in general.[7]

In his fellowship application, he wrote of his wish to conduct "Researches over (sic) the bases of mathematics and of the general theory of sets; specially Hilbert's theory of uncontradictoriness.... [investigations which] have the purpose of clearing up the nature of antinomies of the general theory of sets, and thereby to securely establish the classical foundations of mathematics. Such researches render it possible to explain critically the doubts which have arisen in mathematics".[8]

The doubts referred to emerged largely with Cantor's (1845–1918) work on infinite sets at the turn of the century. Beginning in 1882, in response to several longstanding paradoxes surrounding the concept of infinity, such as Zeno's account of Achilles and the tortoise, Cantor offered what was essentially a revision of the traditional notion of counting. He developed the concept of a transfinite number to describe different kinds of infinite sets. The key concepts were countable, or denumerable, infinity, which corresponds to the set of natural numbers, for example; and noncountable infinity, corresponding to the cardinality of the reals. Related to this was the so-called continuum hypothesis, or Cantor's suggestion that there existed no transfinite number the cardinality of which lay between these two numbers. In the course of this work, Cantor revealed many strange features of infinite sets, such as the fact that the cardinality of the set of natural numbers is the same as that of one of its proper subsets, the even numbers. Further antinomies were pointed out by Bertrand Russell.[9]

---

[7] This is explained by von Neumann himself in his 1947 essay, "The Mathematician".

[8] Fellowship application, von Neumann, Apr. 11, 1926, I.E.B., Series 1, Subseries 3, Box 55, Folder 896 John L. Newmann (sic) 1926–1938.

[9] These typically sprang from the claims to inclusivity or exclusivity inherent in set theory. One that became popular because of Russell is the following: In a certain village, the barber shaves *all* those men, and only those men, who do not shave themselves. Who shaves the barber? Any answer given is contradictory. See Kramer (1982), pp. 594ff.

Another controversy centered on what became known as the axiom of choice, which was made explicit and endorsed by Zermelo in the first decade of the century. The axiom said that it was possible to do something that was implicitly involved in a great many existing mathematical proofs, namely, select an element from each of an infinite number of sets, even if it was impossible to specify the means by which this might be done "in practice" – that is, in finite time. The axiom met with a storm of protest from Borel and others, in the *Mathematische Annalen* of 1904 (see Kline 1980, p. 210). Borel argued that because such a selection process could not be made precise, the axiom lay outside the pale of mathematics. Brouwer objected because he rejected the use of infinite sets *tout court*. Those rising to Zermelo's defence, such as Hadamard, did so on the pragmatic grounds that, even if a selection procedure could not be made explicit, the axiom was reasonable insofar as it made possible the development of mathematics. Hilbert, too, saw it as an unobjectionable requirement, basic to mathematical inference.

These debates were vigorous and, at times, hostile, with mathematicians responding in different ways, often showing philosophical beliefs to be inextricably linked to differences in attitude and taste. There were four broad stances on foundational questions, which, at the cost of some simplification, can be termed Intuitionism, Logicism, Formalism, and Axiomatic Set Theory.

Of Logicism, which stemmed from the work of Leibniz, Dedekind, and Frege, and received its definitive expression in the *Principia Mathematica* of Russell and Whitehead (1910–13), suffice it to say that it sought to show that all of existing mathematics could be constructed from the simple bases of logic. Beginning with the basic logical concepts such as proposition, propositional function, negation, conjunction, disjunction, and implication, it added the axioms of logic. Thus, for example: If q is true, then p or q is true; The assertion of p and the assertion $p \subset q$ permit the assertion of q. From these terms and axioms, the logicists deduced the theorems of logic, which include the principle of the excluded middle. They then proceeded to the logic of propositional functions, such as "x is a Social Democrat". With this, one moves to the discussion of sets, insofar as the preceding propositional function constitutes a definition of the set of Social Democrats. To rule out the paradoxes created by the use of sets that are members of themselves, Russell and Whitehead introduced the theory of types, with individual objects being of type 0; a set of objects being of type 1; a set of sets being of type 2, and so on. The theory of types dictates that if one says a belongs to b, then b must be of higher type than a. By this means, one no longer can speak of sets that are members of themselves, and the paradoxes associated with same are avoided. The statement "All rules have exceptions" no longer can be applied to itself.[10]

The theory of types leads to complex difficulties when one tries to build mathematics in accordance with it.[11] To overcome these, Russell and Whitehead introduced various supplementary axioms, all of which provoked the criticism of other mathematicians during the interwar period, including Poincaré and Weyl. Faced with such criticism, Russell admitted the arbitrariness of certain axioms, but indicated their necessity in building mathematics out of logic. However, the criticism was persistent. If mathematics was reducible to simply logic, then how did new mathematical knowledge emerge? What ensured that the concepts produced by the creative imagination would always fit in the grooves dictated by the postulates of logic? How could this reductionist view of mathematics be reconciled with the latter's seemingly uncanny ability to describe such an enormous and varied range of physical experience? In time, Russell was moved by the steady criticism. By 1926, the year von Neumann moved to Göttingen, in the second edition of *Principia Mathematica*, the doubtful axioms were being used only when necessary in particular proofs, not as general principles. By 1937, Russell rejected them entirely as logical truths, effectively admitting that in its original form, the Logicist programme had failed.

Of Brouwer's Intuitionism, we shall have more to say in connection with Karl Menger in Vienna. For the moment, suffice it to say that the Intuitionism developed by Brouwer, and also Heyting and Weyl, sought to retain only those parts of mathematics based on the "intuitively" plausible natural numbers, and to reject the employment of "nonconstructive" proofs in infinite set theory – that is, proofs that establish existence by contradiction rather than by actually constructing the object the existence of which is being proved. This was based on their rejection of the law of the excluded middle as applied to infinite sets.

Hilbert's Formalism was developed in response to the antinomies in set theory and the intuitionist critique. In 1900, the year after the publication of

---

[10] As we shall see later, it was this discussion that stimulated Oskar Morgenstern in the 1930s in his criticism of the logical paradoxes inherent in the treatment of knowledge and foresight in economic theory – and it was in terms of the theory of types of Russell and Whitehead, *inter alia*, that he groped for a better conception of the economic agent.

[11] The irrational numbers are defined in terms of the rationals, and so are of higher type. The rationals, in turn, are of higher type than the natural numbers, in terms of which they are defined. One cannot, therefore, assert a theorem about the entire set of reals because what applies to one type does not automatically apply to another: one must make assertions about the different types separately.

his *Foundations of Geometry*, with its axiomatic treatment of the Euclidean system, Hilbert gave an important speech to the International Congress in Paris, in which he laid out a series of twenty-three particular problems that mathematicians had still not resolved, and which required urgent attention. In addition to pointing to conjectures in particular areas of mathematics that still required proof, Hilbert emphasised general foundational aspects such as consistency and completeness, stressing "the conviction (which every mathematician shares, but which no one has as yet supported by a proof) that every definite mathematical problem must necessarily be susceptible of an exact settlement, either in the form of an actual answer to the question asked, or by the proof of the impossibility of its solution and therewith the necessary failure of all attempts" (quoted in Reid 1970, p. 82). In 1903, Herman Weyl took a position at Göttingen, and he was followed by Zermelo. At the same time, independently, Russell brought up the antinomies of set theory. In 1904, at the International Congress in Heidelberg, in response to the growing unease provoked by the foundational paradoxes, Hilbert proposed that "proof itself should be made an object of mathematical investigation".

In the following decade, Hilbert was taken up with, not foundations, but the need to provide orderly axiomatic treatments of very diverse physical theories. Developments at this time included Hertz's proving the existence of electromagnetic waves; Roentgen, X-rays; the Curies, radioactivity; Thomson, the electron; and Planck's developing quantum theory. Hilbert turned to the mathematics of space of infinitely many dimensions, or Hilbert space, and developed the theory of integral equations, applying it to the kinetic theory of gases. His colleague, Minkowski, worked on the geometrization of Einstein's special theory of relativity.

By the end of World War I, however, with the foundations debate looming large in the eye of the mathematical public, Hilbert was increasingly drawn into the controversy. By 1922, the debate had so risen in pitch that he turned away from physics – he assigned an assistant, Lothar Nordheim, to survey the activity of Heisenberg, Einstein, and Bohr – and concentrated on responding to Brouwer in particular.

By the early 1920s, Hilbert had declared war on Brouwer. According to Hilbert, if accepted, intuitionist strictures would require sacrificing a large part of extant mathematics, such as pure existence proofs and Cantor's theory of sets, a prospect that appalled him. It was as if, to adapt the architectural analogy so favoured by Hilbert, Brouwer were rummaging darkly in the cellar, worrying about the quality of the edifice's foundations, while the Göttingen mathematician pointed to the structure above, to the beautiful plays of light on the upper floors.

Hilbert's response to Intuitionism was to shift the emphasis. Instead of focusing upon debates concerning the acceptability of particular proof procedures, such as nonconstructive methods, he suggested focusing on the demonstration of consistency, *respecting Intuitionist or constructive methods only*. As Weyl puts it, with Hilbert, "the question of truth [was] then shifted into the question of consistency" (1944, quoted in Kramer (1982), p. 630).

Hilbert laid out his programme in a talk given in 1925 to the Westphalian Mathematical Society.[12] He stressed the distinction between infinity as an empirical reality and infinity as a mathematical object, or the use of symbols signifying infinite sets in mathematical proof procedures. Developments in science, he said, increasingly pointed towards the unlikelihood of the actual existence of infinite sets of anything. Contrary to the old adage that "*natura non facit saltus*", nature did, in fact, seem to make jumps, with the existence of the electron and the quantum of energy both helping displace earlier beliefs in the infinitely small and in continuity. In mathematics, however, the concept of infinity had played a supremely important role in the generation of new fields, which, for Hilbert, was synonymous with the success of mathematics. Indeed, Cantor's theory of sets itself was one of the most remarkable, he said, "the finest product of mathematical genius and one of the supreme achievements of purely intellectual human activity" (p. 188). There had to be some way, therefore, of preserving the use of infinite sets, yet avoiding such paradoxes as had been revealed earlier by Russell. Hilbert's solution was to appeal not to any intuitive notion of the validity of mathematical concepts of the infinite, but to show that the use of such concepts was at least consistent, or did not give rise to contradictions. Further, in order to avoid circularity, and in order to deflate Brouwer's admonition concerning nonconstructive methods, such a consistency proof should rely on finitary methods only. Thus, reference to infinity in classical mathematics would at least have been justified, in the sense of being shown to be consistent, on uncontroversial finitistic grounds. Hilbert's key assistants in the elaboration of this project as of 1925 were his students Wilhelm Ackermann, Paul Bernays, and John von Neumann.

Also associated with Göttingen was the final stance on foundational questions, the axiomatic set theory of Hilbert's colleague, Zermelo, and another German mathematician, Jacob Fraenkel. Like the logicists, they were interested in constructing a theory of sets that would rule out its contradictions and antinomies. To do this, they would have to describe sets and their properties in terms more rigorous than Cantor's. However, rather

---

[12] See Hilbert (1926).

than approach sets through the exceedingly complicated logical apparatus proposed by Russell and the logicists, Zermelo and Fraenkel went straight to the axioms of set theory. Zermelo constructed the first axiomatisation of the theory in 1908 and it was improved by Fraenkel in 1922. Their system allowed for infinite sets, for the axiom of choice, but placed various restrictions on the properties, so that an all-inclusive set could not be considered, and avoided paradoxes to the extent that nobody was able to deduce any within the system.

By 1926, when he went to Göttingen, von Neumann was well steeped in foundations and Zermelo–Fraenkel set theory, and had already published six related papers. Of an early set theory paper by the sixteen-year-old von Neumann, Fraenkel himself reports:

Around 1922–23, being then professor at Marburg University, I received from Professor Erhard Schmidt, Berlin ... a long manuscript of an author unknown to me, Johann von Neumann, with the title *Die Axiomatisierung der Mengenlehre*, this being his eventual doctor dissertation which appeared in the *Zeitschrift* only in 1928 ... I was asked to express my view since it seemed incomprehensible. I don't maintain that I understood everything, but enough to see that this was an outstanding work, and to recognize *ex ungue leonem* [from the claw, the lion]. While answering in this sense, I invited the young scholar to visit me in Marburg, and discussed things with him, strongly advising him to prepare the ground for the understanding of so technical an essay by a more informal essay which should stress the new access to the problem and its fundamental consequences. He wrote such an essay under the title *Eine Axiomatisierung der Mengerlehre* and I published it in 1925 ... [13]

In that 1925 paper ("Eine Axiomatisierung der Mengenlehre", *J. Reine Angew. Math.*, Vol. 154, pp. 219–40), von Neumann took a position in the foundations debate. He objected to Brouwer's and Weyl's willingness to sacrifice much of mathematics and set theory, and to the logicist attempt to build mathematics on the questionable axiom of reducibility. He was more sympathetic to the axiomatic approach of Zermelo and Fraenkel, which, he said, replaced vagueness with rigour, even if the axioms chosen were somewhat arbitrary. His own system, later laid out formally in the 1928 paper ("Die Axiomatisierung der Mengenlehre", *Math. Zeit.*, Vol. 27, pp. 669–752), was, with its single page of axioms, the most succinct developed to date, and it formed the basis for the system later developed by Gödel and Bernays towards the end of the 1930s. [14]

---

[13] Letter, Fraenkel to Stan Ulam, undated, quoted in Ulam (1958), p. 10, n. 3.
[14] As Ulam points out, what was earlier "An Axiomatisation ..." had now become "*The* Axiomatisation ...".

Another paper on set theory ("Über die Definition durch..." *Math. Ann.*), in 1928, proves the possibility of definition by transfinite induction. Von Neumann demonstrated the significance of the axioms for the elimination of the paradoxes of set theory, proving that a set does not lead to contradictions if and only if its cardinality is not the same as the cardinality of all things, and that this, furthermore, implied the axiom of choice (see Goldstine and Wigner 1957). A von Neumann paper on Hilbert's proof theory ("Zur Hilbertschen Beweistheorie", *Math. Zeit.*, 1927) improved a previous proof of freedom from contradiction by Ackermann, giving a finitary proof for a subsystem of classical analysis. Here, von Neumann conjectured that the same method might be used to prove the consistency of all analysis, a conjecture famously proved false by Gödel in Vienna a few years later. Gödel's work, when presented at the famous conference in Königsberg in 1931, was viewed as dealing a death blow to Hilbert's formalist hopes.

Discontinuity in Physics

As noted, by 1926, Hilbert's colleagues were contending with a proliferation of developments in physics. Heisenberg's new theory of quantum mechanics had been shown by Max Born to be explicable in terms of matrix methods. Schrödinger, at Zurich, had constructed a wave mechanics which, although it led to the same results as Heisenberg, proceeded from an entirely different basis. The two were soon mathematically reconciled by Courant, using Hilbert's earlier work on integral equations and Hilbert space. As was clear to even lay observers, including, as we shall see, a young Oskar Morgenstern then reading Whitehead at Harvard, these conceptual changes in physics resonated deeply in mathematics. In physics, the theory of relativity cast doubt upon many concepts that were central to classical mechanics, such as absolute space and time, and simultaneity, and quantum theory threw mechanistic determinism into question, by demonstrating the impossibility of knowing simultaneously both the position and velocity of a particle, necessary to predicting its future evolution. The views of classical mathematicians, such as Poincaré and Volterra, who believed in the underlying continuity of physical events and the possibility of their representation by the infinitesimal calculus and the theory of differential equations, were being questioned. The first phase of quantum physics had indicated the "discontinuous character of all micro-events" with energy states of complex structures such as atoms and molecules being seen to consist of a set of discrete values, and to "jump" from one state to another. The second phase, centered on Born's interpretation of the wave equation, suggested that the basic laws of physics were probabilistic laws, allowing for only statistical predictions. Classical determinism, which had been central to Western scientific culture for more than a century, had been shattered.

Working initially with Hilbert's assistant, Nordheim, von Neumann entered mathematical physics in this, the second phase, beginning with the axiomatisation of Heisenberg's work, and elaborating a mathematical basis for quantum mechanics in Hilbert space. His seminal work in mathematical physics appeared in the form of four articles in 1927, three in 1929, and their condensation into a 1932 book, *Mathematische Grundlagen der Quantenmechanik* (*Mathematical Foundations of Quantum Mechanics*) (see 1927a, b, c, d; 1929a, b, c; 1955a [1932]).[15]

Unlike another commentator, Van Hove, Stan Ulam emphasizes the importance of von Neumann's 1927 paper with Nordheim and Hilbert (1927d). Although the axiomatisation was incomplete and was improved in (1927b), the paper's two aims were, first, to cast mathematical physics in terms of probability relationships rather than the functional relationships of classical mechanics, and, second, to maintain a Hilbertian separation of the mathematical formalism from its physical interpretation, thereby freeing up the mathematics and restoring, as Ulam says, its generative role on a conceptual level (*ibid*, p. 20).

In another paper the same year (1927c), he pushed this probabilistic interpretation in physics further, introducing an idea that remained important in the literature – that of a statistical matrix describing an ensemble of systems of different quantum states (see Feyerabend 1955). He also broached the question of how to construct a mathematical formalism that would provide an adequate theoretical description of the observation process in quantum mechanics, in which the relationship between the observer and the subatomic phenomenon being observed took on special importance. When a measuring instrument was coupled to a very small particle, the very act of measuring affected the object, the characteristics of which were being measured, thereby introducing a certain ambiguity or uncertainty into the process. One could not be sure that the atom coupled with the instrument behaved similarly to the atom on its own. Here, the ambiguity took the particular form of the uncertainty principle, formulated by Heisenberg, which said that, as a matter of principle, not just of empirical fact, one could not know both the position and the momentum of the atom. The surer one was about one, the less sure one could be about the other. At the time, the leading interpretation of quantum theory was that represented by

[15] See Feyerabend (1955), Heims (1980), Ulam (1958), Van Hove (1958), Pinch (1977).

the complementarity principle of Niels Bohr and the Copenhagen School. Bohr attributed this feature of complementarity to other pairs of variables in quantum physics: the energy of an atom and the time at which it has that energy; the wavelike and particlelike properties of the electron or of light. He even expected that the principle of complementarity was some kind of universal, extending to other realms of nature, such as the biological and psychological. As Steve Heims points out, the theory also assumed that the measurements read on the measuring instrument were "objective public facts", independent of the particular person doing the reading. Lastly, it assumed that the observer had no objective view of the isolated atom itself but saw only the combination of atom plus instrument.

In the paper in question, von Neumann gave a formulation in terms of an algebra of operators in Hilbert space, and a philosophical interpretation that was different from Bohr's. Von Neumann's formalism divided the world into the system under observation, the measuring instrument, and the observer. The last remained outside the theory, and von Neumann pointed out that how one chose to describe the observer – where one draws the line between consciousness, eyes, nervous system, brain, and equipment external to the body – was arbitrary. In his system, the observer could reduce a superposition – that is to say, change the state of the system merely by observing it – but the question of intersubjective agreement – that is, whether two observing subjects would reach similar observational conclusions, was left unresolved. We had only the single knowing subject and the world that he alone observes and affects. Unlike previous work, von Neumann's formalism was fully axiomatic and logically rigorous, given the particular assumption that measurement was made infinitely fast, rather than in finite time. This strong assumption von Neumann acknowledges to be incorrect, but he argues that it is essential to preserving the formalism and should not matter in the end. The philosophical details, therefore, are not spelled out. The emphasis is on the mathematical achievement, with the implicit idea that perfecting the formalism is more conducive to progress than is attending to the epistemological details. As Heims correctly puts it, the "algebra interested von Neumann, but not the metaphysics" (*ibid*, p. 133).

Von Neumann's 1932 book assembled and expanded these themes. One issue he emphasised was that of indeterminacy, the so-called "hidden variables" question. Following the work of Heisenberg, Schrödinger, Dirac, and Born, there was common acceptance of what became known as the Born interpretation – namely, that quantum states were statistical in nature. The question was then raised whether there might be a set of "hidden variables" operating below the probabilistic surface that determine the state of the system in a classically causal manner (see Pinch 1977, pp. 184–85; Feyerabend 1957–58). Von Neumann showed in his book that such an interpretation was impossible. He defined the hidden variables in terms of ensembles that were dispersion-free (that is, without statistical spread), and postulated five axioms from which, via a matrix, the statistical results of quantum theory could be derived. He then showed that the "hidden variables" could not exist because it was impossible to derive dispersion-free ensembles that could duplicate the statistical results of the quantum theory: "It is therefore not, as is often assumed, a question of a re-interpretation of quantum mechanics, – the present system of quantum mechanics would have to be objectively false, in order that another description of the elementary processes than the statistical ones be possible" (von Neumann quoted in Pinch, *op cit*, p. 185).

Von Neumann's proof of the impossibility of a hidden variables explanation held sway amongst physicists until challenged by Böhm in the early 1950s. It was another step in the break with classical determinism, further affirmation that subatomic reality was inherently statistical.[16]

Conclusion

In 1926, the 23-year-old von Neumann arrived at Göttingen, with several papers already completed in set theory and foundations, and joined Hilbert in his formalist project to prove the consistency of mathematics. In time, he became involved in providing foundations for quantum mechanics, work that was in keeping with the axiomatic approach developed by Hilbert, beginning at the turn of the century. At Göttingen, as we shall now see, von Neumann also found a ready audience for another of his interests, the mathematics of chess and games.

---

[16] See Böhm (1952). Pinch, *op cit*, describes how many readers, even specialists, could not understand the von Neumann proof. Its authority lay not so much in its mathematical validity as in his reputation as a powerful mathematician. It was this socially constructed authority, more than the legitimacy of the mathematics per se, that made for the considerable resistance encountered by Böhm when he challenged the proof in 1952.

FOUR

"The Futile Search for the Perfect Formula"

Von Neumann's Minimax Theorem

The Infinite Chessboard

Von Neumann's Hungarian teachers had been long interested in the mathematics of games. In 1905, in the columns of *KöMaL*, the mathematics magazine, a short paper by one Jószef Weisz entitled "On the Determination of Game Differences", dealt with a game that was not one of pure chance. Throughout the 1920s, *KöMaL* founder Daniel Arány published papers examining how the probabilities of winning in games of pure chance varied with the number of players.[1] The Eötvös Competition of 1926, conceived by Dénes König, contained a question concerning the solution to a system of two equations, the answer to which was equivalent to proving that, on an infinite chessboard, any square can be reached by a knight via a sequence of appropriate moves.[2] Like many of his colleagues, von Neumann was a chessplayer. Indeed, one of the first things he and his friend Willy Fellner did upon moving to Zurich in 1925 was to join the *Schachgesellschaft Zurich*, one of the oldest chess clubs in the world and a landmark in international chess circles.[3] It therefore is easy to understand why, in the mid-1920s, König, Lászlo Kalmár, and von Neumann took an interest in Zermelo's set-theoretic analysis of chess.[4]

Citing discussions with von Neumann, both König and Kalmár sought to refine Zermelo's 1913 paper. In his paper, "On a Method of Conclusion from

---

[1] See Weisz (1905); Arány, (1924), (1927) (1928) (1929a, b) (1933a, b). I thank László Filep for drawing the Arány and Weisz papers to my attention during my visit to the University of Nyrieghaza.
[2] See Kürschák (1963), pp. 104–06.
[3] Communication with the author from Mr. Richard Forster of the *Schachgesellschaft Zurich*, June 30, 2007.
[4] In 1926, König was aged 42, Kalmár 21, and von Neumann 23.

the Finite to the Infinite", König (1927) followed von Neumann's suggestion to apply to chess a lemma from set theory in order to prove the conjecture that the number of moves within which a player in a winning position can force a win is finite. In order to do this, König invokes the use of an infinite board, but with the usual thirty-two pieces. He also addresses two respects in which Zermelo's earlier proof was incomplete, the most important of which was that it had not been proved that a player in a winning position was *always* able to force a win in a number of moves less than the number of positions in the game. In his proof of this, König again cites discussions with von Neumann. In his paper, "On the Theory of Abstract Games", Kalmár (1928/29) also cites discussions with von Neumann, and, generalizing the work of Zermelo and König, shows that if it is possible in a game to force a win, then this can be done without the recurrence of any position.

It was at this time that von Neumann prepared an altogether more ambitious treatment of the two-person, zero-sum game, which he presented at Göttingen in December 1926, within a few months of his arriving there. Although von Neumann's paper obviously owed something to the Zermelo-type chess discussions with his Hungarian colleagues, it is more easily related, conceptually, to the approaches of Lasker, discussed earlier, and French mathematician Émile Borel.[5]

In Praise of Gambling

By the mid-1920s, Émile Borel was in his mid-fifties and sat at the pinnacle of French mathematics, holding a chair at the Sorbonne in probability, an area, incidentally, that was less well-developed in Hungary. His career path had provided a model of French educational achievement.[6] Following an 1894 doctoral thesis on the theory of functions, Borel had made several important contributions, including work on the theories of measure and of divergent series and an elementary proof of Picard's Theorem, which mathematicians had apparently been seeking for more than seventeen years. His work in this early period culminated in the beginning of a series on the theory of functions, which he edited and to which he himself contributed five volumes. Under his directorship, some fifty volumes of these "Borel Tracts" subsequently appeared.

During the first decade of the twentieth century, his interests shifted towards probability theory. He also became something of a popularizer of

5 Von Neumann (1928b).
6 See Collingwood (1959), Fréchet (1965).

mathematics and science, in 1906 founding the *Revue du Mois*, a magazine
to which he contributed articles of scientific, philosophical, and sociological
interest. He also edited a series of popular books, including one about flight,
l'Aviation, and another about the role of chance in everyday life, *le Hasard*.
After World War I, he entered public life, becoming a member of Parliament
for twelve years, all the while continuing to write in mathematics. In 1921,
he began to edit and contribute to the monumental series of monographs,
Traité du Calcul des Probabilités et de ses Applications.[7]

For much of his career, Borel's home in Paris was the site of an important
salon, where a group of French scientists, intellectuals, and public figures
gathered. They included physicists Jan Perrin and Pierre and Marie Curie,
writer Charles Péguy, politicians Paul Painlevé and Léon Blum, and poet
Paul Valéry. Borel also knew psychologist Alfred Binet and was familiar
with some of his work. Indeed, one of the first things he did in the newly
founded *Revue du Mois* was to challenge a study by the psychologist that
claimed to show that intelligence was correlated with the quality of subjects'
handwriting.[8]

Familiarity with Binet notwithstanding, Borel's work in the mathematics
of games did not originate in an interest in chess. Indeed, in this matter,
he appears to have been influenced by Henri Poincaré's claim that chess
was not a proper mathematical object, because it could be played only on
a chessboard of $8 \times 8 = 64$ squares. In chess, said Poincaré, the value 8
was essential, with no possibility of generalization to a board of n^2 squares.
Even on König's infinite chessboard, so to speak, there were still only sixteen
pieces per team. On the other hand, said Borel, by Poincaré's criterion, card
games were indeed mathematical objects because they were played with $4n$
cards, with n typically varying between eight and thirteen but there being
nothing in principle preventing n from assuming any integer value.[9] It was
thus as a player of bridge and card games that Borel the probabilist began to
analyse games of strategy, and this worldly experience as a player coloured
his view of the power of mathematics in this domain.

---

[7] Émile Borel and Paul Painlevé (1910); Borel (1914); Borel et al (1938).
[8] See Binet (2004, orig. 1906). Borel apparently countered Binet by conducting a similar
     experiment with the same writings, but in typewritten form, to conclude that it was the
     content, not the graphological quality, that was correlated with intelligence. See Borel
     (1908), which discusses another experiment by Binet, designed to investigate how photo-
     graphic views of children's hands affected surmises regarding their gender and intelligence.
     This article was published in Binet's own journal, *L'Année Psychologique*.
[9] See Borel et al (1938), p. 39.

Whether or not Borel read Emanuel Lasker's speculations on the science of struggle, there are striking similarities in certain places. For one, if Lasker had suggested in *Kampf* that the perfect strategist was one who tailored his actions so as to take account of any randomness in their effect, Borel went a step further, recommending the *deliberate* use of probabilistic play. This he did in a series of notes written throughout the 1920s, in which he presented the notion of a strategy in precise form and the principle of random play, and investigated the range of two-person games in which they could be employed profitably.

Borel asks us to consider a game "in which the winnings depend on both chance and the skill of the players", unlike such games as dice in which skill does not influence the outcome.[10] Defining a "method of play" as "a code that determines for every possible circumstance. . . . what the person should do", Borel asks "whether it is possible to determine a method of play better than all others". Considering the particular case in which the number of strategies is three, Borel shows that each player can choose probabilities that ensure an even chance of victory.[11]

In Borel (1924a), he extends this analysis, slightly modified, to the case of five strategies – that is, $n = 5$, in which he shows that "nothing essentially new happens compared to the case where there are three manners

---

[10] Émile Borel (1921). Three of Borel's five notes appeared in the Académie's *Comptes Rendus*. Because some of Borel's observations are repeated from one paper to the next, I consider the three most important to be the aforementioned (1921), (1924a), and (1927).

[11] Borel, "La théorie du jeu" (cit. n. 84), on p. 97. He considers a game with two players A and B, who choose strategy ("method") C_j and C_k, respectively. Each has the same set of n strategies available. Given the strategies chosen, the entries in the matrix represent not payoffs but A's probability of winning the game. The numbers a_{ik} and a_{ki} are contained between $-1/2$ and $+1/2$, and satisfy $a_{ik} + a_{ki} = 0$. Also, $a_{ii} = 0$. His examples are confined to games that are symmetric and fair, in the sense that the expectation for each player is zero.

	Player B		
	C_1	C_2	C_n
	C_1 $1/2 + a_{11}$	$1/2 + a_{12}$	$\ldots\ldots 1/2 + a_{1n}$
Player A	C_2 $1/2 + a_{21}$	$1/2 + a_{22}$	$\ldots\ldots 1/2 + a_{2n}$
	C_n $1/2 + a_{n1}$	$1/2 + a_{n2}$	$\ldots\ldots 1/2 + a_{nn}$

Players are assumed to automatically cast aside "bad" strategies – that is, methods of play that guarantee a probability of winning of less than half. Having done this, the question is how the remaining strategies might be employed in the best manner possible. Borel suggests that a player can act "in an advantageous manner by varying his play" – that is, C_k is played with probability x_k by A and y_k by B, where $\sum_{K=1}^{n} x_k = 1 = \sum_{k=1}^{n} y_k$. Given this, A's expectation of winning is $\sum_{i=1}^{n} \sum_{K=1}^{n} 1/2 x_i y_k = 1/2 + a$, where $a = \sum_{i=1}^{n} \sum_{K=1}^{n} a_{ik} x_i y_k$ and B's probability of winning is thus $1/2 - a$.

of playing": that is, each player can ensure an expected payoff of zero, and Borel wonders whether this is likely to hold for *n* arbitrarily large.[12] This, he conjectures, is improbable. Three years later, however, in another note presented to the Académie, he reports that what has held for three and five strategies seems also to hold for seven, and it therefore would "be interesting either to demonstrate that it is unsolvable in general or to give a particular solution".[13]

In various places in these papers of the 1920s, Borel considers applications. Considering the finite game "paper, scissors, stone", he shows in detail how the calculation of the optimal mixed strategies depends on relative payoffs. For example, if the payoff to *A* for a particular strategy is relatively large, then the probability attached to it in the optimal mixed strategy will be correspondingly low: otherwise, *B* could gain by anticipating *A*'s emphasis on the favored strategy.[14] In a manner very similar to Lasker, Borel notes that the "problems of probability and analysis that one might raise concerning the art of war or of economic and financial speculation, are not without analogy to the problems concerning games".[15]

In a 1924 review of John Maynard Keynes' (1921) *A Treatise on Probability*, Borel was quite explicit about the possibility of a new science. It was in the context of his criticisms of the subjective interpretation of probability proposed by Keynes. Borel said there were situations in which the very attempt to make a probability judgement altered the probability one was trying to evaluate – for example, in betting on the result of an election, in which case the size of bets placed can influence the probability of a

[12] Borel, "Sur les jeux où l'hasard . . .", p. 114.
[13] Borel, "Sur le système de formes linéaires . . . ", p. 117. Borel's analysis is confined to games with odd numbers of strategies because of his use of determinants of skew symmetric matrices to calculate the optimal mixed strategy.
[14] Borel also introduces the infinite game, in which strategies are drawn from a continuum, and shows how the continuous analogue of player *A*'s expected payoff may be expressed as a Stieltjes integral:

$$a = \int_{-\infty}^{\infty} \int_{-\infty}^{\infty} f(C_A, C_B) d\phi_A(C_A) d\phi_B(C_B)$$

where $f()$ is the function relating *A*'s payoff to the strategies chosen, and $\phi_A()$ and $\phi_B()$ are *A*'s and *B*'s respective cumulative distribution functions over strategy space. In Borel's example, each player chooses three real numbers summing to 1, the winner being the one with two choices of greater value than the opponent's. This is what was later to become known as a "game on the unit square", and was taken up in more detail in 1938 by Borel's student Jean Ville. See Ville, "Sur la théorie générale des jeux où intervient l'habileté des joueurs" (1938) in Borel, *Traité* (cit. n. 81), pp. 105–13.
[15] Borel, "La théorie du jeu", p. 10.

candidate's success. In such circumstances, it is difficult to ascribe a precise number to the probability of success attributed by a gambler to a particular candidate:

The problem can be put in a form that is both simple and yet which remains complex enough to preserve its entire difficulty, if we consider the game of poker, where each player bets on his hand against that of the opponent. If the opponent proposes a large bet, this tends to indicate that he has a very good hand, unless he is bluffing; thus, the very fact that the bet is made alters the probability of the judgment on which the bet is based.

The further study of certain games will perhaps lead to the creation of a new chapter in the theory of probability, a theory the origins of which go back to the study of the simplest games of chance; it will be a new science, where psychology will be no less useful than mathematics.[16]

This insistence on the importance of psychology would become a hallmark of Borel's, and it distinguished him from von Neumann. Constantly, Borel reminds his reader of the limited extent to which matters of human psychology can be clarified by the use of mathematics. Games, he says, like problems of war and economics, are in reality highly complex, so that mathematical calculation can be, at best, a supplement to strategic cunning. The only advice the mathematician can give the player "in the absence of psychological information" is to vary his play in such a way that the probabilities remain invisible to his opponent. Repeatedly, Borel insists on the limitations of mathematical analysis in aiding the comprehension of such games. If the player of strategic games must be a good master of combinations, he says, it is no less true that he must be a good psychologist, too. Borel contends that although theoretical solutions may require a player to play probabilistically, the challenge for the player of real games is to be able to discern the way in which the opponent is playing – the probabilities he is

---

[16] Émile Borel, "A propos d'un traité des probabilités", *Revue philosophique*, 1924, *98*: 321–26, reprinted in *Oeuvres de Émile Borel* (cit. n.viii), Vol. 4, pp. 2169–84. Borel goes on to say that this new theory will add to older theories without modifying them. "The theory of value and the law of supply and demand are not changed by a fact such as the following: I am unable to distinguish true from false diamonds and yet would like to buy jewellery today; before the display of a seemingly honest shop, I notice a jewel marked 500 francs and I decide not to buy it because I believe it to be a forgery and the price appears too high; nonetheless, I enter the shop, and the jeweller, after a moment, offers me the jewel I had noticed; I then notice that it is, in fact, marked 5000 francs; I thus conclude that the stones are real, and I decide to buy".

using, so to speak. This requires psychological skill, which is why treatises on card games such as bridge are quite inadequate to the task of teaching superior play. The same problems arise in the "art of war", where, again, "knowledge of the psychology of the adversary" is necessary. Borel continued, long thereafter, to insist on the practical limitations of mathematics given the psychological complexity of real games. Indeed, as we shall see, his most eloquent expression of this was provoked by his encounter with von Neumann's contribution.

The latter entered the picture in May 1928, when he sent a note to Borel in Paris providing an answer to the French mathematician's question concerning the existence of a "best" way to play in the general two-person, zero-sum case. Von Neumann claimed to have been working independently on the matter and to have proved a theorem two years previously.[17]

From Struggle to Equilibrium

The first thing that strikes one about von Neumann's paper is its generality. Citing chess, baccarat, roulette, and poker as examples, he goes beyond everything done previously, laying out a theory of the generic strategic game: "n players S_1, S_2,...,S_n are playing a given game of strategy, G. How must one of the participants, S_m, play in order to achieve a most advantageous result?" The problem, he says, is well known, and "there is hardly a situation in daily life into which this problem does not enter". Yet, the meaning of this question is not unambiguous, he says. For, as soon as there is more than one player, the fate of each "depends not only his own actions but also on those of others, and their behavior is motivated by the same selfish interests as the behavior of the first player. We feel that the situation is inherently circular". Therefore, we must formulate the problem clearly. "What, exactly, is a game of strategy?" he asks. "A great many different things come under this heading, anything from roulette to chess, from baccarat to bridge. And after all, any event – given the external conditions and the participants in the situation (provided the latter are acting of their own free will) – may be regarded as a game of strategy if one looks at the effect it has on the participants. What elements do all these things have in common?"[18]

[17] Von Neumann (1928a).
[18] *Op cit*, p. 13. Here, in a footnote, von Neumann writes that this is the main problem of "classical economics: how is the absolutely selfish 'homo economicus' going to act under given external circumstances?" (p. 13, fn. 2).

He then gives a rather loose definition of a game of strategy, as a series of events, each of which may have a finite number of distinct results. In some cases, the outcome depends on chance – that is, those in which the probabilities are known – in others, it depends on the free choices of the players. For each event, it is known which player affects the outcome, and what information he has with respect to the previous decisions of other players. When the outcome of all events is known, then the payments amongst players can be calculated.

Having formally specified the "rules of the game" – that is, the manner in which the player's result is related to both draws (chance moves) and steps (deliberate moves) – von Neumann undertakes to simplify everything by *collapsing* each possible sequence of chance and deliberate moves to a single strategy, to be chosen by the player, once only, at the beginning of the game. Then, he simplifies even further, integrating the effect of the chance moves on the player's payoff, $f(.)$, by replacing the latter with his *expected* result $g(.)$. We now have, says von Neumann, an even more schematized and simplified basic strategic game:

Each of the players $S_1, S_2, \ldots, S_n$ chooses a number, S_m choosing one of the numbers $1, 2, \ldots, \sum_m$ ($m = 1, 2, \ldots, n$). Each player must make his decision without being informed about the choices of the other participants. After having made their choices $x_1, x_2, \ldots, x_n$ ($x_m = 1, 2, \ldots, \sum_m$, $m = 1, 2, \ldots, n$) the players receive the following amounts respectively: $g_1 (x_1, x_2, \ldots, x_n), g_2 (x_1, x_2, \ldots, x_n), \ldots, g_n (x_1, x_2, \ldots, x_n)$ (where identically $g_1 + g_2 + \cdots + g_n = 0$).[19]

The rules of the game, he says, have now been reduced to a form containing only the essential characteristics of the game, without loss of generality. "Nothing is left of a 'game of chance'", and "everything takes place as if each of the players has his eye on the expected value only".[20] We are now ready to pursue the matter of how to play.

In the two-person case, which represents the bulk of the paper, the players independently choose amongst the numbers $1, 2, \ldots, \Sigma_1$ and $1, 2, \ldots, \Sigma_2$ respectively, and then receive the sums $g(x, y)$ and $-g(x, y)$ respectively. The Laskerian rhetoric of "struggle" is present throughout. "It is easy to picture the forces struggling with each other in such a two-person game. The value of $g(x, y)$ is being tugged at from two sides, by S_1 who wants to maximize it, and by S_2 who wants to minimize it. S_1 controls the variable x, S_2 controls the variable y. What will happen?"[21]

---

19 *Ibid*, p. 20.
20 *Ibid*, p. 21.
21 *Ibid*, p. 21.

By choosing x appropriately, S_1 can guarantee himself at least $\text{Max}_x \text{Min}_y$ $g(x, y)$, irrespective of what S_2 does. Similarly, S_2 can ensure, irrespective of S_1, that $g(x, y)$ reaches no more than $\text{Min}_y \text{Max}_x g(x, y)$. Is it the case that $\text{Max}_x \text{Min}_y g(x, y) = \text{Min}_y \text{Max}_x g(x, y) = M$? Although, in general, Max_x $\text{Min}_y g(x, y) \leq \text{Min}_y \text{Max}_x g(x, y)$, it is not generally true that the equality holds. Thus it does not hold, for example, in the simplest example (tossing pennies) where

$$\Sigma_1 = \Sigma_2 = 2, \quad g(1, 1) = 1, \quad g(1, 2) = -1$$
$$g(2, 1) = -1, \quad g(2, 2) = 1$$

where MaxMin $= -1$ and MinMax $= 1$. Nor does it hold in the game of morra or in paper, stone, scissors.

If, however, rather than choosing a strategy directly, from $1, 2, \ldots, \Sigma_1$, player S_1 instead specifies Σ_1 probabilities $\zeta_1, \zeta_2, \ldots, \zeta\Sigma_1$ and draws his strategy "from an urn containing these numbers with these probabilities", then equality of the two expressions can be ensured. This may look like a restriction of the player's free will, von Neumann says, but it is not. If the player really wants to choose a particular strategy, he can attach to it a probability of 1, but in choosing probabilistically in general he can protect himself against his adversary's "finding him out".[22] Not even S_1 himself knows what he is going to choose! S_2 can do likewise, choosing Σ_2 probabilities $\eta_1, \eta_2, \ldots, \eta\Sigma_2$ and proceeding similarly.

Letting $\zeta = (\zeta_1, \zeta_2, \ldots, \zeta\Sigma_1)$ and $\eta = (\eta_1, \eta_2, \ldots, \eta\Sigma_2)$, S_1's expected value becomes the bilinear form:

$$h(\zeta, \eta) = \sum_{p=1}^{\Sigma_1} \sum_{q=1}^{\Sigma_2} g(p, q)\zeta_p\eta_q$$

and S_2's is $-h(\zeta, \eta)$. The considerations applied earlier to $g(.)$ now can be applied to $h(.)$. In particular, is it the case that $\text{Max}\zeta \text{ Min}\eta \ h(\zeta, \eta) = \text{Min}\eta \text{ Max}\zeta \ h(\zeta, \eta)$?

Anticipating the result, let $\text{Max}\zeta \text{ Min}\eta \ h(\zeta, \eta) = \text{Min}\eta \text{ Max}\zeta \ h(\zeta, \eta) = M$. Now let ϑ be the set of all ζ for which $\text{Min}\eta \ h(\zeta, \eta)$ assumes its maximal value M, and let B be the set of all η for which $\text{Max}\zeta h(\zeta, \eta)$ assumes its minimal value M. Given S_1's and S_2's choice of probability distributions ζ and η respectively, the play has the value M and $-M$ to the respective players. In a "fair" game, $M = -M = 0$: this holds for tossing pennies and morra, mentioned earlier. In a "symmetric" game, the players have the same roles – that is, interchanging ζ and η, we get $h(\zeta, \eta) = - h(\eta, \zeta)$. Thus,

---

[22] *Ibid*, p. 23.

even though the explicitly probabilistic aspects of the game were earlier eliminated by introducing expected values and discarding 'draws', chance has now reappeared spontaneously, in the need for each player to apply a probability distribution to his set of strategies.

Von Neumann then proceeds to the mathematically central part of the paper, the six-page proof that $\text{Max}_\zeta \text{ Min}_\eta \text{ } h(\zeta, \eta) = \text{Min}_\eta \text{ Max}_\zeta \text{ } h(\zeta, \eta)$. He accomplishes this by invoking a painstaking and difficult argument based on the lower- and upper-semicontinuity of the functions bounding the two elements of the saddlepoint. Here, the tone of the paper distinctly changes, and one is struck, not by the simplifying clarity that has characterized the treatment up to this point, but rather the sheer bulldozing power with which von Neumann pursues the proof. It is an early testament to the remark, often made subsequently of him, that elegance in proof frequently gave way to something resembling brute force, in which he showed no fear of pursuing tangential arguments and taking the difficult, sometimes contorted, route. As his only doctoral student, Israel Halperin, said of him: he was "a magician, a magician in the sense that he took what was given and simply forced the conclusions logically out of it, whether it was algebra, geometry, or whatever. He had some way of forcing out the results that made him different from the rest of the people".[23] Or, as Rózsa Péter put it, perhaps less reassuringly, von Neumann proved what he wanted to prove.[24]

Danish historian of mathematics, Tinne Kjeldsen, recently provided an excellent account of the aforementioned proof.[25] Not only does she present it in detail, but she corrects the widespread misunderstanding that it involved the use of an extension of the Brouwer fixed-point theorem. Von Neumann's proof, she shows, did not involve the use of a fixed-point theorem at all.

The mistaken perception that the 1928 minimax proof relied upon the Brouwer theorem can be traced to von Neumann himself. In his later equilibrium growth model of 1937, which does rely on a fixed-point theorem, he pointed out the formal connection between the solution of the growth

[23] Halperin Interview, The Princeton Mathematics Community in the 1930s, Transcript Number 18 (PMC18). See also Paul Halmos (1973).

[24] Similarly, of von Neumann's work on lattice theory, Harvard's Garrett Birkhoff wrote: "a truly remarkable feat of logical analysis and ingenuity...Anyone wishing to get an unforgettable impression of the razor edge of von Neumann's mind, need merely try to pursue this chain of exact reasoning for himself – realizing that often five pages of it were written down before breakfast...". Garrett Birkhoff (1958), quoted in Heims (1980) p. 171. A prominent American mathematician in the interwar period, Birkhoff became a friend of von Neumann after the latter moved to the United States.

[25] Kjeldsen, T.H. (2001).

model and the structure of the 1928 minimax problem. It was a passing remark, expressed with surprise: "The question whether our problem has a solution is *oddly connected* with that of a problem occurring in the Theory of Games dealt with elsewhere" (1937, p. 5, italics mine). Subsequent readings, however, beginning with Kuhn and Tucker (1958), by emphasizing this retrospectively noted formal similarity, and the possibility of solving the minimax problem by fixed-point methods, gradually gave rise to the perception that the two models were based on fixed-point techniques.[26] For example, in their otherwise excellent history of the "Invisible Hand", Ingrao and Israel (1991) state that von Neumann "made decisive use of Brouwer's theorem" (p. 187), that the proof involves seeking the solution to a "problem of algebraic character of which von Neumann demonstrates the close connection with fixed-point theorems and especially with Brouwer's theorem" (p. 211). More recently, Mirowski (2002) writes that von Neumann "admitted later that he used the same fixed point technique for the expanding economy model as he did for the minimax theorem" (p. 113).

The effect of Kjeldsen's careful reading is to remind us that, on the foot of retrospectively noted *formal* connections, we have been too ready to attribute a smooth unity to von Neumann's efforts in economics, retrospectively seeing his forays into game theory and general equilibrium as somehow being of a piece, and thereby eliding the historical particularities of each.[27] In 1926, von Neumann was interested in games and there is no evidence that he saw, at the time, any connection between them and models of economic growth. With his theorem proved, he has shown that, in the generic two-person, zero-sum game, there exists an optimal way to play, possibly requiring a player to choose probabilistically amongst the set of strategies, and ensuring that, on average, victories and defeats counterbalance each other.

As if in response to Borel, von Neumann writes that the existence of an equilibrium demonstrates that "it makes no difference which of the two players is the better psychologist, the game is so insensitive that the result is always the same".[28] Later in the paper, he promises a publication that will contain numerical examples of such two-person games as baccarat and a simplified poker, the "agreement of the results [of which] with the well-known rules of thumb of the games (for example, proof of the necessity

---

[26] See Harold Kuhn and Albert Tucker (1958).
[27] For a rather similar argument concerning the historical interpretation of the Nash equilibrium, see Leonard (1994).
[28] Von Neumann, "Zur Theorie der Gesellschaftsspiele", p. 23.

to 'bluff' in poker) may be regarded as an empirical corroboration of the results of our theory".[29] There was nothing mysterious about bluffing: it was simply rational play. Nowhere does von Neumann discuss the prosaic difficulties of how a player might actually know the correct mixed strategy or how the psychological evaluation of one's opponent might affect one's approach to a game. With existence proved, the game had been collapsed, reduced to its essential skeleton, and any psychological complications consigned to the periphery. The disinterest in psychology remained a characteristic of von Neumann's, including later, when his attention shifted from the chessboard and poker to the realm of politics.

At the point in the paper at which probability makes its appearance in the form of the possibly optimal mixed strategy, von Neumann writes:

Although... chance was eliminated from the games of strategy under consideration... it has now made a spontaneous appearance... The dependence on chance (the "statistical" element) is such an intrinsic part of the game itself (if not of the world) that there is no need to introduce it artificially by way of the rules of the game: even if the formal rules contain no trace of it, it still will assert itself (*ibid*, p. 26).

To the extent that this passing remark about the importance of chance in "the world" was a reference to the statistical aspect of physics, it has provoked the suggestion that von Neumann's analysis of games may be related in some essential manner to his work in quantum mechanics. Mirowski (2002) insists strongly upon this connection, pointing to the fact that the bilinear forms in the game paper can be found in von Neumann's treatment of the Heisenberg Interpretation and suggesting that the place of probability in the game comes from Niels Bohr's Copenhagen Interpretation. The idea that "each observer influences the results of all other observers", Mirowski says, arose in quantum mechanics, and would become "insistent" in game theory (p. 109).

Both *a priori*, and based on the evidence presented, I find this argument to be quite implausible. Not only is the idea that one player influences the results of the other as old as the history of chess, but in the mid-1920s when von Neumann turned to games, there was, as we have already seen, an established literature in existence, beginning with Lasker and continuing with Borel, in which at least probability, and, in the case of Borel, recourse to randomization, already were central.[30] It is not because von Neumann

---

[29] *Ibid*, p. 42.
[30] Having said that, I find Mirowski's argumentation to be less than clear, and I would strongly recommend that the reader confront and evaluate it himself.

makes a passing reference to the statistical nature of the world that one can convincingly conclude that quantum mechanics was necessary to what he was doing. What von Neumann did was to put mathematical structure on the ideas with which Lasker and Borel had been toying, and show that an equilibrium existed.

We may never know to what extent he had read Lasker or Borel in detail. In a priority debate raised in the 1950s by Borel's protégé, Maurice Fréchet, von Neumann replied, quite starchily, that his own work was independent and that Borel's work had come to his attention only as he was writing up his own paper. Interestingly, he does not deny familiarity with Borel but claims that there was nothing worth reporting in game theory until the theorem had been proved (see Fréchet 1953). My own feeling is that, in 1926, von Neumann *must* have had some inkling, even if only through hearsay at Göttingen, of how Borel had framed the problem and of what he was trying to do.

More plausible is Mirowski's endorsement of the idea that Hilbert's axiomatic method was important for von Neumann's game analysis. As Corry has shown, Hilbert was very keen to develop an axiomatic treatment of a range of disparate fields, including insurance mathematics. Von Neumann's heroic "taming" of the strategic game would surely have appealed to the Master. Yet was it a properly *axiomatic* treatment? Certainly, von Neumann provides a theoretical mathematical model of the strategic game, but he does not do so by postulating an explicit system of axioms, the independence, consistency, and completeness of which can then be investigated. In fact, he would not do any of this until fifteen years later at Princeton, when he returned to game theory after a long hiatus. Only then, in the early 1940s, would he explicitly invoke Hilbert and provide a properly axiomatic presentation of the concept of the "game".

With the tension of the central proof dissipated, he draws the paper to a close with preliminary considerations of the three-person, zero-sum game. Here, he says, certain complications are essential to the matter and cannot be overlooked. Von Neumann asks, can we find for each of the players the value of the game, w_1, w_2, w_3? For these values to be satisfactory, it must be the case that no two players are able, together, to secure a value exceeding the sum of their individual values. Letting $M_{1,2}$ be the amount that the coalition of players S_1 and S_2 can secure, we must have:

$$w_1 + w_2 \geq M_{1,2} \qquad w_1 + w_3 \geq M_{1,3} \qquad w_2 + w_3 \geq M_{2,3}$$

$$w_1 + w_2 + w_3 = 0$$

This is possible if and only if $M_{1,2} + M_{1,3} + M_{2,3} \geq 0$.

However, there are many games for which this is not true – that is, in which it is impossible to provide individual values w_i, because two players can collaborate and "rob" the third, thereby doing better in coalition than they could have individually. If S_1 succeeds in entering a coalition, he can expect to receive $1/2 \, (M_{1,2} + M_{1,3} - M_{2,3})$; if he fails, he will receive $- M_{2,3}$. Because the rules of the game have nothing to say about which coalition will be formed, von Neumann suggests that we regard the probability that S_1 enters a coalition as $2/3$. His basic expected value, v_1, is thus $1/3 \, (M_{1,2} + M_{1,3} - 2M_{2,3})$. In general, the three-person, zero-sum game can be divided into two types. In the first, $D = M_{1,2} + M_{1,3} + M_{2,3} = 0$, and S_1's basic value is $- M_{2,3}$. This type, like the two-person game, is strictly determined. In the second, $D > 0$, and S_1's basic value is $- M_{2,3} + 1/3 \, D$. In this type of game, the multiple possibilities of coalition make the three-person game qualitatively different from the two-person one: it is symmetric, but not strictly determined. The outcome of any such negotiations depend on factors regarding which the coalitional values can shed little light *a priori*. Here, as von Neumann puts it, the "actual game strategy of the individual player recedes into the background", and a new element enters, which is "entirely foreign to the stereotyped and well-balanced two-person game: struggle".[31] He concludes with the suggestion that a similar approach could be taken with games of four and, ultimately, any number of players, the result of which would be a "satisfactory general theory" of all such games.

Before leaving this paper, which appeared in the prestigious *Mathematische Annalen*, we may ask what influence the Hilbert programme in metamathematics – the other source upon which Mirowski (2002) appears to lean – had upon it: "It has now become widely accepted that von Neumann's early work on game theory was prompted by questions thrown up by Hilbert's program of the formalization of mathematics" (p. 108).

Certainly, the manner in which Hilbert shifted the terms of the foundations debate – away from discussing the acceptability of particular axioms, such as the axiom of choice, and towards the examination of consistency using finite methods – led to the evocation of the game metaphor in this connection. Hermann Weyl saw Hilbert as trying to "save" mathematics by radically reinterpreting it: transforming it "in principle from a system of intuitive results into a game with formulas that proceeds according to fixed rules" (Weyl quoted in Kline, p. 252). At the aforementioned Königsberg conference in 1931, von Neumann himself spoke of Hilbert's Formalism as "an internally closed procedure which operates according to

---

[31] Von Neumann, "Zur Theorie der Gesellschaftspiele", p. 38.

fixed rules known to all mathematicians and which consists basically in constructing successively certain combinations of primitive symbols, which are considered 'correct,' or 'proved' . . . a combinatorial game played with the primitive symbols" (von Neumann 1984 [1931], p. 62). There is no doubt, therefore, that, precisely because it involved a kind of shift to the surface, to the manipulation of signs, Hilbert's approach to the foundational debate was likened to the playing of a game.

Beyond the appearance of the word "game" in the two contexts, however, it is not at all clear, nor does Mirowski explain, how the technical, metamathematical problems with which Hilbert proof theory was concerned, *viz.* consistency proofs using constructive methods, might be related to the mathematical problem of modelling strategic games. As with the quantum mechanics argument, I am inclined to attach greater importance to the historical fact that there already existed in the mid-1920s a rich literature on games, to which von Neumann's work plausibly can be related.

More subtle, and in my opinion more persuasive, is the suggestion that, by virtue of his particular cast of mind, which excelled in the abstract, gamelike manipulation involved in foundations and set theory, von Neumann was naturally drawn towards the analysis of chess and parlour games. Support for this can be found in Stan Ulam's comments on the Hungarian's creative style, based upon his early papers on set theory and foundations. Amongst mathematicians, Ulam distinguishes between two broad kinds: those who reason by "seeing" and those who reason "abstractly". The former create concepts and new proofs in a visual or tactile fashion, "seeing" in their mind the manipulation of mathematical objects. For example, the basic operations in set theory, such as union and intersection, are particularly easily imagined in this way. According to Ulam, the other kind reason, not visually but more formally, proceeding through the application of abstract rules.

In Ulam's opinion, von Neumann's early papers in set theory and foundation show him to be definitely of the latter, abstract, type, observing a sharp distinction "between the formal symbols and the game played with them, on one hand, and the interpretation of their meanings, on the other". Ulam (1958, p. 12) writes, "The foregoing distinction is somewhat like that between a mental picture of the physical chess board and a mental picture of a sequence of moves on it, written down in algebraic notation", with von Neumann being inclined towards the latter. It is easy to see why someone of this mindset would have been attracted to a thoroughly abstract treatment of chess that said nothing about the actual moves upon the board. Ulam's comments are also in keeping with Borel's assessment of von Neumann as altogether too abstract for his liking.

The "Futile Search for a Perfect Formula"

Having presented von Neumann's minimax note to the Paris Académie in 1928, Borel appears to have ignored the main paper for a number of years. Not until 1936, when giving a talk at a mathematical congress in Oslo, was he reminded of it by someone in the audience, which prompted Borel to say that he hadn't had the time to study it carefully.[32] Indeed, the best that can be said about Borel in relationship to von Neumann in the 1930s is that he never embraced the Hungarian's contribution. On the few occasions he was to write about it, it was to add qualifications regarding its usefulness for the consideration of real games.

The culmination of Borel's work in this period was his 1938 volume on the analysis of games of chance, based on university courses given in 1936–37, and written up by his student Jean Ville. The recent Depression, on which Borel had written economic articles in the early 1930s, allowed him to draw connections between parlour games and the economic world:

[Economic] phenomena are caused, on the one hand, by material causes, which have concrete manifestations, such as the valuation of existing stocks, and, on the other hand, by causes dependent on the human will. Economic theories that take account only of causes of the first kind give rise to developments that are interesting, but of practically little value. And economists may be reprimanded in a manner similar to the way meteorologists are criticised: just as the latter are excellent at scientifically explaining yesterday's weather rather than forecasting that of next week, economists are better at producing theories of something that has just happened than they are at prescribing measures to be taken to ensure that tomorrow's economic life remains normal. In order to treat economic questions satisfactorily, room must be made for probability and psychology: the study of games of chance, and of psychology will provide a useful basis for such inquiry.[33]

Later in the book, he returns to the parallel between games and problems of strategy or economics, analysing the problems of two enemies allocating their opposing forces across several battlefields, two merchants deciding how to apply discounts in different markets, or two entrepreneurs bidding for the same contracts.[34]

---

[32] Émile Borel (1936). There, he continues, possibly with the slightest hint of displeasure: "von Neumann cited . . . a note I had published in 1927 in the *Comptes rendus de l'Académie des Sciences* de Paris, but was unaware of the third edition of my *Éléments de la théorie des probabilités* (Paris: Hermann, 1924), a work in which I develop, for certain symmetric games, in particular the Japanese game of paper, stone and scissors, considerations quite analagous to his (he calls this game baccara du bagne)" (p. 1173).

[33] Borel et al. (1938), pp. X–XI.

[34] *Ibid*, pp. 86–87.

In that 1938 volume, the attention paid to the von Neumann proof is curiously marginal, it being the subject of a note, not by Borel himself, but by his student Jean Ville, in which the latter provides an elementary proof of the minimax theorem. All of this is quite congruent with Borel's having prodded Ville to take a look at "that theorem", following the Oslo meeting.[35] Following Ville's proof, Borel offers some telling remarks in commentary. "It appears essential for me to indicate, however, to prevent all misunderstanding, that the practical applications of this theorem to the actual playing of games of chance is, for a long time, unlikely to become a reality". Actual games are exceedingly complicated, Borel says, and even if one could simplify a game to the point at which such calculations were possible, the advantage of playing according to von Neumann's prescription would be had only on average, after a great many rounds. Even if one could draw on experienced players to locate reasonable strategies, the number of variables remaining was still so great as to make the task of writing the equations "absolutely insurmountable".[36] Borel says games are interesting precisely because they perpetually evolve. Sometimes, they even move in cycles. No sooner has agreement been reached concerning a good way to play than players take advantage of that consensus by introducing novel approaches, only to later find themselves returning to older ways of playing. He also says that even when ideal play involves the use of probabilities, it is very difficult not to follow some regularity when actually playing. In bridge, for example, probabilistic play intended to defeat one's opponent may well mislead one's partner also! Borel concludes, "All these remarks are obvious to anyone with some experience in games. Perhaps they will make clear, to those uninterested in games, how enjoyable games are as leisurely distraction, at the same time showing to those who would wish to turn games into an occupation, how futile is the search for a perfect formula which is forever likely to elude us".[37]

Conclusion

Von Neumann's seminal game paper was part of a rich contemporaneous discussion of the mathematics of chess and parlour games in the first three decades of the century, involving diverse contributors, from Lasker and Zermelo to König, Kalmár, and Borel. It was a multifaceted literature,

[35] *Ibid*, pp. 105–13.
[36] *Ibid*, p. 115.
[37] *Ibid*, p. 117.

embracing Lasker's philosophical probing of the place of struggle in business and war; Zermelo and the Hungarians' set-theoretic analyses of chess; and Borel's own attempt to create a novel form of social inquiry, blending probability and psychology. Von Neumann's contribution cut through the manifold complexity of game situations, turning "the game" into a mathematical object. The relevant feature of the two-person game was that it was characterized by a straightforward equilibrium. For larger games, in which "struggle" was important, von Neumann provided only an outline, but he expressed the belief that such games could be described in mathematical form.

By all accounts, this work was a minor interest, overshadowed by von Neumann's other contributions in set theory and foundations and quantum mechanics. In the period from 1925 until his emigration in 1931, these contributions sealed von Neumann's reputation and opened doors to him amongst physicists and mathematicians, all the way from Budapest to Copenhagen. When, many years later, von Neumann's widow, Klari, sought to explain the vehemence with which he turned against Europe during the 1930s and the war, she insisted on the depth of his attachment to the world that was being destroyed.

In Berlin there was Max Planck, von Laue, Einstein; Sommerfeld had his Institute in Munich, Weyl and Schrodinger were moving back and forth between Zurich, Berlin and Göttingen. In Göttingen there was Klein, Hilbert and Courant; in Kopenhagen, Harald and Niels Bohr. Of the younger men, Heisenberg, Wolfgang Pauli, Paul Dirac and Enrico Fermi were intensely and fruitfully active; Theodor von Karman, Szilard, Teller and Wigner, of the Hungarian contingent, were all very much in evidence already... Within relatively short geographic distances, there was an easy and natural communication both between the younger group and the older men who had already world-wide reputations. Scientific discussions and friendships grew together – there were no secret documents or restricted data – it must have been near-paradise for the academically minded... For a young man, with his keen mind and impatient drive, the experience was unforgettable – an apprenticeship under such favorable conditions does not happen very often. No wonder that he had such deep roots in Germany and such love and admiration for the spirit of urgent search for knowledge that existed there, before the Nazis appeared and disrupted everything.[38]

With the 1928 paper and the note to the French academy, von Neumann moved on; for more than a decade, he put game theory, Borel, and all that

[38] Von Neumann-Eckhart, Unpublished Draft Autobiography, p. 11. This draft and some letters are located in the papers of the late Klari von Neumann-Eckhart, which are in the possession of Professor Marina von Neumann Whitman, Ann Arbor, Michigan.

completely aside. Indeed, he wouldn't learn of the existence of Borel's 1938 book until the early 1940s, when the volume was brought to his attention by Viennese economist, Oskar Morgenstern, he having stumbled across it by accident in the library of the Institute for Advanced Study at Princeton. By then, the Hungarian's interest in game theory had already been reawakened, but it was not by chess, nor, arguably, even by economics. By 1938, for von Neumann it was politics that mattered.

Plate 1. Emanuel Lasker. *Credit:* Courtesy of the Lasker Gesellschaft, Berlin.

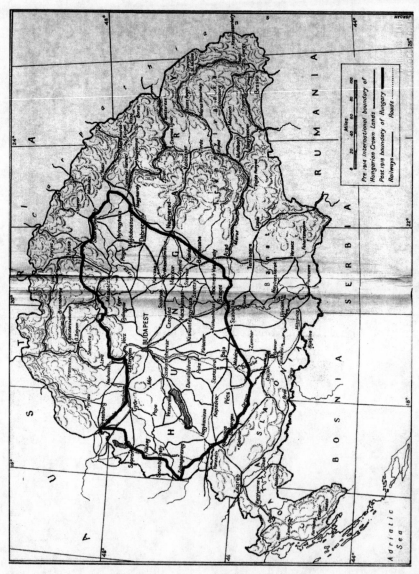

Plate 2. Reduction of Hungarian territory as of 1919. *Credit:* From Michael Karolyi (1956), *Memoirs of Michael Karolyi*, London: Jonathan Cape.

Plate 3. Bolyai Institute of Mathematics, Szeged, 1928. Back row: Frigyes Riesz, Béla Kerékjartó, Alfred Haar, Denés König, Rudolf Ortvay. Middle row: Jószef Kürschák, George Birkhoff (visiting), O. Kellogg (visiting), Lipót Fejèr. Front row: Tibor Radó, István Lipka, László Kalmár, Otto Szász. *Credit:* Courtesy of Dr. Peter Szabó, Szeged.

Plate 4. Willy Fellner, Emery Reves, and von Neumann, Zurich, 1926. *Credit:* Courtesy of Marina von Neumann Whitman.

Plate 5. Wedding breakfast for von Neumann's cousin Lily. Seated, from left: von Neumann; the newlyweds; Mariette von Neumann, friend of the newlyweds; Lily's mother-in-law; Lily's father and mother; and Lily's father-in-law. *Credit:* Courtesy of Marina von Neumann Whitman.

PART TWO

OSKAR MORGENSTERN AND INTERWAR VIENNA

FIVE

Equilibrium on Trial

The Austrian Interwar Critics

[At] the beginning of a game of chess, the result is neither subjectively known to the players nor objectively settled ... each move in a game of chess is dependent upon previous ones and the final result upon all the moves that have gone before ...

In essence, there is an immanent, more or less disguised, fiction at the heart of mathematical equilibrium theories: that is, *they bind together, in simultaneous equations, non-simultaneous magnitudes operative in genetic-causal sequence as if these existed together at the same time.* A state of affairs is *synchronized* in the 'static' approach, whereas in reality we are dealing with a *process.*
Hans Mayer (1932), pp. 91–92, emphasis in original

Introduction

While the Hungarian mathematicians tackled chess in a formal manner, deliberately circumventing the game's psychological complexity and inherent unpredictability, amongst the Viennese economists, the chess metaphor was being put to different use. They seized upon it as an exemplar of complexity.

To the extent that a game of chess involved interdependent steps, taken through time and cumulative in their importance, it was, in the eyes of Oskar Morgenstern's teachers, the quintessential embodiment of a time-bound activity with an unpredictable outcome. If, therefore, one saw the essence of economic exchange, not in the mechanical and instantaneous achievement of an equilibrium, but as a process of uncertain outome in which the passage of time was crucial, then the joint search for a result at the chessboard offered a powerful analogy with which to counter the mechanistic view. This was the understanding of the game metaphor that best characterized the thought-world of Oskar Morgenstern in the late 1920s, and that fact is illustrative of both the gulf then separating him from the Hungarian mathematicians and the distance he would travel in becoming co-author of *The Theory of Games.*

Morgenstern: The Beginnings

We know a little about Morgenstern in the years leading up to 1922 and his entry to the University of Vienna. His father had come to Austria from Görlitz, in Silesia, having apparently lost the family fortune on poor business ventures. In Vienna he worked for Julius Meinl and Co., the coffee and tea importer, and the family was modestly comfortable. At school, Morgenstern was strong in German; less so in Latin and mathematics. With Vienna ravaged by tuberculosis in the postwar years, those who could afford it sent their children away to breathe better air in the Austrian provinces and abroad. Thus, Morgenstern spent the summers of 1921 and 1922 in the company of relatives in Sweden, where he gained a passing knowledge of the language.[1]

He came of age in a shrunken, tense Vienna. At the *Gymnasium*, like many of his fellows, he entertained strong German nationalist feelings, writing youthful articles denouncing foreign cultural influences.[2] At the University, amidst overcrowding and anxiety about the future, these strains were apparent, leading Morgenstern's contemporary, Friedrich Hayek, to later write that interwar Vienna was "not intelligible without the Jewish problem": "Which was not a problem simply of Christians and Jews but a very large middle group in between the two, partly of baptized Jews, partly of Christians who had made friends with the Jews; and there was close contact between the purely Christian group and the mixed group, and again between the mixed and the Jewish group, but not between the two extremes".[3]

The dissolution of the Empire had seen an inflow of Jewish population from the Eastern provinces, the return of veterans to higher education, and the arrival of acute economic hardship. The sense of loss, the economic

[1]   Silk (1977). See also Rellstab (1991), (1992a); Shubik (1989).
[2]   "[W]hat would Europe be, what would the world be without the German people and its achievements! Germany can be satisfied with itself, there is no need to borrow from foreign peoples, who force it to further intellectual enslavement, but from its simply inexhaustible resources it is able to give to all those who want to receive. And yet, Germany does borrow: theatre and poetry are signs of a depressed mind. This has to be fought. All the foreign dirt which will wipe out healthy ideas, has to be removed with violence". In "Werktätiges Deutschland", OMDU, Box 1, Correspondence 1918–1924, M-N, quoted in Rellstab (1991), p. 5.
[3]   Hayek (1994), p. 60.

hardship, and the sight of the numerous orthodox Eastern Jews that had flooded the city after the war all contributed to a hardening of attitudes. An anti-semitic wave found support amongst university students of both Pan-German and Catholic Austrian nationalist inclination, and many belonged to fraternity organisations such as the *Deutsche Studentenschaft* (German Student Organization), which had been founded in 1919 with the express aim of diminishing the "racially alien" presence.[4] Such groups also had the support of a great many professors and teachers, and the tacit approval of sections of the University authorities, who granted them the use of academic facilities. There were protests in support of a "Numerus Clausus", or limitation on the number of Jewish professors and students, such as had been adopted by the Hungarians.

Morgenstern entered the University of Vienna to pursue a *Dr. rer. pol.*, which was nominally a degree in political science, although it allowed him to concentrate in economics. As was traditionally the case in the German-speaking universities, this field was taught under the auspices of the Law Faculty. Whereas students were exposed to a good deal of economic thought, philosophy, and political and economic history, they received instruction in neither mathematics nor the sciences.[5]

At the University, the three most important figures to Morgenstern were Othmar Spann, Hans Mayer, and Ludwig von Mises. Spann and Mayer held two of the professorships in economics in the Faculty of Law, the third being held by Count Ferdinand Degenfeld-Schönburg. Mises was a university lecturer without a chair. Morgenstern's three mentors, Spann, Mayer, and Mises, were influential figures, in time regarded by the Rockefeller Foundation as the "Prima Donnas" of Viennese economics.

The Preacher

Although Othmar Spann had originally studied the methodologically individualistic economics of Carl Menger, by the time Morgenstern encountered him, he had left that behind, moving over to Romanticism and idealistic philosophy. After an initial exposure to Carl Menger's teaching, Spann became active in Zurich, under Herkner, in the social reform campaign of the *Verein für Socialpolitik*. Following several years in Frankfurt, he received

---

[4] On anti-semitism in interwar Vienna in general, see Pauley (1992); Berkley (1988); Oxaal, Pollak, and Botz (eds.) (1987); and Wistrich (ed.) (1992). For a treatment of the matter from the perspective of a philosopher and social scientist, see Malachi Hacohen's very thorough biography of Karl Popper (Hacohen 2001).

[5] For an account of the background to the Austrian School encountered by Friedrich Hayek, see Caldwell (2004), esp. chapters 1–6.

his *Habilitation* in 1908 and then spent the following decade as professor of political economy at Brünn (Brno), before being wounded in the war.[6]

Although the greatest living influence on Spann was Tübingen sociologist and corporatist opponent of Marxism, Albert Schäffle,[7] his social philosophy owed even more to German Romantic Adam Müller (1779–1829). The latter's 1809 treatise, *Die Elemente der Staatskunst*, was a neo-feudal work that portrayed the corporatist state or *Ständestaat* as society's natural form. In this hierarchical arrangement, social harmony would stem from individuals adopting the role for which they were destined by nature.[8] In Austria, ideas of this sort had been well received by the Hapsburg court, and a certain tradition had been maintained by various Austrian Catholic writers who continued to target democracy and individualism. For example, one Baron Karl von Vogelsang (1818–1890) had responded to the Viennese stock-market crash of 1873 with a sustained critique of the capitalist competitive ethic. He argued for the role of the state as a countervailing force against the anonymous forces of the market and proposed that the territorial system of representation be replaced by one in which parliamentary deputies would represent a particular trade or profession, rather than the constituents of a geographical area. Vogelsang was followed by Karl Lueger, the dominant force in Catholic, anti-semitic politics at the turn of the century.

Spann adapted Müller's ideas in the creation of Universalism, a holistic, anti-rationalist view of the social order. According to Spann, there were two ways of understanding individual behaviour: the genetic, or causal, interpretation, and the functional one. Rejecting causal accounts, Spann argued that only a functional explanation could portray social phenomena as parts of a larger totality. Certain elementary dualities, such as man and woman, youth and age, Man and God, he regarded as fundamental to social life. In any of these relations, said Spann, neither part played a role without the other, and the whole was greater than the sum of the parts. Methodologically individualistic social science, he said, ignored the fact that individuals functioned in society, that it was the community that gave behaviour meaning. The Universalism championed by Spann took society as the primary theoretical unit. The social whole was the final cause that gave

---

[6] On Spann, see Haag (1976/77), Johnston (1983 [1972], pp. 311–15), and Hayek (1994, p. 54).

[7] See Spann (1924).

[8] Spann was responsible for the 1922 republication of Müller's 1809 work. For some years, Spann's 1911 *Die Haupttheorien der Volkswirtschaftslehre* (Leipzig: Quelle & Meyer) was *the* textbook on the history of economics and it was responsible for making the forgotten Müller visible in Austrian academic circles. See Haag, *op cit.*

purpose to its subtotalities. Only when the *telos* governing the evolution of society was known could the behaviour of the individual be understoood.

Spann's university lectures to Morgenstern and others were a war against the various "isms": "individualism, atomism, psychologism, Marxism, and other dead sciences".[9] According to Spann, individualism had grown out of a mistaken belief in individual freedom. Furthermore, methodologically individualist theories, in their refusal to recognize that the individual was subordinate to society, had cast the individual into the void, inducing the contemporary feelings of malaise and dislocation.

Although this individualist–universalist distinction had already been explored by Karl Pribram, a prominent sociologist at the University, Spann used it to shape a neo-Romanticist doctrine.[10] By the mid-1920s, having acceeded to the chair of Böhm-Bawerk, who died in 1914, Spann had given up all pretense of being an objective scholar, and had engaged students and others in spreading his ideas. A charismatic speaker, he captured the hearts of many young students, decrying individualism as the enemy of the German nation, and attacking Marxism as a distortion of German idealism. He introduced his students to the philosophy of Fichte, Hegel, and Schelling, and to the mysticism of Novalis and Augustine. He lectured his students: "In the Creation God wills Himself and therefore the world can will only God. That is, the mystical core of all history and of all pairing, and discloses that we have worked not only for ourselves but for the whole, and for God".[11]

For a while, Morgenstern was amongst those swept along by Spann, becoming something of a protégé.[12] "Looking at my papers, work and reading", he wrote in his diary, "one would think I were a student of philosophy and not of economics. But that is no disgrace ... my philosophical knowledge is very fragmentary in comparison to my knowledge of economics".[13] It was a view he would grow to reject entirely.

---

[9] Quoted in Haag, *op cit*, p. 239.

[10] On Pribram, see Johnston, *op cit*, pp. 76–78 and *passim*.

[11] Spann, quoted in Johnston (1972), p. 313. "Emphasising spirituality (*Geist*) and qualitative elements over force of numbers and democratic majorities, Spann's *Der wahre Staat* was a slashing criticism of democracy. Youthful and impetuous, many of them war veterans, Spann's students at the University of Vienna were carried away by the rhetorical boldness of the slight, handsome professor with the burning eyes of the true believer.... [Dreams] of Utopias – even arch-reactionary Utopias – were infinitely more pleasant than a world of defeat, inflation, and gnawing hunger". Haag, *op cit*, p. 237.

[12] "He is really looking after me and he wants to make me, it seems, a Privatdozent. I don't say anything about it, anyhow; I work in a way that prepares me for all eventualities". Morgenstern Diary, Apr. 27, 1923, OMDU, quoted in Rellstab 1991, p. 12.

[13] Morgenstern Diary, Mar. 18, 1923, OMDU, quoted in Rellstab 1991, p. 12.

Spann's ideas were used by rightist forces in German and Austria, and his network of influence extended to the Pan-Germans, the Christian Socials, and the Nazi front-organisation, the *Kampfbund für deutsche Kultur.* "Spannianer" occupied influential positions in government and industry, and in youth movements. Although he did not join the Nazi movement, in November 1923, Spann did praise Hitler's recently attempted *putsch*, and, in 1929, on the centenary of Müller's death, respects were paid by both the Nazi students of the University and the Austrian Catholic newspaper, the *Reichspost.*

Morgenstern's interlude with Spann was intense and shortlived. Around 1923, when he was reading and criticising Marx for Spann's Sunday morning seminar, he encountered Böhm-Bawerk's work and, at the same time, became aware of the towering presence of Carl Menger. The latter had died only in 1921, however, for reasons that were not without a hint of intrigue, had retired much earlier from university life. Later in 1923, Morgenstern took a class in sociology with one of Menger's disciples, von Wieser, who was very critical of Spann in his lectures. He repeatedly emphasised the methodological individualism of Austrian theory and dismissed as dangerous metaphysics anything that smacked of German idealism.[14] By degrees, Morgenstern became aware of the existence of a different perspective on economics, a tradition based at the University itself and stretching back some fifty years. Spann's worldview lost its fascination for him. Critical in this process, and stepping in to fill the void left by the break with Spann, was Wieser's chosen successor, the second of the Prima Donnas, Professor Hans Mayer.[15]

[14] Although his admiration for von Wieser grew in time, Morgenstern initially smarted at his criticism of Spann: "Might he be Jew, or half Jew, or liberal?" he wondered. OMDU, Diary, Sept. 12 1923, translated and quoted in Rellstab *op cit*, p. 13. By the time Wieser died in 1925, however, Morgenstern had published a finely wrought commemorative article in the *American Economic Review*, praising both Wieser's accomplishments in reintroducing into Mengerian economics the influence of costs upon price and, in particular, his later sociological work, as exemplified in his final book *Das Gesetz der Macht* (1926, Vienna: J. Springer). See Morgenstern (1927b).

[15] "Spann thought that through me he would exert an influence on Mayer and get him on the right track – you know, anti-Marxist and anti-everything.... It didn't turn out this way because I got captured by marginal utility theory and Mayer pulled me into his orbit and Wieser's orbit": Morgenstern, quoted in Craver (1986a), p. 10. His alienation from the *Spannkreis* was both difficult and liberating for the young Morgenstern: "I see again how awfully lonely I am and how I miss the relationship with several like-minded people. I go everywhere alone and I have been used to it for ever. That's why I feel some kind of release to be freed from the Spann circle" (Morgenstern Diary, Nov. 31 1924, OMDU, quoted in Rellstab, *op cit*, p. 28).

The Critic

Because of the disgrace he later brought upon himself, in 1938, Hans Mayer is the neglected figure in the history of Austrian economics in the interwar period.[16] Having taught at Freiburg in Switzerland, he was decorated in the war, and then taught at Prague and Graz before returning to Vienna, thanks to his protector, Wieser. Like Spann, towards whom he bore great antipathy, Mayer assembled a group around him and ran a seminar, which included his assistants Paul Rosenstein-Rodan, Alexander Gerschenkron, and Morgenstern. He was the first editor of the *Zeitschrift für Nationalökonomie* and became President of the Economic Society (*Nationalökonomische Gesellschaft*) in 1928.

A theorist in the tradition of Menger and von Wieser, Mayer was interested in the imputation problem.[17] However, amongst the issues to which considerable discussion was devoted in his seminar in the late 1920s were those of the incorporation of time into equilibrium theory, the psychological bases of marginalist economics, and the place of mathematics in economic analysis. Regardless of what later was said of Mayer,[18] his critical writings here are an integral part of the late Austrian perspective, and helpful in understanding the development of Morgenstern as critic.

Mayer saw himself as an Austrian bulwark against Marginalist orthodoxy. As early as 1911, in a review of Schumpeter's *Das Wesen und der Hauptinhalt der theoretischen Nationalökonomie*, he criticized the use of the differential

---

[16] Given Schumpeter's erudite celebration of figures infinitely more obscure – not to mention his own provenance – his omission of Mayer from the *History of Economic Analysis* in 1954 can only have been deliberate. Vaughn (1994), too, has very little to say about Mayer, and in Cubeddu (1993), Mayer does not even appear in the Index. Caldwell (2004), deals with Hayek's 1930 critique of static equilibrium and his interest in causal-genetic theory, but gives little attention to Mayer. On the latter, see Mahr (1956), Weber (1961), and Craver (1986a), pp. 10–13. For a harsher view, see Mises (1978) pp. 94–99.

[17] This was the problem of identifying the proportions of the value of the final product attributable to the factors of production used in its creation. See Mayer (1928), an article regarded by Craver (1986a) as representing the end of the Austrian phase of this discussion. She claims that further formal development was impeded by mathematical weakness.

[18] Of Mayer, Mises (1978) writes harshly: "Mayer was the favorite pupil of Wieser. He knew the works of Wieser and also those of Böhm-Bawerk and Menger. He, himself, was totally without critical faculty, never manifested independent thought, and basically never comprehended what it was all about in economics. The awareness of his sterility and incapability depressed him badly, and made him unstable and malicious. He occupied his time with an open fight against Professor Spann and with mischievous intrigues against me. His lectures were miserable, his seminar not much better. I need not be proud of the fact that the students, young doctors and the numerous foreigners who studied in Vienna for one or two semesters, preferred my instruction" (p. 94). This view may well have coloured the opinion of Mayer held by the Austrian school in North America.

calculus in economics. He said that although the possibility of infinitesimal change was plausible in the measurement of time and space, it was inapplicable in the consideration of economic quantities: what sense did it make to speak of the satisfaction yielded by an infinitesimal portion of, say, a shoe? He stated that in the economic realm, reasoning in mathematical terms, made possible through the adoption of methods that had proven themselves in the natural sciences, marked a surrender to pure "form".[19] By the mid-1920s, the supposed sterility of the use of mathematics and the excessive simplification involved in reducing subjective action in time to a static, mathematical description had become essential motifs in Mayer's work. In his seminar meetings, he took his students through the early work of Cournot, the Lausanne School, and Jevons. One might say he was putting the concept of static equilibrium on trial.[20]

According to Mayer, Jevons, in his intention to explain the process of price formation, exemplified the causal–genetic type of inquiry that the Austrian school wanted to promote. However, in succumbing to representation by the mechanical concept of equilibrium, and in particular by the law of equal marginal utility, Jevons had fallen short of a truly dynamic, causal theory. Mayer said, "Everything developed by Jevons up to the law of the equalization of marginal utility completely coincided with what the Austrians developed with their pure focus on causal theory... But the Austrians did not put forward the law of a uniform level of marginal utility; they expressly rejected it".[21]

Mayer regarded this "Law of the Level of Marginal Utility" as empirically unfounded. It was based on the assumptions that the individual's utility functions for each good were continuous and that these utility functions existed "alongside one another" (1932, p. 78) so to speak, at the *outset* of the choice problem, so that a change in income could be distributed amongst *all* possible wants. Mayer said similar assumptions underlay the indifference curve representations of Edgeworth, Fisher, and Pareto. Mayer disputed the claim that an economic subject would distribute an increase in income amongst all possible uses in such a way as to maintain the equality of weighted marginal utilities: "Experience shows that any fall in income is not

---

[19] See Mayer (1911).
[20] See Mayer (1932). On genetic-causal equilibrium, see Fossati (1957), pp. 43–45.
[21] Mayer (1932), p. 75. There was also the inevitable shot at Spann: the inadequacies of value theories had nothing to do with "that distinction between 'individualist' and 'universalist' theories which may be appropriate for a judgemental (normative) attitude to things but which becomes a meaningless, artificial, illusory opposition if applied to theoretical work whose only aim is to elucidate processes of empirical reality" (p. 56).

at all followed by a decrease in all branches of consumption: that in such a case the quantity of some goods remains constant, that of others is reduced in varying degrees, more sophisticated items are replaced with cruder ones, and some kinds of goods are given up altogether. *The derivation of the law of marginal utility level from external experience must therefore be regarded as completely unsuccessful*" (ibid, pp. 79–80, emphasis in original).

Mayer said mental facts, or self-observation, sufficed to establish two things. First, want satisfactions of different kinds were interdependent and, second, they were linked, not in the simultaneous manner suggested by indifference curves, but in a dynamic, causal, fashion: the relationship between wants and their satisfaction was not one of general, mutual dependence as suggested by the indifference mapping, but one in which new wants *emerged* in time, depending on the degree to which existing desires had been satisfied. It was thus invalid to assume that all wants were present at the beginning of the "problem". The "postulate of the law of equal marginal utility . . . becomes impossible in the real world of the psyche" (p. 81). Given that it was the basis on which the theory of general equilibrium depended, if "this fundamental law of the equalization of the level of marginal utility did not hold, the whole theoretical system of equilibrium prices would lose its main support" (p. 77).

According to Mayer, conventional equilibrium theory made it appear irrational to satisfy a particular want completely, except in the trivial case in which satiation was reached for all goods simultaneously. It did not allow for the fact that a particular want could die out at the point at which new wants appeared. The quantity and sequence of previous consumption was critical in determining the configuration of wants faced by the economic subject at any given moment.

It is thus quite clear that the law of marginal utility level is an impossibility. It assumes a synchronization of needs satisfaction in the different branches at a level simultaneously applying to all . . . whereas the essence of things is a *sequence*. It is as if one were to express the experience of aesthetic value on hearing a melody – an experience determined by successive experiences of individual notes – in terms of the aesthetic value of the simultaneous harmonization of all notes making up the melody (p. 83).

Mayer pursued this theme of "inappropriate collapse" in the critique of Jevons' equations of exchange, in which he claimed that the assumption that the price ratio was *constant* during the exchange process was invalid. The English economist realised fully that the assumed succession of minor exchanges by which equilibrium was reached, coming one after the other, would alter the determinants of utility and thus the exchange relations. However, Mayer claimed, he deliberately elided this problem of dynamics by resorting to the static law of indifference, according to which any two units of the same good coming onto the market at the same time must fetch the same price. In the case in which two individuals form the entire market, one could not assume that the marginal exchange ratio would be the same as the previous partial exchange ratios, said Mayer. Instead of inquiring into the process by which price ratios emerged genetically from the sequence of exchanges, Jevons had resorted to a static picture in which were given not only the utility scales but, indeed, that which it was the theory's task to explain – the price ratio. Mayer argued Jevons had jumped to the conclusion – the representation of an already attained equilibrium – without considering the process. It was similar to assuming that the result of a game of chess could be known before the game had even been played, said Mayer.[22]

Mayer also criticized Walras' 1874 *Elements of Pure Economics*, which laid out the simultaneous equations that described an exchange economy, the roots of which were the prices and quantities that, when in equilibrium, cleared all markets. Of this, Mayer wrote: "No element is given before any other: there is no one-way causal connection between them, but they all mutually determine one another; they relate to one another in *all-round, reversible dependence*, in 'general interdependence,' as variable elements of a closed system . . . a simultaneous system excluding time and causality" (pp. 59–60).

In Mayer's view, the adoption of the mechanical metaphor of equilibrium by both Jevons and Walras had the effect of inducing one to see exchange in static terms and to gloss over the problem of price formation.

The central problem therefore remains unsolved: how are the exchange ratios for the first exchange acts determined – those ratios from which the transfer totals . . . themselves result? . . . [Q]uite apart from the unreality of the law of marginal utility level, this is in fact no solution: it is a *hypothesis* that, by analogy with the laws of

---

[22] "*Only at the end of the process*, then do the exchange totals (x, y) ensue through summation of the partial quantities of goods transferred in the individual phases of the barter. They are not known beforehand to the parties to the exchange – if they were, the parties would immediately barter the full totals and there would no point in the succession of small partial exchanges; nor do they become definitive before the conclusion of the last partial exchange – just as, at the beginning of a game of chess, the result is neither subjectively known to the players nor objectively settled. For the final totals depend upon the exchange ratios which have come into force in the individual, successive, partial exchanges, just as each move in a game of chess is dependent upon previous ones and the final result upon all the moves that have gone before" ([1932], p. 91).

mechanics by which the collision between several bodies of different velocity and weight must eventually lead to a predictable state of equilibrium, so in the case of dealings between different economic subjects, a definite equilibrium corresponding to the law of marginal utility level and therefore definite exchange ratios must establish themselves after a complicated and, in its details, *unexplained* process has run its course... (*op cit*, pp. 93–94).[23]

The inappropriateness of mechanical explanations of exchange, the importance of action and reaction in time, the centrality of genesis and process: these were the issues with which Morgenstern grappled, along with Hayek, Machlup, Haberler, Rosenstein-Rodan, Gerschenkron, Steffie Braun, and the others in Mayer's seminar. From Mayer, Morgenstern inherited not only an early suspicion regarding the use of mathematics in economics but a scepticism about the ability of marginalist economics to deal with time. In these attitudes, he was reinforced further by Mayer's archrival, the third of the Prima Donnas, Ludwig von Mises.

The Ideologue

The appointments of Spann and Mayer were a source of controversy, largely because of the presence of two other prominent candidates, Schumpeter and Mises. A generation ahead of Morgenstern and a man of great accomplishment, Schumpeter apparently was passed over because of his actions as Finance Minister in the first coalition government of the Austrian Republic, when he sullied his reputation by refusing to compensate holders of government bonds for the postwar inflation of 1919.[24] By the time Morgenstern was a student, Schumpeter had already left the country and was well established at Harvard.

The case of Mises, who occupied a curious position in Viennese economics, was a little more intricate.[25] He had studied at the University of Vienna in the first decade of the century, completing several historical studies under Carl Grünberg, and then taught at various independent academies and institutes before being appointed *Privatdozent* at the University in 1913. Several years later, he became Professor *Extraordinarius*, or Associate, but he did not go any further in academia and his greatest disappointment was his failure to

---

[23] And again on Jevons, "The mechanical analogy and the simultaneous equations based upon it are thus inapplicable to the problem of price formation" ([1932], p. 95).

[24] See Craver (1986a). On Schumpeter, see Morgenstern's 1951a obituary, Allen (1990), and Swedberg (1991).

[25] On Mises, see his 1978 *Notes and Recollections*, which was written in 1940 upon his arrival in the United States; Hayek's "Foreword" in Mises 1922, pp. xix–xxiv; and the essays in Hayek (1992).

be appointed to a chair in economics, particularly in 1923 when von Wieser moved aside to leave room for Mayer. Why Mises had been overlooked in that case was later attributed by Hayek to the fact that he was Jewish and, politically, an outspoken classical liberal. Although the first alone would have constituted a barrier to a university appointment, it was not insurmountable if one had the support of other Jewish members of the faculty. However, in contrast to the majority of the faculty, most of the Jewish professors were social democratic, left-leaning progressives and therefore were antipathetic to Mises.[26] In addition, Mises was given to a certain brusqueness, which handicapped him in academic politics. Denied the recognition he believed he deserved, Mises remained antagonistic towards Mayer.

Mises nonetheless exerted a great influence on those who entered his sphere. An official at the Vienna Chamber of Commerce, he gave an occasional class or seminar at the University, and, from the beginning of the 1920s, conducted a fortnightly evening *Privatseminar* at his professional office. Its clientele overlapped with that of the Mayer group and included, amongst others, Haberler, Hayek, Felix Kaufmann, Fritz Machlup, Rosenstein-Rodan, Karl Schlesinger, and Alfred Schütz.[27]

As a theorist, Mises made his early reputation as a monetary expert. In the first decade of the century, he wrote several essays on Austro-Hungarian monetary policy and emerged as an independent thinker, unafraid to oppose powerful interests. Before World War I, for example, he argued for the legalisation of gold payments by the Austro-Hungarian Central Bank, despite great opposition from the Austrian monetary authorities.[28] He followed this with a 1912 book, *The Theory of Money and Credit*, in which he broke with the implicit Austrian view that money was neutral or stood apart from *real* economic activity of exchanging goods and services. Mises wanted to cover ground that Menger and Böhm-Bawerk had neglected and he placed new limitations, beyond anything occurring in Menger's *Grundsätze*, on what could be said about "value". There were acts of valuation of individual objects, he claimed, but it was not possible to

---

[26] A "Jewish intellectual who justified capitalism appeared to most as some sort of monstrosity, something unnatural, which could not be categorised and with which one did not know how to deal. His undisputed knowledge of the subject was impressive, and one could not avoid consulting him in critical economic situations, but rarely was his advice understood and followed", Hayek (1992), p. 157. See also Hayek (1994), p. 59.

[27] On the Mises seminar, see Mises (1978), pp. 93–100; Craver (1986a); Hayek (1992); and the account by Haberler in Mises (1980), pp. 276–78.

[28] Mises later hinted that the real reasons underlying the opposition of the bank management was a corrupt scheme of secret accounts and side payments. See Mises (1978), pp. 43–53.

calculate or measure value, and any imputation of the value of a total supply from that of a part, or vice versa, was illegitimate. This sharply disaggregative view left no place for the equations of exchange of Irving Fisher or Gustav Cassel: it made no sense to speak of a constant relationship between the quantity of money and *the* price level. To consider the effects of changes in the quantity of money, one had to do a step-by-step analysis, explicitly allowing for time and the effects, with varying lags, on different economic quantities.

Regarding the use of mathematics in economic theory, Mises was aware of Menger's and Böhm-Bawerk's views, but he pushed the matter much further. Mathematical representation, he said, was at best a concise way of portraying ideas already formed prior to the use of formalism: it could never add anything by way of theoretical insight. When used inappropriately, it could induce one to see economic activity in a mechanical manner, to search for "equilibrium" when, in reality, there was no equilibrium to be found, when overcoming "frictions" was what economic action was all about.

If no condition is considered 'normal', if we are aware that the concept of a 'static equilibrium' is alien to life and action which we study, and that this concept is merely a mental picture we use in order to comprehend abstractly human action through the idea of a state of nonaction, then we must recognize that we always study motion, but never a state of equilibrium. All of mathematical economics with its beautiful equations and curves is nothing but useless doodling. The equations and curves must be preceded by nonmathematical considerations; setting up equations does not enhance our knowledge. Because there are no constant relations in the field of human action, the equations of mathematical catallactics cannot be made to serve practical problems in the same way the equations of mechanics solve practical problems through the use of data and constants that have been ascertained empirically.[29]

This passage, of course, could have been written by Mayer, and the similarities between his and Mises' views are striking. Each was critical of the static theory of Jevons and Walras, and each regarded as illegitimate the conception and representation of equilibrium theory in terms of the differential calculus.[30] However, so great were the personal divisions between the two men that any common ground remained unexplored.

[29] Mises (1978), p. 58.
[30] "Thus far, the use of mathematical formulations in economics has done more harm than good. The metaphorical character of the relatively more easily visualized concepts and ideas imported into economics from mechanics . . . has been the occasion of much misunderstanding. Only, too often, the criticism to which every analogy must be subjected has been neglected in this case". Mises (1960 [1933]), p. 117.

In terms of the Austrian economic lineage, Mises was closer to Böhm-Bawerk than to Wieser. Böhm-Bawerk was Finance Minister at the time of Mises' involvement in the pre-war monetary debates and had acted as confidante. In the first decade of the century, when Böhm-Bawerk took an academic chair upon retirement from the Ministry, Mises was a prominent participant in his seminar. By the time Morgenstern was a student, this seminar had achieved a reputation of legendary proportions as the site of debates on Marxism, with Böhm-Bawerk and Mises aligned against the Viennese socialists, Otto Bauer and Otto Neurath.[31] After the war, socialism gained ground in Hungary, Germany, and Austria, and the climate of popular opinion tended towards planning, with the arguments of Engels, Kautsky, Neurath and others dominating discussion. In 1920, in the form of his paper, "Economic Calculation in the Socialist Commonwealth", Mises launched a vigorous attack against the arguments for common ownership of the means of production. He followed this with a 1922 book, *Socialism.*

Mises' arguments against planning were simple. Exchange, he said, was central to value determination and relied on both private ownership of the goods to be traded and the freedom to enter exchanges. The socialisation of the means of production, by removing them from the nexus of exchange, would sever the causal link running from valuation of consumption goods to valuation of factors or goods of higher order. The value of the latter would thus have to be ascribed by fiat, by the arbitrary design of a committee, or by Party chief. According to Mises, with money prices denied their role as signals of social value, the process of production through time and the perpetual allocation and reallocation of scarce factors to different uses would become mere "gropings in the dark" (1960 [1933], p. 101). Although a human intelligence could conceivably plan the activities of a *household* through time without the benefit of monetary exchange, it quickly became impossible with an economy of any size:

[T]he mind of one man alone – be it ever so cunning, is too weak to grasp the importance of any single one among the countlessly many goods of a higher order. No single man can ever master all the possibilities of production, innumerable as they are, as to be in a position to make straightaway evident judgments of value without the aid of some system of computation. The distribution among a number of individuals of administrative control over economic goods in a community of men who take part in the labour of producing them, and who are economically

[31] See Böhm-Bawerk (1896); Mises (1978) pp. 40ff.

interested in them, entails a kind of intellectual division of labour, which would not be possible without some system of calculating production and without economy (1920, p. 102).[32]

If, to borrow Mayer's terms, the economy was a crucible, then one had to respect its mysterious process. The removal of factors of production from the market would obliterate their relative prices as signals of value and thus render impossible the rational allocation of factors to economic activities. Price signals were crucial given the limited capacities of the human intelligence when faced with the complexity of the economy. If one added Otto Neurath's proposal to dispense with monetary calculation and replace it with calculation *in natura* to socialisation of the means of production, then nothing short of civilization itself was threatened.

[It] would be impossible to speak of rational production any more. There would be no means of determining what was rational, and hence it is obvious that production could never be directed by economic considerations. . . . Rational conduct would be divorced from the very ground which is its proper domain. Would there, in fact, be any such thing as rational conduct at all, or indeed, such a thing as rationality and logic in thought itself? Historically, human rationality is a development of economic life. Could it then obtain when divorced therefrom? (*ibid*, p. 105).

Like Mayer, Mises insisted upon the complex, dynamic nature of the economy, but towards ends that were different from Mayer's. To Mises, the complexity of the economy ensured, above all, the impossibility of our ever "knowing" it sufficiently well to be able to control it. His critique of equilibrium theory was intricately bound up with his views on political order, and his brand of Austrian economics duly became linked in the public eye to a pronounced liberal politics.

For all their shared theoretical ground, the relationship between Mises and Mayer was one of mutual hostility. Mayer was jealous of Mises' popularity, and matters were not helped by his anti-semitic streak. Personal acrimony affected how they interpreted the Austrian tradition and with whom they chose to align themselves. Each had his interpretation of doctrinal history, which was why Mayer was keen to insist on the distance separating Wieser and himself, on the one hand, and from Mises on the other. Each had his sphere of influence and many students, such as Morgenstern, were participants in both groups.

---

[32] Here, incidentally, in the context of an attack on socialism, we have Mises broaching the theme of the division of knowledge, of which Hayek (1937) was an elaboration.

Conclusion

These were the early influences on the young Morgenstern. It was in this setting that his early sensibilities and outlook were shaped. His scientific emphases were resolutely Austrian: the critique of the notion of static equilibrium, the quest for a theory of process, and a suspicion of the subversive effect of mathematical representation. As we shall see, what is remarkable is the manner in which he would maintain certain elements of this epistemological critique, yet turn towards the mathematicians for guidance.

SIX

Wrestling with Complexity

Wirtschaftsprognose and Beyond

Again Holmes would have calculated that, and therefore would have chosen Dover. Thus, Moriarty would have acted differently. And from so much thinking, there would have come no action, or the less intelligent one would have handed himself to the other at Victoria Station because all the fleeing would have been unnecessary. Examples of that kind could be taken from everywhere. Chess, strategy, etc.
Morgenstern, *Wirtschaftsprognose*, 1928, pp. 98–99

Rockefeller Years

When Morgenstern entered the University of Vienna, the Laura Spelman Rockefeller Memorial had been in existence for several years. Named after John D. Rockefeller's widow, its endowment ran more than $70 million and it had become the main source of Rockefeller support for the social sciences before being absorbed into the Rockefeller Foundation in 1929.[1] The Memorial's favourable disposition towards social science research during the 1920s was attributable largely to its director, Beardsley Ruml. He arrived at Rockefeller with a doctorate in experimental psychology from the University of Chicago and was interested in the construction of a less "abstract and remote" social science that would be of immediate use to the "social engineer".[2]

Ruml established a system of grants, placing emphasis on empirical research and preferring universities that also emphasised teaching. In the United States, the universities of Chicago, Harvard, Columbia, and Iowa State, and the National Bureau of Economic Research were amongst the main beneficiaries. In England, money was given initially to the London

---

[1] On the history of the Rockefeller Foundation, see Craver (1986b), and other works cited therein, including Fosdick (1952) and Bulmer and Bulmer (1981).
[2] Craver, *op cit*, p. 206.

School of Economics and later to Oxford. The Cambridge of Keynes and Robinson, on the other hand, showed little interest in shifting focus in response to Rockefeller, and was passed over. In Continental Europe, Rockefeller tended not to support universities directly, in the belief that European social science was still highly speculative, the universities were often torn by internecine strife, and the risks of wasting money through bureaucracy were high.[3] Consequently, the Foundation devoted its efforts to a programme of fellowships intended to free future teachers "from the traditional conceptual thinking and *a priori* generalizations of the present generation of teachers" and to the funding of institutions that were independent of universities.[4] The choice of Rockefeller Fellows depended on the favour of local advisors such as Gösta Bagge in Sweden and Karl Pribram in Vienna. It was at Mayer's recommendation that von Wieser put in a good word on Morgenstern's behalf to Pribram. In order to qualify for the fellowship, Morgenstern had to accelerate the completion of his first degree – his "doctorate" – with an essay on marginal utility, which he duly dispatched in June 1925.[5] He left Vienna in September 1925 and, over the course of the next three years, visited England, the United States, France, and Italy.[6]

In England, he spent most of his time in London, where he was registered at the London School of Economics (LSE). He improved his English and became familiar with recent work in economics in that country. At the LSE, he met A. L. Bowley and T. E. Gregory and, in frequent meetings at the Political Economy Club and the Athenaeum, F. Y. Edgeworth from Oxford. The latter impressed him greatly. After Edgeworth's death, Morgenstern wrote that his work embodied that quality of "*Wertfreiheit*" so stressed by Max Weber. There was a modesty in Edgeworth's theoretical aims, with none of the extravagances associated with those German "system builders", whom he held in disdain. Edgeworth "came closer to the ideal of the 'pure scientist' than hardly (sic) any other in the social sciences. . . . His theories are generally free from value judgements, to a higher degree . . . than those of many of his contemporaries and he was fully aware of this need in regard to logical properties and necessities" (1927a, p. 477). Edgeworth also helped Morgenstern gain some distance from German economics, for he

---

[3] See Craver, *op cit*, who notes that exceptions here included the *Institut universitaire des hautes études internationales* at Geneva, the Institute of Social Science at the University of Stockholm, and the Institute of Economics and History at the University of Copenhagen.
[4] Quote from Rockefeller official, economist Lawrence Frank, in Craver, *op cit*, p. 209.
[5] I have been unable to trace Morgenstern's *Dr. rer. Pol.* thesis.
[6] See material covering Morgenstern's fellowship activities in OMDU, Box 1, Folder I.-S, Autobiographical Material.

"helped to protect economics in England from those extravagances which are so typical to the development of economics in Germany [where] every beginner believes... that he should create entirely 'new' foundations and a completely new methodology. In England, however, the use of mathematics is more common than elsewhere, a further circumstance to frighten away dilettantes who are only half interested in economics" (*ibid*, pp. 478–79).

In conversations with Morgenstern, Edgeworth showed interest in questions concerning what economic actors know and how they interact with each other, and agreed that the temporal dimension of economic theory, in which the mathematical difficulties were very great, was its most unsatisfactory and difficult aspect. Morgenstern was struck by his dictum: "The most helpful applications of mathematics to economics are those which are short and simple and which employ few symbols; and which aim at throwing a bright light on some small part of the great economic movement rather than representing its endless complexities".[7]

Just before Christmas 1925, Morgenstern left for the United States, in time for the annual meetings of the American Economic Association. He spent the next six months at Columbia University in New York. There, his closest contacts were with Wesley Mitchell and Henry Moore; he also frequently saw Frank Fetter from Princeton. After that came several months at Harvard, where he had contact with Charles Bullock, Warren Persons, and others involved in the analysis of business cycles.

Prediction

Morgenstern's choice of business cycles as a topic of study was continuous with discussions at the time in Vienna. Upon completing his studies in 1921 and following an introduction by Wieser, Hayek was employed as a temporary civil servant under Mises in an office established to administer the settlement of pre-war debts – the *Abrechnungsamt*. As an official of the Chamber of Commerce, Mises was one of the directors of the office, and he quickly became a source of support and guidance for Hayek. Not long afterwards, Hayek travelled to New York, spending a year as research assistant at New York University to one Professor J. Jenks, another occasional

---

[7] Morgenstern 1927a, p. 478. "His magnificent scholarship, his vivid interest in the development of economics in all countries and his outoing friendliness will forever remain unforgettable. He had astonishing knowledge of the most recent German language literature and had very sharp views about the current 'builders of systems' which showed a profound feeling for the situation in Germany. He viewed the strong methodological bent there with suspicion and thought little of its productivity" (*ibid*, pp. 478–79).

visitor to Vienna. He returned from New York, in 1924, with "a new idea of great predictions, the sort of thing which the Harvard economic barometer had developed in the 1920s".[8]

By the late 1920s, the most prominent contributions to the study of the cycle fell into two groups: the attempts by Henry Ludwell Moore, like Jevons before him, to show the existence of a periodic economic cycle generated by noneconomic forces, and the empirical work by both Mitchell, at the National Bureau of Economic Research, and Persons and Bullock, constructing the business barometers at Harvard's Committee for Economic Research.[9]

Critics of Moore's 1914 *Economic Cycles – Their Law and Cause* and his 1923 *Generating Economic Cycles* rejected the idea that the business cycle could be reduced to a single general cause, and his use of periodic cycles and frequency analysis did not catch on. Other, simpler, statistical ideas and methods were developed and the search for an explanation through statistical regularities gave way, in the work of Mitchell, Persons, and others, to descriptive empirical analysis with relatively little emphasis on theory. Mitchell and Persons resembled each other in that they did not use statistical data for the construction of qualitative explanations of the business cycle. They differed in that the former was interested in empirically defining the cycle, whereas the latter sought to represent it. Between the two of them, their empirical approach dominated business cycle work in the 1920s and 1930s.

By 1927, when Morgenstern visited him at Columbia, Mitchell was the preeminent figure in international business cycle research. In his first book, *Business Cycles and their Causes* (1913), he had stated that the various extant theories of the business cycle were to be tested, not by further examining their logic, but by studying the facts they purported to explain.

The point of interest is not the validity of any writer's views, but clear comprehension of the facts. To observe, analyze, and systematize the phenomena of prosperity, crisis, and depression is the chief task. And there is better prospect of rendering service if we attack this task directly than if we take the roundabout way of considering the phenomena with reference to the theories.... Whatever chance there may be

---

[8] Hayek (1994), p. 67. As it happened, in May 1924, on the boat returning towards Vienna from New York, Hayek learned from von Wieser that he had won a Rockefeller Fellowship. He never availed of the scholarship, however, with employment and marriage preventing him doing so after his return to Vienna. *Ibid*, p. 63ff.

[9] See Morgan (1990), chapters 1 and 2.

of bettering the work already done lies in securing data more full and more precise than the data heretofore employed (p. 20).

Mitchell did not reject the need for a theory: he simply emphasised the need for greater accumulation of basic quantitative facts. His book gave a massive range of statistics for four countries – the United States, England, France, and Germany – for the period 1890–1911, covering prices and quantities and other phenomena such as migration patterns. His analysis of these statistics, the central part of the book, was limited to the presentation of averages and to comparison by means of graphs: "A theory of business cycles must therefore be a descriptive analysis of the cumulative changes by which one set of business conditions transforms itself into another set."

According to Mitchell, "the deep-seated difficulty in framing such a theory arises from the fact that although business cycles recur decade after decade, each new cycle presents points of novelty. Business history repeats itself, but always with a difference. This is precisely what is implied by saying that the process of economic activity within which business cycles occur is a process of cumulative change".[10]

Looking at the various theories of the cycle and comparing them – casually, not quantitatively with his statistical data – Mitchell felt that none of the theories was wrong, but none was clearly right either. No theory, he concluded, could ever be perfectly adequate because of the process of cumulative change of the real economic world. Cycles were not regular, but varied in duration and intensity. Every point in every cycle was unique, the outcome of a unique set of circumstances. Extraneous factors such as the weather, war, and policy changes also played a role in giving each cycle its unique character. He pointed to the Harvard business barometers, suggesting that their improvement might help in making predictions and thus in "perfecting social control over the workings of the money economy".[11]

Mitchell was instrumental in establishing the National Bureau of Economic Research in 1921, and he continued his earlier statistical research there. Here, he produced his second book, *Business Cycles: the Problem and its Setting* (1927), which was a survey of the field, not another statistical study. He was more hesitant than ever about attempting to define the term business cycle: "The more intensively we work, the more we realize that this term is a synthetic product of the imagination – a product whose history

[10] Mitchell (1913) p. 449, quoted in Morgan, *op cit*, p. 46.
[11] See Mitchell, *op cit*, pp. 456–57, and p. 588ff.

is characteristic of our ways of learning. Overtaken by a series of strange experiences our predecessors leaped to a broad conception, gave it a name, and began to invent explanations".[12]

The person most responsible for the development of those methods Mitchell cautiously praised – the decomposition of time series and the construction of indicators of business activity with a view to improving forecasts – was Warren Persons. He joined the faculty at Harvard in 1919, brought there by the Committee for Economic Research, newly established in 1917 and run by Charles Bullock. The aim of the Committee was to study the collection and interpretation of economics statistics and, in 1919, it launched the *Review of Economic Statistics* to provide an outlet for work of this kind.[13] Persons was still at Harvard in 1927, when Morgenstern visited, although he left the following year to become a private consultant.

Although Persons, like Mitchell, was aware that theoretical preconceptions preceded the very decision regarding what data should be considered important, his approach to the business cycle was resolutely empirical: the emphasis was on representation rather than explanation. In two papers published in 1919, Persons raised the level of sophistication of the statistical methods used.[14] The second of these combined the derived cyclical data of twenty time series to construct composite indicators of economic activity. In order to combine the various series into well-matched groups, Parsons superimposed them on transparent sheets and asked "observers" to compare them visually. It was clearly somewhat hazardous, as the extent to which any two series seemed to match each other depended on how the superimposition was done – for example, the duration of the lag allowed. Opinions differed regarding the degree of correlation observed. Persons settled on three groupings that were used to construct three indices of the business situation – the Harvard A-B-C curves, representing speculation; physical productivity; and the financial markets. The result was a representation of economic activity constructed with minimal recourse to *a priori* theory. As far as method was concerned, the whole was very dependent

---

[12] Mitchell (1927), p. 2, quoted in Morgan *op cit*, p. 49. "Perhaps some single factor is responsible for all the phenomena. An acceptable explanation of this simple type would constitute the ideal theory of business cycles from the practical, as well as from the scientific, viewpoint. But if there be one cause of business cycles, we cannot make sure of its adequacy as an explanation without knowing what are the phenomena to be explained, and how the single cause produces its complex effects, direct and indirect" (p. 180, quoted in Morgan *op cit*, p. 49).

[13] Renamed the *Review of Economics and Statistics* in 1949.

[14] See Persons (1919a; 1919b).

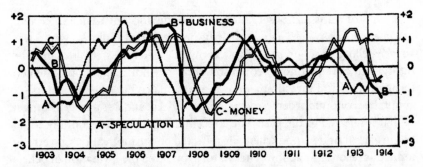

Figure 6.1. Persons' "Index of General Business Conditions", in Wesley Clair Mitchell (1927), *Business Cycles: The Problem and Its Setting*, p. 294. *Credit:* Reprinted with permission of the National Bureau of Economic Research.

on a simple visual approach: nothing more than two-variable correlation techniques was used, so the interdependence amongst three and more variables remained unanalysed, and the question of the theoretical mechanism linking the observed patterns remained open.

However, as Mitchell (*op cit*, pp. 324–26) points out, the main purpose of Persons' three-curve charts was to improve forecasting and aid the social control of the economy (Figure 6.1). Even if the materials were thin and somewhat arbitrarily chosen, it remained a fact that, for the 1903–1914 period, the cyclical fluctuations of business (curve B) followed those of speculation (curve A), with an average lag of eight months. Money rates (curve C) followed both business, with an average lag of four months, and speculation, with a delay of one year.

Vienna's Mises was critical of all such work. To him, Mitchell's emphasis on the "purely empirical" examination of business cycles, unhindered by theoretical preconceptions, smacked of historicism. Mises argued that even if no satisfactory theory of the business cycle yet existed, it was not going to emerge inductively from the consideration of masses of statistical data. He saw Mitchell's form of Institutionalism as an American reincarnation of the German historical method, and both as species of statist apologetics.[15] Then there was the aim of the Harvard group to improve the predictive

---

[15] Mises wrote: "No economist ever dared to assert that interventionism could result in anything else than in disaster and chaos. The advocates of interventionism – foremost among them the Prussian Historical school and the American Institutionalists – were not economists. On the contrary. In order to promote their plans they flatly denied that there is any such thing as economic law. In their opinion governments are free to achieve all they aim at without being restrained by an inexorable regularity in the sequence of economic phenomena . . . they maintain that the State is God" (1922, p. 487). See also Mises (1960) [1933], p. 8.

capacities of statistical observation: the examination of trends in certain barometers was to help predict their future path, and Bullock's group was already selling its services to private companies. To Mises, the idea that the future path of the economy could somehow be known, and approximated through statistical measurement, was anathema. What statistics could say about the economic order was very limited indeed, and was purely historical in that it referred to the past. The conceptual order of the economy could be grasped or understood when one reflected on the motivations governing individual economic action. This, in turn, involved the formation of beliefs and expectations, which were in perpetual flux. The economy, although it could be "understood", could be only inadequately represented by statistics and could never be predicted.[16] More dangerous, in Mises' view, was the short step that lay between believing that one could capture the economy through statistical representation and advocating state intervention to correct anticipated downturns. He felt institutionalist modesty concerning the possibilities of theory had given way to illusions concerning the capacities of empiricism. All of this found an echo in the young Morgenstern's response to the business cycle literature.

Wirtschaftsprognose

Morgenstern subsequently referred to his *habilitation* thesis as the place he first emphasised the theoretical complications arising from the fact that beliefs about others, and thus beliefs about beliefs, can affect economic actions.[17] It is true: it was here that he became interested in this problem, sufficiently so to take it up again in the early 1930s, in a changed context. However, *Wirtschaftsprognose* was much more than the "prelude" to game theory later portrayed by Morgenstern. It was a treatise on the methodology of economic forecasting, in which he determined to prove the futility of all attempts at prediction. At this point, contrary to what he sometimes later implied, he was not appealing to the mathematicians: he was explicitly

---

[16] "[Economic] knowledge is not quantitatively definite", said Mises. "For example, economics is not in a position to say just how great the reduction in demand will be with which consumption will react to a definite quantitative increase in price. For economics, the concrete value judgments of individuals appear only as data. But no other science – not even psychology – can do any more here . . . we here come up against a boundary beyond which all scientific cognition is denied to us. Whoever wants to predict valuations and volitions would have to know the relationship of the world within us to the world outside us" (1933, p. 118).

[17] See Morgenstern (1976) pp. 805–06.

pessimistic about any possibility of treating these questions in a formal manner.

The following inquiry is an attempt to pick up systematically today's very contemporary problem of economic prediction and to lift the discussion out of its state of aphoristic contradictions. An attempt is made to reach more or less final results, first by illustrating and examining all asumptions, and secondly by seeking to answer the question – which is totally different from that of the problem of possibility – of the impacts of prediction and its sense as a tool of rational economic stabilization (1928, p. iii).

Morgenstern set out to show the impossibility of making any complete forecast of the state of the economy given the complexity of the mechanisms that shape economic events. He came against economic prediction from several angles, even contradicting himself in the process.

Thus was it "in principle" impossible to use economic theory and statistics to make economic forecasts. He based this claim on arguments concerning economic data, processes, and actors. The lack of homogeneity and the small size of samples made data wholly inadequate for statistical induction. Contrary to in the natural sciences, the data problems of economics were so significant as to render futile any attempt to apply probability methods. As for economic processes, Morgenstern held that attempts to understand the business cycle based on statistical considerations alone could never facilitate economic forecasting. For that, one needed to look to the underlying processes, of which prices were the surface phenomena. These mechanisms, however, lacked the regularity necessary to make them useful for any kind of prediction: only loose and inexact laws could be discovered. Finally, even when predictions were made, their effect was to create anticipations on the part of consumers, the reactions of whom would only serve to make the original forecast false. Unlike astronomy or medicine, the social sciences had the peculiarity of being able to affect their object of study. The prediction of the astronomer could have no effect on the movement of the stars, but that of the economist could change economic events. Morgenstern's criticism of static theory is thoroughly Misesian: "In the static economy, nobody acts economically any more, and that means that they no longer ascribe value, and that no more acts of choice are made, and no decisions are made, because everything stands still" (*ibid*, p. 7).[18]

---

[18] Although the themes developed by Morgenstern strongly echo those discussed by Mises and Mayer, the thesis contained no direct reference to Mises. Explaining his conflict with Mayer, the latter later wrote that "those who sought to qualify for a university lectureship had to be careful not be known as my students" (1978, p. 95).

However, in the real economy, individuals had a system of orientation points (*Orientierungspunkte*) that included both their knowledge of the laws of nature and their knowledge and beliefs about *other* economic subjects: "every action influences the other actions, and each is reflected in the other. The set of all different actors is similar to a cupola, in which every stone supports the others, and vice versa, and none can stand freely on its own" (pp. 92–93).

Morgenstern insisted this element of interdependence defeated any attempt at economic prediction. Because economic actors would always incorporate such an authoritative forecast into their plans, the prediction could *never* be realized. Any attempt by the government to take account of the public reaction would only lead to a similar attempt by the public. The result would be a game of guessing and outguessing, the outcome of which could never be known in advance. To illustrate the idea, Morgenstern turned to Arthur Conan Doyle:

An analogy would be appropriate here, which is at the same time quite amusing: as Sherlock Holmes, chased by his enemy Moriarty, leaves from London to Dover with a train which stops at an intermediate station, he gets off the train instead of going on to Dover. He saw Moriarty at the train station and, considering him very intelligent, expected that Moriarty would take the faster train, to await him in Dover. This anticipation of Holmes turned out to be right. But what if Moriarty had been even more intelligent and had considered Holmes' capacities even greater, and therefore had predicted Holmes' action? Then he [Moriarty] would obviously only have gone to the intermediate station. Again Holmes would have calculated that, and therefore would have chosen Dover. Thus, Moriarty would have acted differently. And from so much thinking, there would have come no action, or the less intelligent one would have handed himself to the other at Victoria Station because all the fleeing would have been unnecessary. Examples of that kind could be taken from everywhere. Chess, strategy, etc., but there one needs to have special knowledge, which simply makes the examples more difficult (1928, pp. 98–99).

As before with Mayer, the chess metaphor was once again being employed to highlight the *intractability* of sequential interaction, the difficulty of dealing with process. Although the context had changed – before, in Mayer's case, the problem was that of price formation; here, it was that of stabilising a prediction – chess still served as an exemplar of the complex, the unforeseeable.

The "entire modern methodology since Menger–Rickert–Max Weber contains all the arguments against attempts at prediction, but the diehards don't seem to be aware of this", wrote Morgenstern (*ibid*, p. 117). Statistical services such as the Harvard Business Barometer, Moody's, and the Babson and Brookmire agencies could never "help to improve the rationality of the economic system" (p. 122). Their only legitimate function was to collect statistics and make information available, with as little interpretation as possible, especially regarding the future.

In reviews of *Wirtschaftprognose,* Arthur Marget and friend Eve Burns were deft in responding to Morgenstern's nihilism.[19] Marget rebuffed him in his own words: "One feels that Dr. Morgenstern has here given us what, despite his declared intention, can be described only as a satire" (1929, p. 332), and proceeded to strip *Wirtschaftsprognose* bare.[20] Burns, likewise, suggested that his emphasis on complexity was tantamount to abandoning "all hope of creating a usable economic science, or indeed any science at all" (p. 161). Lionel Robbins later read the work and although he admitted that the disastrous experiences of the Depression had shown economic forecasting to be a fiasco, he nonetheless felt that Morgenstern was going too far: "I do not believe . . . that it is necessary to arrive at a scepticism like that which my friend Dr. Oskar Morgenstern has shown in Wirtschaftsprognose. If I had to choose between Dr. Morgenstern and Harvard, I would be very embarrassed. But in order not to suppress all hope of the future of economic science, I would maintain Harvard in being".[21] He went on to say that he shared Morgenstern's "scepticism about the possibility of deriving laws, concerning changes in the heterogeneous data, and I think that statistical services have for their principal object the development of precise and verifiable statistics. Nevertheless, I believe that to disown all power of diagnosis and of prediction is to exhibit a truly superfluous austerity" (Robbins [1938], p. 175). Privately, Morgenstern's friend Haberler also took him to task.[22]

---

[19] See Marget (1929), Burns (1929).

[20] Marget insisted that the Harvard methods did *not* rely on the use of probability techniques, pointing to Persons' own insistence re same; that, even if they had, this in itself was not an argument against prediction *per se*; that there was no *a priori* reason why anticipations should respond to predictions in such a way as to contradict the forecast and that, in Morgenstern's example of the "total prediction", the effect was to *reinforce* the direction of the prediction; that Morgenstern's claim that a prediction, to be useful, must be *exactly* right, was an unrealistic demand; and that his argument against prediction based on the operation of noneconomic factors was, if interpreted logically, an argument against economic science itself.

[21] See Robbins (1938), translated from the French and quoted in O'Brien (1988), p. 172.

[22] "You say 'I wanted to write a formal methodological work. And there I can't mix in empirical research.' To respond to that concretely: I suggest that, for you, the knowledge goal (the impossiblity of prediction) is more important than the method, and that's why you say that the knowledge goal cannot be reached with empirical methods. If you want to concentrate on formal methodological methods . . . you have to choose another topic, one that can be handled that way". (Letter, Haberler to OM, Jan. 2, 1929, OMDU).

In all its confusion, the book reflected the various influences on Morgenstern at the time. It remained faithful to the general Austrian opposition to Historicism and Institutionalism, emphasising the primacy of theory. Although Morgenstern did not mention Mises in the book, it is difficult not to feel his presence in the emphasis on theory, the critique of Institutionalism, and the attack on prediction. As for Mayer, even if the book lacked his finely wrought argumentation, his influence was reflected in the critical approach and the emphasis on time. However, the book was distinctly Morgenstern's. Written in the trenchant yet convoluted style that characterised many of his writings of the next decade, it was his introduction to the "empire of the mind", a realm in which he was now determined to make his mark.[23]

Reading

For all that, it would be easy to overstate the influence the business cycle literature had on Morgenstern. He was more likely to stay up late struggling, not with business barometers, but with the writings in logic and the philosophy of science that had increasingly begun to interest him. In the previous years in Vienna, Spann had no interest in scientific philosophy, regarding it as a sign of the century's spiritual impoverishment. Neither were these issues discussed in any detail with Mayer. In the Mises group, discussions had touched on recent developments in the philosophy of science, but even here, readings were largely confined to the phenomenological writings of Husserl and Alfred Schütz, a seminar participant.

In England, however, things had begun to change. Morgenstern had just left Vienna, where he had been buffeted between conflicting influences and where questions of method were not only foremost but linked to deep personal hostilities. The Weberian counsel that social theory should be "value-free" had been much touted in debate in Vienna, but these discussions took place in a setting ridden with political intrigue. After this, Edgeworth's calm and simplicity were like a breath of fresh air.

Morgenstern found the qualities encountered in Edgeworth again in Allyn Young, whom he met at Harvard early in 1927, the year he was to succeed Edwin Cannan at the University of London. Like Edgeworth, whom

---

[23] "One phrase of your's keeps running through my head: to the effect that making love was something anyone could do whereas the empire of the mind was reserved for the few who were competent to rule there". Thus wrote David Watson, an American biology student met at Columbia, to Morgenstern in Vienna. (Letter, Watson to OM, Apr. 27, 1932, OMDU).

he, too, admired greatly, Young was as much mathematician and statistician as economist, and he was enormously stimulating for the younger set that surrounded him in evening discussions at his home in Cambridge throughout the best part of 1927. There, he emphasised the need for a reconciliation of the Austrian and Lausanne schools, and Morgenstern later said of him that he regarded Vienna as the only source of hope in the German literature.[24]

Around this time Morgenstern began to extend his reading in philosophy of science and mathematics. He read an earlier little book by Russell on mathematical philosophy and tackled Wittgenstein's *Tractatus*.[25] He also broached Herman Weyl's book of the late 1920s on the philosophy of mathematics and science, and during his time in Cambridge, Massachussetts, was lucky enough to meet Whitehead, which drew him to the latter's Lowell Lectures of two years earlier, *Science and the Modern World*.[26]

Whitehead wrote of the dissolution of the scientific cosmology that had obtained for the past three centuries, a scientific materialism that presupposed the "ultimate fact of an irreducible brute matter" (1967, p. 17). This worldview had now become obsolete because, with the development of relativity and quantum theories, science had reached a turning point:

The stable foundations of physics have broken up: also for the first time physiology is asserting itself as an effective body of knowledge, as distinct from a scrap-heap. The old foundations of scientific thought are becoming unintelligible. Time, space, matter, material, ether, electricity, mechanism, organism, configuration, structure, pattern, function, all require reinterpretation. What is the sense of talking about a mechanical explanation when you do not know what you mean by mechanics? (*ibid*, p. 16).

In his essay on "Mathematics in the History of Thought", Whitehead remarked that "the pursuit of mathematics is a divine madness of the human

---

[24] See Morgenstern (1929), pp. 487–88.
[25] Russell (1919); Wittgenstein (1921).
[26] Weyl (1927); Whitehead (1967, orig. 1925). "[Sunday] evening at the Whiteheads, who had an open house . . . . . . [W]e talked very long and extensively, for example, about Kant, whom he regards highly, and about his stance on Husserl['s position], which he thinks is very much related to his own, then about Dewey, Lewis Mumford, and Emerson, etc. . . . He holds a private seminar every Friday; now on logic (Russell, Wittgenstein, Mill are on the upcoming program). He invited me to participate and naturally I accepted immediately" (OMDU, Mar. 10, 1927). Then, "Tonight at Whitehead's seminar, about Mill's doctrine of judgement. Quite good, but not overwhelming. At least, I learnt much. It showed that between Whitehead and the phenomenologists there is much relation. Next time, Russell is going to be discussed" (*ibid*, Mar. 11, 1927). At subsequent meetings, Wittgenstein and Keynes were the topics of discussion.

spirit, a refuge from the goading urgency of contingent happenings". The essay in question emphasised the human achievement represented by abstract mathematics and the astonishing capacity the latter had shown in applications to various parts of empirical science. Mathematics in the seventeenth century provided the "background of imaginative thought with which the men of science approached the observation of nature. Galileo produced formulae, Descartes produced formulae, Huyghens produced formulae, Newton produced formulae" (p. 30). Their analyses of periodicity, which gave birth to modern physics, were made possible only because mathematicians had already worked out the abstract notions surrounding periodicity. Whitehead wrote, "the paradox is now fully established that the utmost abstractions are the true weapons with which to control our thought of concrete fact" (p. 32). He drew the parallel between the two periods in which mathematics had had its greatest influence on general thought – that from Pythagoras in the sixth century B.C. to Plato, and then in the seventeenth and eighteenth centuries. In both periods, he said, the general categories of thought fell apart, with waves of religious enthusiasm, a turning inwards to the self to find enlightenment, and a movement towards the reconstruction of traditional ways. Now, he said, we had reached such a point again:

[T]he temporary submergence of the mathematical mentality from the time of Rousseau onwards appears already to be at an end. We are entering upon an age of reconstruction, in religion, in science, and in political thought. Such ages, if they are to avoid mere ignorant oscillation between extremes, must seek truth in its ultimate depths. There can be no vision of this depth of truth apart from a philosophy which takes full account of those ultimate abstractions, whose interconnections it is the business of mathematics to explore (p. 34).

Morgenstern would soon find himself trying to reconcile Whitehead's argument for the mathematical analysis of ultimate abstractions with Mayer's and Mises' arguments against it.

From the United States, he returned, temporarily, to Vienna, where his future was taking form. Mayer indicated that he wanted him to take over the editorship of the *Zeitschrift für Nationalökonomie* when his fellowship ended. Mises, who had been instrumental in setting up a small Institute for Trade Cycle Research that same year, appointing Hayek as director, intimated that there would probably be some work there when Morgenstern returned to Vienna. The following year, he spent a few months each in Paris and Rome, where he met the French and Italian economists, and completed the revisions to his thesis, all the while reading further in mathematics and philosophy.[27]

Once back in Vienna, he reintegrated himself into familiar circles. Meetings of Mises' *Privatseminar* continued to be held fortnightly at his Chamber of Commerce office. They met at seven in the evening, later retiring to the café. If it had always been important, this seminar now outstripped anything that went on at the University, and there were frequent visitors from abroad, including Robbins from the LSE, Howard Ellis from the United States, and John Van Sickle, who was connected with the Rockefeller Foundation. In Morgenstern's absence, the tension beween Spann and Mayer had risen. Spann had actually succeeded in blocking Mayer's plan for Morgenstern's *Habilitation*, which was thus delayed for a year, arguing that Mayer constantly supported "Jews" such as Morgenstern. As for Mayer, Morgenstern found him wasting his time in suspicion and speculation. His slothfulness delayed Morgenstern's publications and he became increasingly impatient with him. The conflict between Mayer and Mises continued unabated.[28]

Morgenstern became involved in the *Nationalökonomische Gesellschaft*, or Economic Society, which had been reconstituted in 1928, largely through Hayek's efforts, in a bid to bridge the gap between Mises and Mayer. The latter was made President and the meetings were intended to draw on the common pool of participants. They met in rather formal meetings at the offices of the National Bankers Association, and continued the discussion into the early hours at the Café Künstler. Morgenstern was also involved

---

[27] At Cannes, Morgenstern wrote in his diary of receiving Hilbert's "*Theoretical Logic*" (OMDU, Diary, May 27, 1928). This was presumably Hilbert and Ackermann's 1928 *Grundzüge der theoretischen Logik*, or Principles of Theoretical Logic (Berlin: Springer). In Rome, he referred to the impending visit of several mathematicians: "The programme of mathematicians has arrived. *Very* good: Hilbert, Weyl!" (OMDU, Diary, Aug. 28, 1928).

[28] See Letter, Elster to OM, Apr. 4, 1928, OMDU. A few months later, in Morgenstern's diary: "Yesterday I met Schams in the Café. . . . We should complain about Mayer. Why doesn't Mayer work? He doesn't write, doesn't read, doesn't finish the second volume, he doesn't act on behalf of the journal, but he sits for hours in the café and talks about Spann. We are all of the same opinion of him. We could do much more if we could fill the journal. But it stays at the planning stage, and one gets nowhere. I become green and yellow with anger about the useless journal . . .". (OMDU, Diary, Dec. 22, 1928). Five months later: "Hayek made a good suggestion: Mayer should challenge Spann, along with their followers, to a public debate on some easily comprehensible item. That would be excellent, because Spann would probably appear stupid, or he wouldn't come at all, which would reveal his true colours. But, Mayer will not want to, as usual" (OMDU, Diary, Apr. 19, 1929). See also Craver *op cit*, p. 8, for similar opinions in Gerschenkron and Hayek.

in the *Geistkreis*, which Hayek and Herbert Fürth, preceding him in their defection from the Spann circle, had formed in 1921. It was an eclectic group of students of economics, history, law, and psychoanalysis – amongst them Machlup and Schütz – and they discussed widely in economic and political philosophy, literature, art, and music.[29] As he slipped back into familiar surroundings after three years of travel, Morgenstern felt as though he had grown in some respects. In others, however, it was as though he had never been away.[30]

Around Mises, discussions had moved away from the problems of socialist planning and more towards questions of method he would soon present in *Epistemological Problems of Economics* (1933). No less than before, however, Mises managed to give matters a political slant and all of his deliberations, whether on the difficulty of dealing with time in economic theory or on the naturalness of "frictions" in human action, invariably issued in the endorsement of a nonmathematical economics based on the rationality of individual action, with all public intervention or attempts at collective action being deemed irrational. In time, Morgenstern grew suspicious of this seamless fusion of economic theory and liberalism and found himself searching for a way to respond to it.

Conclusion

At the close of the 1920s, although he had no clear plan for theoretical achievement, Morgenstern found himself questioning aspects of what he had been taught at the foot of Edgeworth and Whitehead and via other readings in philsophy and mathematics. In time, Mises would come in for criticism. As for Mayer, although Morgenstern would soon distance himself from him in the context of Viennese power struggles, he would retain the essence of the Mayerian epistemological critique, though by embracing – not shunning – mathematics and logic. Although Morgenstern's theoretical concerns were still those of the "Austrian" economist – the difficulties of

---

[29] Craver, *op cit*, p. 16, also mentions Felix Kaufmann and Max Mintz, amongst others.
[30] "Monday there was the 'Mises Seminar.' Hayek read a chapter from his book; pretty good, but surely not original. Afterward an unpleasant discussion in this arrogant circle of Jews ... " (OMDU, Diary, Jan. 17, 1929, Box 12, quoted in Rellstab 1991, p. 26). "Yesterday, there was also the Geistkreis; before that, Hayek was here for dinner. Fürth spoke about the renewal of the law. Mintz seems to know a lot, but is dislikeable. I was the only pure Aryan (out of 8!), Hayek is probably only 1/2 or 2/3. This is uncomfortable" (OMDU, Diary, Apr. 25, 1929, quoted in Rellstab, *op cit*, p. 26).

accounting adequately for time, uncertainty, expectations, and the inadequacy of mechanical structures to the description of social processes – he was, in certain respects, becoming a reluctant Austrian, trying to reconcile his mentors' interests with a style of scientific argument that they had rejected. A key influence in the course of the 1930s was that of the son of the founder of the Austrian School, mathematician Karl Menger.

SEVEN

Ethics and the Excluded Middle

Karl Menger and Social Science in Interwar Vienna

Introduction

In the autumn of 1926, the twenty-four-year-old Viennese mathematician Karl Menger was on a two-year postdoctoral stay at the University of Amsterdam, as assistant to L. E. J. Brouwer, when he wrote to his girlfriend back in Vienna:

The Düsseldorf conference was, as far the mathematics section is concerned, so uninteresting and insignificant that I took the remaining 10 days to go to Paris for my intellectual refreshment. That, on the other hand, is quite fantastic... I liked Paris so much that I would like to return if it is at all possible. Just in passing, and to annoy you, I also visited Mondrian the painter, who paints only squares. He showed me photos of all his pictures; the development is quite interesting. Also his studio, whose walls are covered with large rectangles in different colours and sizes. I liked it very much....

From there I returned to Amsterdam after all. I am lecturing on the calculus of variations. Personally, I am occupied by geometry of all kinds, furthermore by epistemology. I hope I'll get the energy to put together my views about the problem of truth. In the last weeks, I have had so many ideas that I don't have any time at all to write them down, and run away every evening to distract myself... in order not to overwork. Apart from that, I curse the fact that I am not in Vienna but rather here. I can't get used to living here, and I will try my best to leave here forever in the month of June.[1]

Menger had completed a doctorate two years previously at the University of Vienna, under the direction of mathematician and socialist Hans Hahn.[2]

---

[1] Letter, Menger to Hilda "Mitzi" Axamit, undated, Karl Menger Papers, Illinois Institute of Technology (hereafter KMIT). These papers have since been arranged and archived in the Economists' Papers Project at the Special Collections Section of the Duke University Library.

[2] For autobiographical detail on Menger, see the postscript to the English translation of Menger (1934d); the historical notes in Menger (1979), *passim*; and Menger (1994). Partial

Some sense of his personality is to be had from the aforementioned letter, with its hint of distaste for Germany, diversity of interests, and preference for modern art. Then, there is the admission that he was not at all happy in Amsterdam. Indeed, within a year, Menger was back in his beloved Vienna, where, in the following decade, he gave full rein to his broad interests – in particular, finding himself thrown into the role of intellectual intermediator between the worlds of mathematics and social science. He brought to Viennese economics an increased concern for logical rigour and greater recourse to the use of mathematics. As such, he was highly influential upon Morgenstern.

Menger is interesting for the subtle complexity of his story. Beyond doing creative work in mathematics proper, he wrote on the philosophy of mathematics, on economics and ethics, and, in a politically electric Vienna, he was not without social committments. In what follows, I portray his contribution to social science as part of a complex of concerns, embracing not only social theory but also the contemporaneous debates on the foundations of mathematics and political conflict in the Austrian First Republic. Menger's theoretical engagement with the social order is part of an intense, personal struggle with questions of foundation and legitimacy in diverse realms.

Breaking with Brouwer

By the time Menger was born in 1902, his father's seminal work had, of course, been recognized for more than thirty years as a milestone in the development of modern economics. In his 1871 *Grundsätze der Volkswirtschaftslehre*, Menger Sr. had moved beyond the Classical political economy of Smith, Ricardo, and J. S. Mill in several respects – in particular, in his attribution of the "value" of a good to the individual's subjective estimation of its utility, rather than the labour expended in its manufacture, and an associated shift of emphasis away from the exploration of economic growth towards the analysis of individual choice and resource allocation. The young Menger had a deep understanding of his father's work, even as an entering university student, and he was not yet twenty when he wrote an erudite introduction to a posthumously published second edition of the paternal *magnum opus*.[3] The son was early and well connected to the

treatments of Menger in relation to social science may be found in Bassett (1987); Cornides (1983); Ingrao and Israel (1991); Punzo (1989, 1991, and 1994); and Weintraub (1985).
[3] See Menger, Carl (1923).

economists of his father's Vienna and to the city's academic community in general.

At university, he started out, not in economics, but in physics, and shortly thereafter was won over to mathematics in a 1921 seminar on curve theory, given by the aforementioned Hans Hahn. In those meetings, in which Hahn showed how the various extant mathematical definitions of the seemingly intuitively obvious concept of a curve were all inadequate, Menger made his mark by providing a precise definition of a curve, based on the dimensionality of the set defined by the intersection of the curve and the neighbourhood of any point on it. Before he could publish his results, however, he contracted tuberculosis and entered a sanatorium that Autumn in the mountains of Styria. Prior to this, he informed Hahn of his results, and deposited them in a sealed letter with the Vienna Academy of Sciences. Then, for three semesters, in a mountaintop clinic, he recovered slowly from the "Viennese disease", immersed in literature, reading philosophy, and, as his strength returned, doing mathematics.

It was at this time that he came across earlier work on topology by Dutch mathematician Brouwer, including attempts at the definition of dimension. Although it was related to his own work, Menger saw it as being quite different, and said so in his first published article, which he submitted from the sanatorium in early 1922.[4] Upon returning to the city in early 1923, frail but cured, he learnt of another "virtually unknown" 1913 paper by Brouwer in *Crelles Journal*, which offered a definition of dimension close to his own but failed to develop a systematic theory on the basis of it. He also became aware of the 1922 results on curves and dimension by Russian mathematician, Pavel Urysohn, which did, in fact, offer a systematic theory, almost equivalent to his own. The latter Menger acknowledged in the published version of his own work. He then chose to use a recently won Rockefeller Fellowship for a sojourn with Brouwer in Amsterdam, prompted, however, not so much by their common interest in dimension theory as by his recent reading in an area evoked in our earlier account of von Neumann – namely, Brouwer's "intuitionist" writings on the philosophy of mathematics.

Brouwer's turn towards extralinguistic intuition as a basis on which to ground legitimate mathematics was rooted in his suspicion of language as an instrument of domination. Not one to whom social intercourse came easily, Brouwer had retreated to the physical and psychological isolation of his hut on the moor at Laren during the first decade of the century, where

---

[4] Menger, Karl (1923).

he constructed a critique of language and a philosophy of mathematics. Human society, he said, was a "dark force enslaving the individual, forcing on him not only a pattern of behaviour, a moral code, but even a pattern of thinking", and language was the primary means by which such authority was exerted. Brouwer's early writings were pessimistic as to the possibilities for human communication, and championed an austere individualism: "Self-liberation of man will gradually move him away from society; it will teach him to seek the night, the heath and solitude".[5] Mathematics and language, he claimed, occupied different spheres, with the former depending on a primal, extralinguistic, *urintuition*.[6] This conception of mathematical intuition had its roots in the work of the nineteenth-century mathematician Kronecker, famous for his already-cited dictum that only the whole numbers had been created by God; the rest was simply the invention of man. Intuitionism sought to retain only those parts of mathematics based on the "intuitively" plausible natural numbers and to reject the employment of "nonconstructive" proofs in set theory – that is, proofs in infinite set theory that establish existence by contradiction rather than by actually constructing the object the existence of which is being proved. This was based on their rejection of the law of the excluded middle as applied to infinite sets.[7]

Around 1912, when Brouwer had secured a chair at Amsterdam and was in less difficult circumstances, there emerged a certain hope regarding the possibility of developing a language adequate to the expression of mathematical concepts. This "interlude" lasted for a good fifteen years, and Brouwer

[5] Brouwer (1905a), p. 512. On the same page, he wrote that "self-liberation for the woman will lead her in sweeping moves of flight to acts against the laws of the land, against current morality; for these are directed against God, serve to perpetuate life as it is, the realm of the devil" (*ibid*). See also Brouwer (1905b).

[6] See W. P. Van Stigt (1981). For the authoritative biography of Brouwer, see Van Daalen (1999) and (2005).

[7] The law of the excluded middle states: either p or ~p. This can be seen to hold for finite sets because the members of the set can be "examined" to see whether or not they have the property p. In the case of infinite sets, however, the Intuitionists claimed, because the members of the set cannot be examined exhaustively, the law of the excluded middle cannot be invoked. They thus rejected existence proofs based on contradiction because they proceed in the following manner: the claim that a certain property holds for some member(s) of an infinite set is proven by showing that, were the property *not* to hold, a contradiction would arise; therefore, by the law of the excluded middle, the property must hold. Because of this invocation of the law of the excluded middle, the Intuitionists regarded such proofs as "meaningless", an abuse of language, and accepted only proofs that were "constructive" – that is, that either produced the existent directly or indicated how it could be done in a calculation involving a finite number of steps.

was instrumental in launching the Signific Movement, devoted to the development of an improved international language. By the mid-twenties, however, he had relapsed into pessimism, falling out with fellow members of the Movement and his colleagues in the international community. Thus Menger arrived in Amsterdam when Brouwer was in an unsettled phase and, although they got along very well for the first few months, it is not hard to see why the young progressive from Red Vienna and the austere individualist might have had trouble finding common ground.

Within a few weeks of arriving in Amsterdam, Menger became assistant to Brouwer, reading manuscripts to be refereed for *Mathematische Annalen* and the *Amsterdam Proceedings*, and then, in the autumn of 1925, teaching a university course on curve and dimension theories. In time, he grew to know Brouwer better. He later recalled his intense, highly strung personality, strict vegetarianism, and heightened sensitivity towards other mathematicians' opinion of his work. Menger shared Brouwer's deep appreciation of Beethoven, but they failed to contact on art: "In painting he seemed to favor the style of the Renaissance and was fond of portraits of that period. He looked with obvious distaste at the expressionistic graphic with social themes by the Dutch and Flemish artists of the 1920s (which I greatly admired) – even at Masereel's marvellous novels in woodcuts. They lacked the kind of beauty he seemed to expect from art". The artists admired by Menger included Gerd Arntz, Frans Masereel, and Peter Alma, modern graphic artists who explored themes of social inequality and exploitation. Alma and Arntz were linked to Menger's acquaintances in the Vienna Circle through the political activities of Otto Neurath, and, in fact, it was Alma who directed Menger towards Mondrian's studio on Rue du Départ in Paris.[8]

Gradually, Menger began to feel estranged from Brouwer. He read his 1905 booklet, *Life, Art and Mysticism*, which he thought mystical and mindful of the writings of Austrian misogynist contemporary, Otto Weininger: "In his political views, Brouwer was decidedly antisocialist, far to the right and of a militant Teutonic chauvinism. . . . from Milan he once wrote me that its inhabitants are no less Teutonic than, for example, Munich's". Menger noted that although the mathematician had one violently antisemitic close friend, he himself showed no evidence of anti-semitism. His strong antipathy towards Einstein, Menger felt, was mainly because of the latter's opposition to German nationalism.[9] Menger also became aware of

---

[8] See Postcard, Alma to Menger, Sept. 27, 1926, KMIT.
[9] Quotations from Menger (1979), pp. 242–44. The booklet by Brouwer that Menger read was his 1905 *Leven, Kunst en Mystiek*. On Otto Weininger, author of *Geschlecht und*

Brouwer's quarrelsomeness and his petty, legalistic involvement in academic and political controversies, especially with the French.[10]

In early 1926, their relationship deteriorated further when Brouwer began to insist that the papers on dimension theory submitted to him contain references to his own, Urysohn's, and Menger's work, arguing that this would help safeguard the rights of the deceased Russian. To Menger's chagrin, however, Brouwer's preoccupation with references soon "became an obsession" with his own place in the history of dimension theory, with conversations about new mathematical results quickly degenerating into searches through footnotes and references. By the autumn of 1926, the Dutch mathematician's obsessions were taking a more disturbing form, with Brouwer apparently tampering with Urysohn's republished papers and Menger's own manuscripts, and aggressively resisting all entreaties by the young mathematician. We can thus better understand why Menger wrote to his girlfriend that he was keen to leave Amsterdam. The conservative politics, the intuitionist philosophy: nothing was quite right. He was thus relieved when Hahn wrote him in early 1927 concerning an appointment as *Professor Extraordinarius* at Vienna, a post made vacant through the departure to Prague of Kurt Reidemeister, geometer, socialist, and participant in the *Wienerkreis*. Menger seized the opportunity and, within a few months, was back in the Austrian capital.

Red Vienna: Politics and Philosophy

His return to Vienna coincided with the first tremor in the fragile social order of postwar Austria. The First Republic had emerged from the war as the beleaguered rump of the Austro-Hungarian Empire and Vienna, formerly the political and cultural capital of a Hapsburg monarchy of fifty-four million people, was now the struggling centre of a impoverished, depleted country. The years 1918–1920 were marked by famine, disease, and inflation, and government requisitions of food for the city from the countryside added to already existing tensions between Vienna and rural Austria, increasing the suspicion with which the capital was regarded in

---

Karakter: Eine prinzipielle Untersuchung (Vienna, 1903), (trans. *Sex and Character*), for an introduction and further references see Johnston (1983), pp. 158–62; and Janik and Toulmin (1973), pp. 71–74.

[10] See Menger (1979), pp. 243–45. For example, Brouwer was at war with the *Union Mathématique Internationale*, which, after World War I, had restricted participation in its international mathematical congresses (Strasbourg 1920, Toronto 1924) to countries belonging to the *Conseil International des Recherches*, thus ruling out Germany.

subsequent years: "From the beginning the country was divided into two camps – Vienna and scattered industrial enclaves against the largely agrarian and Catholic provinces – whose hostility was not simply political. It expressed itself in hate mongering by the Catholic church, the far-from-silent partner of political reaction, for whom socialism was the Antichrist".[11] In the capital, outbreaks of tuberculosis and influenza continued throughout the 1920s, unemployment was high even during economic recovery later in the decade, and, all the while, Vienna gradually gave way to Berlin in cultural and intellectual terms.

Politically, the new Austria was labyrinthine.[12] The Christian Social Party represented the interests of Catholic, conservative, agrarian nationalism, and was led through the 1920s by the priest-chancellor Monsignor Ignaz Seipel, and from 1932 by Engelbert Dollfuss. The other large party, the Social Democrats, had the support of the working class in Vienna and other cities, and was run by a group of Viennese progressive intellectuals including Otto Bauer, Karl Renner, and Julius Deutsch. At the outset, the socialists were committed to eventual *Anschluss* with Germany, which was also under Social Democratic rule. This was an important point of difference with the Christian Socials, committed as the latter were to maintaining an independent, Catholic Austria. To the right of the Christian Socials lay the paramilitary, largely Austrian nationalist, *Heimwehr*, and a Nazi party, which sought unification with Germany, but for reasons very different from the socialists'. Although initially small, the Austrian Nazis gained in prominence after the demise of the German Social Democrats, and their methods extended to bombing and other acts of terrorism from the late 1920s through Hitler's rise to power in 1933. The paramilitary groups on the Austrian-nationalist right included the *Frontkämpfen*, whereas, to the left lay the *Schutzbund*, military wing of the socialists. Matters were further complicated by the ethnic factor: amongst the population of the First Republic there were Austrians, Germans, Czechs, Hungarians, Rumanians, Poles, and, intersecting with several of these groups, a large population of Jews. Amongst the latter, one could distinguish the long-established, well-integrated families of the Viennese *haute-bourgeoisie* from the recent influx of Eastern Jews from Poland and Rumania who had fled the Russians during the war. As I have already evoked in my discussion of Morgenstern,

[11] Gruber (1991) p. 10.
[12] For detailed accounts of political developments, see Rabinbach (1983); and Rabinbach (ed.) (1985). Gruber (*op cit*) is a sympathetic account of the cultural and social achievements of the Viennese Social Democrats. Useful "popular" commentaries on the period include Clare (1982) and Hofmann (1988), chapters 3 and 4.

the arrival of this Orthodox, and thus visible, minority from the East was important in the resurgence of anti-semitism in the interwar period.

For nineteen months until June 1920, the Social Democrats and the Christian Socials had managed to form a governing coalition, but, as Rabinbach (1983) points out, its collapse effectively marked the end of the socialist ascendancy in national politics. Thereafter, significant Socialist presence was confined to the municipality of Vienna, where the postwar extension of the franchise had yielded the Social Democrats a solid majority by 1923. As if to compensate for their relative incapacity at the national level, it was in the domain of municipal politics that the Socialists exercised power and enacted redistributive measures. In the 1920s, Red Vienna was the site of the socialist experiment implemented by Bauer and his colleagues, the Viennese search for a "third way", intermediate between Bolshevik Communism to the left and parliamentary reformism to the right. Under socialist mayors Reumann and then Seitz, many reforms were carried out, particularly in the areas of housing, health care, and education. The socialist programme of *Bildung* was aimed not so much at the immediate, massive, appropriation of economic resources as at the creation of a socialist mentality in preparation for a future, poorly specified, revolution. The programme was essentially one of cultural transformation, extending from the organisation of communal living quarters, through educational reform, to instruction in the most private reaches of workers' lives – *viz.* birth control, personal hygiene, and social mores. Helmut Gruber has suggested that this emphasis on cultural transformation was, in itself, a form of compensation for impotence in "real politics". It was also a reflection of the cultural gap that existed between the intellectual elite at the head of the party and the uneducated, urban proletariat of the rank-and-file.[13]

The architects of the socialist policies came from the intelligentsia of which some members of the Vienna Circle and the young Menger were part. His mentor Hahn was an active socialist, and Hahn's brother-in-law, Otto Neurath, was even closer to the leaders of the Social Democrats. Thus, Menger, although notoriously restrained in these matters, speaks of the achievements of the Vienna socialists with clear admiration.[14] It was also the

---

[13] Rabinbach (1983), p. 26; Gruber (1981), *passim.*
[14] Hofmann (1988), p. 189. See also Stadler (1991) in Uebel (ed.) (1991); Menger (1994), pp. 1–2 and *passim.* As a member of the Viennese social elite and as a Catholic, however, Menger was also sensitive to the fact that these achievements had beeen funded through finance minister Hugo Breitner's penal taxation of wealthier Vienna, and that some of the changes, such as the secular programme of school reform of education minister Otto Glöckel, were a direct challenge to the power of Seipel's Church.

case that many of the intellectuals behind the Social Democratic movement were Viennese Jews and that the party, as inheritor of the Austrian liberal tradition, could rely on the support of much of the Jewish professional and commercial interests of the capital. This contributed further to the perception of a gap between the urban, "Godless" socialists of Vienna and the conservative, Catholic interests of provincial Austria. Throughout the 1920s, tensions rose.

Matters came to a head on July 14, 1927, when a jury acquitted three *Frontkämpfen* youths in the murder of a *Schutzbund* member, and the judgement provoked an immediate and massive popular reaction. On July 15, thousands of protesting workers descended on Vienna along the main arteries leading to the *Ringstrasse*, and the result was a riot in which several policemen were battered to death, the *Justizpallas* set on fire, and some eighty-five protesters killed by the security forces. These events permanently affected political relations in the First Republic and, thus, when Menger returned from Amsterdam a few months later, it was to a tense, unstable Vienna.

Particularly shaken by these events were the Social Democratic intellectuals, including their spokesmen in the Vienna Circle, Neurath and Hahn. The Vienna Circle's roots lay in the *Verein Ernst Mach*, a philosophical discussion group started in the early twenties by Hahn and Neurath, along with Moritz Schlick and physicist Philipp Frank, all of whom were devoted to perpetuating Machian empiricism in the face of a general shift towards metaphysics in German-speaking philosophical circles.[15] In 1923, through Hahn's intervention, Schlick moved from Prague to Vienna to take the Mach-Boltzmann chair in the philosophy of the inductive sciences, and in time the group expanded to include Victor Kraft, Kurt Reidemeister, Rudolf Carnap, Herbert Feigl, and Friedrich Waismann. By 1929, with their bold manifesto, the Mach group became more clearly identified as the *Schlick Kreis*, or Vienna Circle. Most histories of the group focus somewhat narrowly on their philosophy, but there was an important political dimension to the turn towards logical empiricism.[16] It is a constitutive refrain in the writings of Neurath and Hahn and is key to understanding Menger.

[15] See Uebel (ed.) (1991), especially Friedrich Stadler's "Aspects of the Social Background and Position of the Vienna Circle at the University of Vienna", pp. 51–77. Also Victor Kraft (1953). Rudolf Haller, in "The First Vienna Circle", in Uebel (ed.), *op cit*, pp. 95–108, points out the importance of pre-War discussions among Hahn, Neurath, and Frank in Vienna in 1907–1912. Haller says it was Hahn's organisation of this group, "whose members were also very close in their political convictions", that allows us to regard him as the true founder of the Vienna Circle (p. 108).

[16] For example, see Uebel (ed.) (1991) and Reisch (2005).

Otto Neurath[17]

Neurath's views on science, philosophy, politics, and art were interconnected. His aim was the creation of a new, modern form of life: a collectivist transformation of the social order, predicated variously upon the methodological unification of science and the use, for propaganda purposes, of modern art, in which the visual clarity achieved by disposing of ornament would be harnessed towards political ends. If theology and metaphysics had served to console the workers, true science would alert and unify them: "the better the proletariat grasps the social engineering relations of our order and surveys its own chances, the more successfully it can fight.... For the proletarian front the technique of the struggle and the interests of propaganda coincide with high esteem for science and the overcoming of metaphysics".[18] This commingling of philosophy and politics is evident in the Vienna Circle's manifesto, "The Scientific Conception of the World", presented by Neurath in 1929 to the First Congress for the Epistemology of Science, at Prague, where he demanded the "purification" of the social sciences along the lines of that achieved in physics, and equated the new scientific outlook with the advancement of a collectivist form of life.[19] In an age of increased mechanization and modernization of production, concepts such as the "folk spirit" were to be abandoned in favour of an empiricist, antimetaphysical, social science, the object of study of which would be simply "people, things, and their arrangement".[20]

In his 1931 book, "Empirical Sociology", Neurath described further this "Physicalist Sociology" or "Sociology on a Materialist Basis", a neo-Marxist sociology that employed only such concepts as were ultimately reducible to spatiotemporal terms, and pointed inexorably towards central planning. Neurath advocated a strict behaviourism in which humans were to be

---

[17] Like Hahn, Neurath emerged from World War I with strong socialist convictions. Having studied at Heidelberg, he worked at the War Ministry in Vienna and then directed the Museum for War Economy in Leipzig, which marked the beginning of his activities in visual education. When the museum's operation collapsed in 1918, he joined the Social Democratic Party and was heavily involved in the administration of the short-lived Bavarian 'Soviet Republic' of 1919. When this movement was crushed, Neurath was imprisoned and then released through the intervention of Max Weber and the Austrian government. On Neurath, see Cartwright et al (1996); Uebel (ed.) (1991); Rosier (1987); and Zolo (1990). By Neurath himself, see the essays in Neurath (1973) and (1983).

[18] See "Personal Life and Class Struggle" in Neurath (1973), pp. 249–98. Quotations from pp. 296–97.

[19] See "Scientific Conception" in Neurath, *op cit*, pp. 299–318. Quotations from pp. 304–05.

[20] *Ibid*, p. 315.

studied as one might study a colony of ants. One considered the economic geography of the group, their physical surroundings – what Neurath termed the life terrain – and then examined the living standards secured by different forms of socioeconomic arrangement, different "social orders". The "metaphysical" concept of utility, which, significantly, was central to the economic theory of Menger's father, was to be dispensed with by considering *physical* bundles of consumption and their associated standards of life: "If in one case a man sits hungry and crying in a dirty, little hole, and in another the same man has a friendly smile and eats happily in a bright villa, we should say that in the first case, the living standard is less favourable than in the second".[21] In short, one did not need a theory of (invisible) utility in order to understand the conditions of the working class. Neurath went further with this emphasis on the material, rejecting any link between money values and true living standards. The task of a progressive sociology, he said, was the clarification, and then transformation, of the social order: "The theory of social structure is essential for any social engineer, which means anyone who participates as collaborator in the planned organization of all social formations. Prediction of the coming social structure and of the functioning of a given social structure is then at the center of a planned way of life".[22] Such prediction and transformation required that the people be politicised and educated – which was where Neurath's Museum came in.

In 1924, he set up the Social and Economic Museum of Vienna (SEMV), with funding from the Viennese municipal government, some trade unions, and social insurance funds.[23] Using the "Vienna Method" of pictorial statistics, this centre exhibited statistical information on social and economic change to the workers of Vienna. Pictorial symbols were used to overcome literacy barriers and stimulate the interest of the uneducated, who would probably have never set foot inside a museum otherwise. By demonstrating clearly to the Viennese working class that infant mortality rates were falling in the poor ghettos but still lagging behind that in the wealthy enclaves, or that the Social Democratic municipal government had made great strides in provision of housing and education, the Museum's pictorial statistics were both a constituent element of Neurath's empirical

---

[21] Neurath (1931b), p. 401.

[22] *Op cit*, p. 403. Anti-socialist Friedrich Hayek's rejection of "positivism" in the social sciences was intimately tied up with his political differences with Neurath. See Hayek (1994), p. 50.

[23] See Müller (1991) and Stadler (1991). Also "Museums of the Future" and the other essays in Neurath (1973).

sociology and endorsement of a particular politics. The most important of the SEMV's informative graphic art came from the chisel of Gerd Arntz, Neurath's chief designer from 1928 and one of the artists to whom Menger had pointed when contrasting his own tastes with Brouwer's. Like Alma and Masereel, Arntz used simple forms, in his case black and white woodcuts and linocuts, to protest against socioeconomic conditions, and this appealed to the sensibilities of Menger and many socially progressive moderns.

But this work also points to the delicate position in which the young mathematician found himself at the end of the 1920s. He was sympathetic to progressive reform, but opposed to Neurath's radical change. He was close to the agnostic, Jewish socialists of his milieu, yet, as one of Catholic background, understood something of the suspicions of Austrian, clerical, conservatism. He deeply understood and was sympathetic to his father's utility-based economics, which Neurath dismissed as bourgeois metaphysics. Menger, in short, found himself torn between conflicting loyalties and contradictory worldviews, drawn into an imbroglio requiring clarification. Chastened by his experience with Brouwer and now back in a politically charged Vienna, he became increasingly sensitive to the conflation of science and politics, whether that took the form of Brouwer's declarations concerning "meaningful" mathematics or Neurath's dictates as to what constituted an acceptable, proletarian philosophy. For his own peace of mind, Menger was drawn into a bid to clarify the shifting boundaries amongst the scientific, the political, and the aesthetic. This required that he come to terms with the ideas of his revered Hahn, whose marshalling of the philosophical and the political was even more subtle than Neurath's.

Hans Hahn

Like Neurath, Hahn presented the promotion of a "scientific view of the world" as both a philosophical and political endeavour.[24] At the Prague Congress in 1929, addressing the audience before Neurath's delivery of the Circle's Manifesto, Hahn spoke about "The Significance of the Scientific World View, Especially for Mathematics and Physics". He laid down the

---

[24] Having taught at the University of Czernowitz, Hahn went on to fight on the Italian front in World War I, where he was badly wounded. He then taught at Bonn and returned in 1921 to his native Vienna, to a chair in mathematics. His early mathematical work was on the calculus of variations, and on real and set functions. After the War, he published volume I of his *Theorie der Reellen Funktionen* and continued to work on these topics and the theory of integrals. On Hahn, see Karl Menger (1935); Menger's "Introduction" in Hahn (1980); Popper (1992), p. 40 and (1995); and Sigmund (1995a) and (1995b).

basic propositions of the logical empiricism of the Vienna Circle, opening
with a leap of faith: "[We] *confess our faith* in the methods of the exact sciences, especially mathematics and physics, faith in careful logical inference
(as opposed to bold flights of ideas, mystical intuition, and emotive comprehension), faith in the patient observation of phenomena, isolated as much
as possible, no matter how negligible and insignificant they may appear in
themselves (as opposed to the poetic, imaginative attempt to grasp wholes
and complexes, as significant and as all-encompassing as possible)...".[25]
Experience or observation, he said, were the only means of knowing the
facts that make up the world, and all thought was nothing but tautological
transformation. The scientific view of the world had the joint tasks of critically examining the statements of science, exposing the meaninglessness
of metaphysical propositions, and thereby undermining idealistic philosophy's lofty claims to authority. Knowledge had to be constituted through
observation and the tautological transformation of thought through logic,
something which left room for neither metaphysics nor theology. Like
Neurath, Hahn emphasised that social, cultural, and artistic factors were
not simply being harnessed to buttress the new philosophy, they were inseparable from it. The scientific world view was the "true expression of the
time", corresponding to new structures and organisations, a new rational-
isation in industry, and a new objectivity in architecture and the applied
arts.[26] Thus did Hahn subtly intertwine his philosophical arguments with
arguments for a philosophy.

In a 1930 popular pamphlet called "Superfluous Entities or Occam's
Razor", Hahn further aligned the scientific view of the world with a socialist
outlook. A new scientific philosophy was needed to break the stranglehold
that idealistic philosophy and religion had had on the people. In refusing to
face up squarely to the empirical world and in entertaining the belief that
there existed some reality beyond that attainable by the senses, metaphysics,
or "world-*denying* philosophy", had acted as a political opium, "consoling
the masses".[27] Hahn equated the campaign for logical empiricism with the
struggle for social reform and suggested that it was no accident that the
England of Occam, Locke, and Hume was also the nation that "gave
the world democracy". Occam's Razor, the thirteenth-century scholastic
imperative that "One must not assume more entities than are completely
necessary", is appropriated by Hahn, the modern agnostic, who equates its
achievements in philosophy with those of the axe in politics – in a brilliant

[25]   Hahn (1930–31) in (1980), p. 30.
[26]   *Ibid*, p. 30.
[27]   Hahn (1930), p. 1.

stroke, so to speak. Cutting out the unnecessary in philosophy, he says, is continuous with purging politics of the divine right of kings and other forms of undemocratic superstition: "the land that saw the beheading of a king also witnessed the execution of metaphysics. For all those other-worldly entities of metaphysics... the gods and demons of the religions, and the kings and princes of the earth – all must share a common fate".[28]

By the early 1930s, therefore, Menger was faced with conflicting world-views. Brouwer's politics gravitated around an austere antisocialism, according to which liberation was to be found by turning one's back on society, resorting to a primordial intuition to overcome the constraints of language. In Brouwer, Menger observed, the approach seemed to trans-late into an inability to get along with other mathematicians, dogmatism in philosophy, and a willingness to engage in subterfuge and distort the histor-ical record. The political stance Menger faced in Vienna was quite different. Although the Circle philosophers began with a starting point similar to Brouwer's – in which language again was the culprit and its abuse, through metaphysics and religion, permitted the construction of repressive forms of authority – to Hahn and Neurath, liberation lay in denying the "world-denying philosophy" and constructing instead a philosophy that would promote a collectivist mentality. Brouwer had turned his back on society; Hahn and Neurath were obsessed with the socialist ideal.

Of course, these worldviews did not present themselves to Menger as forms of life to be calmly evaluated. It was relatively painless for him to break with Brouwer, but his relationship with the Viennese, and Hahn in particular, was very different. The latter was his mentor, now colleague, and had continued to act as intermediator in the interminable priority debates with Brouwer on the theory of curves and dimension. The personal stakes were important for Menger, so he had to tread delicately. What followed, before politics took over entirely, was that he launched into a condemnation of Brouwer and then, altogether more subtly, put some distance between himself and the Circle.

From Mathematical to Social Order

Against Intuitionism

In 1928, Brouwer was invited to Vienna to give two lectures, the first of which was a talk on "Mathematics, Science, and Language". Here, he

---

[28] Hahn (1930), p. 4. Hahn's lecturing abilities stimulated many, and it was partly through Hahn that Karl Popper came in contact with the Vienna Circle. See Popper (1992), pp. 39–40 and (1995).

outlined the philosophical view with which Menger was familiar, speaking about mathematical *contemplation* ("*mathematische Betrachtung*"), the primordial phenomenon of the intellect ("*Urphänomen*"), and other concepts which Menger found to be "completely obscure".[29] Brouwer found little warmth in the reception accorded to him by the Viennese. His philosophy was part of a campaign to proscribe parts of mathematics, he was bellicose and, to Menger and his colleagues, his attacks were based on an ultimately religious or mystical authority the opposition against which the Circle owed its very constitution. Brouwer, they felt, had to be countered.

In 1930, in a paper titled "On Intuitionism" that he might more appropriately have called "Against Intuitionism", Menger delivered a scathing critique of Brouwer, condemning intuitionist judgements of what is and is not "meaningful" or "valid" in mathematics.[30] He said the Intuitionists, who were not in agreement or, often, even individually clear as to what constituted "constructive" mathematics, regarded other developments as meaningless, a stance he declared to be "devoid of cognitive content": "For what matters in mathematics and logic is not which axioms and rules of inference are chosen, but rather what is derived from them". The idea that individual mathematicians should make judgements regarding what is plausible or meaningless is a fact "of interest for the biographies of the mathematicians . . . and perhaps for the history of mathematics, but they are not relevant for mathematics and logic". Menger advocated an *implicationist* position, in which all that mattered were the mathematical statements and the explicitly stated rules for their transformation: "Attempts to found the acceptance or rejection of propositions or transformation rules on intuition are ultimately empty words".[31] Menger thus rejected Intuitionism by redefining what constituted mathematics proper. Legitimate mathematics involved first-order statements and their transformation by clearly stated rules; it did *not* involve second-order claims as to what was meaningful or acceptable. The latter belonged outside mathematics, to the realms of psychology, biography, and history. Menger implied that even if as wide a range of mathematical practices as possible was to be tolerated, not everything could be tolerated in mathematics.

His position here not only cut him off from Brouwer, but gradually carried him away from the Circle, even opening up a rift between him

---

[29] See Menger, "Wittgenstein, Brouwer, and the Circle" in (1994), pp. 131–32.
[30] See Menger (1930).
[31] Menger (1930) in (1979), p. 57.

and Hahn.[32] Important here was the adoption by several Circle members of the Wittgensteinian terms "meaningless" and "tautological" to describe mathematics because mathematics, as a self-contained body of formalism, did not say anything *about* the world. Menger was miffed at the apparently cavalier manner in which some philosophers, especially Waismann, seemed to construe this technical term as a dismissal of mathematics, as though theorems were "merely" tautological. More important, Menger rejected the assumption, implied by several Circle members, that logic, as a body of rules by which mathematics was created, was somehow unique. Wittgenstein's concept of tautology was clear, Menger felt, but it was confined to the simplest elements of logic and was inadequate to capture mathematical practice in all its diversity. Yet some philosophers seemed to persist in the notion that, because mathematics was tautologous, it could somehow be reduced to the operation of a single, absolute logic.[33] To Menger's chagrin, even his admired Hahn seemed to subscribe to the simplistic view that such a set of rules was unique. Although less troubling for Menger personally, Carnap, too, continually alluded to *the* logic. All of this was to Menger's consternation, for, in his arguments against Intuitionism, he had explicitly argued that mathematics involved a twofold freedom: in the choice of basic propositions *and* the logical rules for their transformation. He was painfully aware that his view went against the grain in the Circle and he took solace in the quiet assent evinced by Kurt Gödel.[34]

In an attempt to persuade the Circle that logic could take various forms, in 1930, Menger invited the Warsaw logician, Alfred Tarski, to give a series of talks to his Mathematical Colloquium.[35] For a number of years, Menger had been in contact with several Polish logicians, including Bronislaw Knaster,

---

[32] Unlike those concerning Brouwer, Menger's remarks on Vienna colleagues are discreet and allusive, and we are left to piece together the clues trailing in footnotes, letters, and obiter dicta.

[33] See "Introducing Logical Tolerance" in Menger (1979), pp. 11–16. Menger writes that he seriously questioned the uniqueness of language and logic in Circle discussions, but met with no enthusiasm: "Schlick and Waismann at that time refused to take my skepticism seriously: Hahn was not favorably disposed; Carnap, too, at first shook his head; Neurath was not very interested in the question, while Kraft remained silent; and Husserl's disciple, my personal friend F. Kaufmann . . . was dead set against the idea" (*Ibid*, p. 12).

[34] See Postscript to the translation of Menger (1934d), p. 111.

[35] Menger had begun organising the Colloquium in 1928, running it along the lines of the Schlick group. Regulars included Gödel, Georg Nöbeling, and Abraham Wald; outside visitors such as John von Neumann. The proceedings were published as *Ergebnisse eines Mathematischen Colloquiums*: see Menger (ed.) (1928–1936). Volumes 1 through 5 and 7 were co-edited with G. Nöbeling and K. Gödel. The collected *Ergebnisse* are currently being republished by Professor Karl Sigmund of the University of Vienna.

Kazimierz Kuratowski, Leon Chistwek, and Jan Lukasiewicz, and he was familiar with their work. Tarski gave three talks, and Menger, by special effort, managed to persuade members of the Circle to attend one of the more philosophical presentations. Amongst the topics discussed by Tarski was Lukasiewicz's three-valued logic, in which propositions may be classed as True, False, or Uncertain, the third value being the excluded middle of the traditional two-valued system. He also showed how logical operations could be described using a parenthesis-free notation, in contradiction to remarks by Wittgenstein in the *Tractatus* concerning the indispensability of parentheses in logic. Bit by bit, Menger acted to counter what he perceived to be the Circle's narrow views of logic.

By the time he was writing his second paper on the foundations topic, even greater economic disarray and political fragmentation existed in Vienna. Securing teaching or research positions had become difficult for many of Vienna's younger academic talents and, in an atmosphere of growing anti-semitism, not least in conservative University circles, the Jews amongst them found it even more difficult. Menger came up with the idea of raising research money by organising a series of scientific lectures in 1932, directed towards the Viennese intelligentsia, on "Crisis and Reconstruction in the Exact Sciences". These were an outstanding success, with a full audience paying ticket prices similar to those for the Vienna State Opera. The series was closed with a talk by Menger on "The New Logic", in which he spoke to educated Vienna about the debates on the foundations of mathematics. Once again, he criticised Intuitionism sharply, and described in more sympathetic terms Hilbert's competing bid to establish the consistency of classical mathematics. He then seized the opportunity to announce, with a flourish, the results of work by Vienna's own Gödel, whose proofs of incompleteness and undecidability had put an end to Hilbert's project, demonstrating in particular that "a universal logic (such as Leibniz dreamed of) which, proceeding from certain principles, makes possible the decision of all conceivable questions, cannot exist".

He returned to his earlier insistence on the banishment of value judgements from mathematics and, in an allusion no doubt lost on the lawyers of Vienna, took a jab at the vegetarian Brouwer. Menger said that the assertion that certain methods of inference are based on intuition whilst others are meaningless belongs to neither logic nor mathematics but is reflective of psychology or taste: it reminds one of "the dietetic rules enunciated by certain philosophical schools".[36] Mathematics is like music, he says, which

---

[36] Menger (1933), pp. 37–38.

nobody is ever required to defend or justify. Once Intuitionism is purged of normative assertions, it can be seen to follow a system of inference according to certain rules, just as classical mathematics does. He closed with a call for tolerance in a world that had been shown by Gödel to be even less certain than hitherto believed.

What interests the mathematician and all that he does is to derive propositions by methods which can be chosen in various ways but must be listed, from initial propositions which can be chosen in various ways but must be listed. And to my mind all that mathematics and logic can say about this activity of mathematicians (which neither needs justification (*Begründung*) nor can be justified) lies in this simple statement of fact.[37]

He labelled this stance Logical Tolerance, and the attitude not only divided him from Brouwer, but drew him gently away from Hahn on logic and Neurath on the unity of science. Whereas Hahn endorsed Neurath's scientific monism, Menger could not do so. Scientific practice neither could nor should be reduced to a set of codified procedures, any more than mathematics could be reduced to a single set of legitimate practices or logic limited to a dichotomous system of truth and falsity.[38] Thus emerged the motif of tolerance that is key to understanding Menger throughout this period. It was the attitudinal thread by which he guided himself through the murky 1930s' debates in philosophy and mathematics. It also sustained him as he was drawn to the world of politics, to foundational questions of a different kind, where the "exclusion of the middle" was taking on a darker connotation.

The Politics of the Excluded Middle

Following the events of July 1927, relations between the Social Democrats and the ruling Christian Socials declined. The true political limitedness of Bauer's socialist party became clear and they compensated with their

---

[37] *Ibid*, p. 40, italics in original.
[38] Hahn published his 1933 essay, "Logik, Mathematik und Naturkennen" in Neurath's Unified Science series, *Einheitswissenschaft*. Menger, however, from the outset, refused to endorse the unity of science program, with the extant correspondence in the 1934–37 period showing him fobbing off Neurath's several entreaties for support. (See Letters KMIT and also Box 267, Wienerkreis Archives, Rijksarchief in Noord-Holland, Haarlem, Netherlands). This distancing – there was no radical break – began after Neurath's 1929 delivery of the politically pointed Circle Manifesto in Prague, upon which Menger requested that he be listed as somebody sympathetic to, rather than belonging to, the Circle.

"anticipatory socialism": the social programmes, the proletarian culture, the public festivals and parades. The Right, on the other hand, was energized by the conflict and the second half of 1928 saw the appearance of confrontations and skirmishes between the *Schutzbund* and Prince Ernst Rüdiger von Stahremberg's *Heimwehr*, as the latter organised rallies in traditional socialist strongholds such as Wiener Neustadt. Because it helped in the suppression of the "enemies of Jesus Christ", the illegal *Heimwehr* was ignored by Monsignor Seipel. Internal jostling within the paramilitary movements of the Austrian Right also saw some of them flirt with the Nazis, for whom support in certain quarters was rising. In October 1929, the Wall Street crash announced the onset of the Depression and in 1931, Austria's leading bank, the overextended *Allgemeine Österreichische Boden-Creditanstalt*, collapsed. Until 1933, Austrian production slumped and unemployment rose, adding to the numbers in the Nazi party and further hardening anti-semitism. In the late 1920s, Alpine resorts had begun barring Jewish visitors or limiting the duration of their stay. Nazi gangs in Vienna targeted Jewish businesses and clubs, and anti-semitic rhetoric was aimed at the leaders of the Social Democrats. At the University of Vienna, there were periodic scuffles between Germanic student groups and Jewish students, increased protests regarding the strong Jewish presence amongst faculty and students, and, in 1930, the Pan-Germanic rector, Gleispach, made an abortive attempt to limit Jewish enrollment. In 1932, while Menger was lecturing on Logical Tolerance to the city's liberal Jews, a Nazi campaign poster in Vienna read "When Jewish Blood Will Squirt from the Knife" and fifteen Hitlerites were elected to the city government. Later that year, Joseph Göbbels addressed a rally at a crowded Engelman sports arena.[39]

When Seipel died in 1932, he was replaced by a new Christian-Social Chancellor, Engelbert Dollfuss, who aligned himself with von Stahremberg and sought the support of Mussolini in trying to restore political stability to Austria. The support of the Italian dictator and the Austrian prince was necessary to Dollfuss in resisting the Austrian Nazis, who had begun a campaign of terror aimed at destabilising the government, but von Stahremberg and the Austrian Chancellor also bore a common enmity towards the Social Democrats, the elimination of whom Mussolini actively

---

[39] During this period, Hahn and Menger themselves were targets of anti-semitic attacks at the university. Even though Menger himself was not Jewish, he kept bad company: some of his closest colleagues, such as Hahn and Wald, were Jewish, and the Vienna Circle was regarded with suspicion by the largely conservative, nationalist university authorities.

encouraged. Dollfuss's first step in this regard was to outlaw the socialist paramilitary wing, the *Schutzbund*, in 1932. With Hitler's rise to power in early 1933, confrontation between the Dollfuss régime and the Austrian Nazis rose further. Dollfuss's strategy in September was to suspend the national parliament, in an infringement of the Austrian constitution, effectively assuming dictatorial power. He now had complete freedom to crush the Social Democrats so as to better resist Hitler, the irony of which was not lost on Menger.[40] The showdown between the socialists and the allied government troops and *Heimwehr* occurred in February 1934. The three-day offensive began as raids on *Schutzbund* offices and culminated in artillery attacks on workers' cafés and occupied apartment complexes such as the Karl Marx Hof, the fortresslike aspect of which had long made them suspect to the Right. Martial law was proclaimed throughout and more than a thousand people died in the fighting mainly in Vienna, Upper Austria and Styria. Immediately afterwards, fifteen Social Democrat leaders were summarily hanged or shot.

Thus, while Menger, the public figure, could stand confidently before the lawyers and merchants of Vienna and call for logical tolerance, in private he was struggling with the strain and his correspondence bears witness to this rising preoccupation with the political and social order.[41] His concerns included Brouwer the conservative – vehement, intolerant, with his mystical appeals to Intuition, and Hahn and his comrades – impassioned, but strangely unconcerned with the multiplicity of ways of reasoning, ways of seeing the world . . . In different ways, Menger had come to terms with each of them but now there was this mounting political tension, with arguments over purity and legitimacy, of the kind he had fought in mathematics, engulfing the social domain. Hitler was making threats; Dollfuss was menacing the Social Democrats; Mussolini was abetting in the background; Jews were being threatened in the streets and slandered at the University. Where were the moderate voices? Even Schlick was caught up in the panic, as the appearance of a newspaper "extra" every few hours made concentration difficult and added to the tension.

Menger himself spoke of the strain in November 1933, when he wrote Schlick that he had begun writing a "little book" that he wanted him to read. Several weeks later, from the foot of the Rax in the Austrian Alps, where he had retreated to revise the manuscript, he confided to Schlick his fears about how the book would be interpreted. He was concerned that it

---

[40] See Postscript to the English translation of Menger (1934d), p. 114.
[41] For example, see Letter, Nöbeling to Menger, Oct. 9, 1933, KMIT.

might be misinterpreted and cause unpleasantness, but he would stand by his ideas. Yet he did not want others to come to harm because of it, or to have his views attributed to others. "Nor do I wish that my book, through its proximity to Neurath's, be associated with left radicalism, which absolutely does not correspond to my views. I naturally hope that I am seeing the whole thing too darkly".[42] The developments culminating in the events of February 1934 weighed heavily on Menger, as on all Viennese. Scientific work became difficult and letters of support came in from Princeton's Oswald Veblen and other colleagues abroad.[43] In mid-March, the graves of the dead were still fresh and the shell holes in the Karl Marx Hof still gaping open when Menger presented his manuscript to Carnap, saying that he had "half crossed over" to the philosophers: "For the work that has been occupying me so intensively in the last months was nothing less than the writing of a philosophical book, and I would be extremely happy were it to get your approval".[44]

---

[42] Letter, Menger to Schlick, Dec. 21, 1933, KMIT. It was largely through Schlick that the book was published.

[43] See, for example, Letter, O. Veblen to Menger, Jan. 6 1934, KMIT. On Feb. 13, the day after the shooting started in Vienna, Maria Knaster, wife of the Polish logician, wrote to Menger, inviting him to flee the violence and come stay with them in Warsaw. Letter, M. Knaster to Menger, Feb. 13, 1934, KMIT.

[44] Letter, Menger to Carnap, Mar. 15, 1934, KMIT. In fact, an epistolary priority battle was fought between Menger and Carnap during this time. Previously, in January, Carnap had written to Menger, sending him the manuscript of his *Logical Syntax of Language*, and requesting his support in obtaining a special Rockefeller Fellowship to do work in mathematics for several months at Princeton or Harvard. He had already solicited the help of others, including Russell and von Neumann, and wanted Menger to speak to the Rockefeller officials in Vienna. Seemingly in response to an earlier letter by Menger, Carnap defended himself against the claim that he was a supporter of Intuitionism: "I believe even now (as you will see from the manuscript) that Intuitionism has a certain historical value; that seems to be your opinion also. As for a few points, which I supported in earlier discussions of Intuitionism, I have in the meantime become sceptical, in the sense of a greater tolerance. Even in the past, I did not consider these points to be fixed but rather discussable. I have always been more careful in this sense in my publications than in oral comments. I cannot remember ever having written that the uncountable or other concepts of classical mathematics rejected by the intuitionists were 'meaningless'" (Letter, Carnap to Menger, Jan. 16, 1934, KMIT). In a February letter, he reported to Menger that he had been turned down by the Rockefeller Foundation. In the March reply, cited earlier, Menger indicated that he had been reading Carnap's manuscript, and challenged him directly on the tolerance principle of syntax elaborated therein. As much for what it says about Menger as the question of priority, it is worth quoting at length:

"A personally more important comment is about your tolerance principle of syntax. In this outstanding paragraph, I see the major point of your book, as far as I can determine from the manuscript, and I am extraordinarily happy about the beautiful conception of this paragraph. On the other hand, I am not quite in agreement with the concluding

Ethics and the Social Order

Published that year, Menger's book opens with a bold statement demarcating his new approach to ethics. Personal value judgements are to be cut out of the analysis: he will apply logic to ethics "without allowing it to be influenced by subjective feelings". Unlike the traditional moral philosophers, he will steer clear of the search for ultimate meanings or essences such as "the concept of morality" or "the principle of virtue". History has shown concepts of good and evil to vary enormously through time and space, says Menger, and he is not interested in offering "a particular moral system with a claim to universal validity".[45] Instead, he offers five unsystematic

bibliography. You say there that the intended attitude should be obvious to most mathematicians without it being in general expressed specifically. As far as I am concerned, I am ready to admit that my explicit formulation of this attitude did not cost me any great effort. On the other hand, how you can designate as obvious an attitude which is diametrically opposed to the most important published works on foundational questions, a view which contradicts Hilbert's epistemology as well the total attitude of all intuitionists, is a complete puzzle to me. It is true that my specifically expressed view was not noticed by philosophers but I do not understand why you, who have made this interpretation your own, should be so dismissive of it. I am personally as convinced of the importance of the tolerance principle of syntax, which I consider as my intellectual property, as of its contradictory character to any research done in this area up to now. That it does not conform to your earlier statements I have already mentioned in my last letter.

And now as to your quoting Gödel in connection to this principle, I probably don't have to tell you how much I appreciate Mr. Gödel since I have expressed this in enough publications and also through actively supporting him (please read, for example, my lecture about the New Logic). However, as far as this tolerance principle is concerned, I was representing in my *publications* the bases of his views at a time when Mr. Gödel can scarcely have even mentioned this orally. To me in any case he only told me very much later that he joined me in what I had written in my publications. In conclusion, I would like to make the following suggestion, to change the concluding bibliography of your first chapter in a way that would be more true to the facts and which would be of advantage to your book. If you would allow me a suggestion, it would be that you should say that the attitude meant here, which is totally counter to all formalist publications as well as all intuitionistic publications has until now been exclusively, and even with greatest emphasis, represented in my publications. This view, as you know from oral statements, is already also supported by Mr. Gödel.

I don't have to tell you how little I appreciate the philosophers you are fighting and how little I am indifferent to what they write or do not write about me. If the case is so totally different as far as your work is concerned, I would ask you to see it as a sign of my high esteem of you. Just at this very moment, I am especially happy that you support my views on the foundations of mathematics, as I must confess to you that I have half crossed over to the philosophers. . . . "

A postcard from Carnap later that month effectively conceded the point (Carnap to Menger, Mar. 21, 1934, KMIT).

[45] Menger (1934d), in translation, p. 2.

"epistemological notes" about "good and evil", all intended to underline his conventionalist stance.

In the first of these, for example, he discusses how individuals, as they mature, develop views of different ethical norms. Each person can ultimately be understood as tripartitioning modes of behaviour into what he *feels* is good, indifferent, or evil; what he *characterizes* by these words; and what he regularly, sometimes, or never *follows*. This is as "deep" as Menger wishes to go: feelings, utterances, and actions are his starting point and he will not attempt to go "beyond" or "below" them in a foundationalist attempt to ground any particular norm or system of norms. In a second note, he highlights the inadequacy of various foundational analyses of moral responsibility, of "what we ought to do". According to Menger, the Tolstoyan response, acting "according to the will of nature", says Menger, receives content only after one has made specific stipulations about the will of nature. Similarly, the directive to "do what is just" is as vague as the recommendation to do what is "good". Kant's categorical imperative – "Always act according to that maxim of which you can wish that it become a general law" – places enormous cognitive demands on the individual of average intelligence trying to determine the consequences of a particular rule, Menger says.[46] How should a person react when another intentionally disregards his duty and consciously disobeys the categorical imperative? There also is ambiguity in the wording of the imperative: what does one mean by "can wish"? In general, says Menger, if the categorical imperative is to lead to any concrete precepts, it requires extra stipulations. Finally, in the precept that one should act in the *interest* of society or the group, the concept of interest is ridden with the same problems as that of 'good', the meaning of which has usually been illustrated through silly circular exchanges of the kind: "What is worthy of love? That which is good. What is good? That is obvious". Menger scornfully says such discussions are even considered to be important achievements of well-known moral philosophers. In another note, he makes it clear that he is interested in a "demystified ethics" and, when accused that, in neglecting to define good and evil, he is skirting around the study of ethics proper, he replies that he is no more interested in debates about what constitutes a valid object for any particular science than he is in any campaign for scientific unity – an oblique aside, of course, towards the unnamed Brouwer and Neurath. He is concerned with the formal side of ethics and the psychological and biological aspects

---

[46] From Kant's *Foundations of the Metaphysics of Morals*, 1785, quoted in Menger (1934d), translation, p. 9.

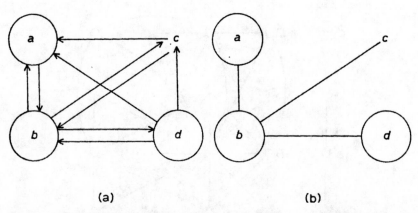

Figure 7.1. Formation of Compatibility Groups. *Credit:* From Karl Menger (1934d), pp. 103–04.

are best left to other specialists. He is interested in the possible structures, the possible geometries, of the social order, but he cautions that "those who intend to evoke from within themselves a self-evident system of norms for ethical behavior must not be permitted to refer to geometry. Rather, it is the person [like me] who establishes the existence of various systems of norms and their possible coexistence, who can speak of the analogy to geometry".[47]

The analytical core of his book, again, takes the form of not a complete system, but a series of independent "logico-mathematical notes on voluntary associations". To give a flavour of the analysis, let us consider the fourth note, in which Menger considers a person's demands on himself and on others, showing how sets of individuals may be partitioned to form groups of mutually compatible members. Assume that each individual can have, or can lack, each of two characteristics: politeness and sensitivity. We thus have four possible types of person: (a) polite, sensitive, (b) polite, insensitive, (c) impolite, sensitive, (d) impolite, insensitive. It is clear that any member of (c) is willing to associate with (a) and (b) – he is sensitive to their politeness or lack thereof – but not with (d) or with other members of (c). On the basis of this, we can construct socially compatible groups. In Figure 7.1, on the left, mutual compatibility is indicated by a double arrow; self-compatibility by a circle; on the simplified right, mutual compatibility is shown by a solid line.

[47] Menger (1934d) in translation, p. 29.

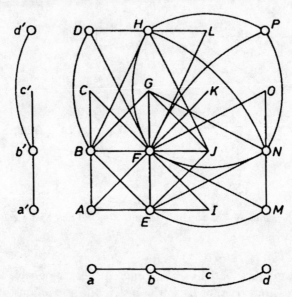

Figure 7.2. Compatibility Groups, given four possible attitudes to each of two different norms. *Credit:* From Karl Menger (1934), p. 123.

Depending on the size of the class of individuals and the number and type of characteristics, one can partition a given class in a number of different ways, and Menger shows how to survey the various total possible partitions. One might have different criteria for choosing amongst them – for example, one might wish to obtain compatibility groups of roughly equal size. He goes on to show why, in this particular case, Kant's categorical imperative, once again, is not a guarantee of harmony. A *simple* person is one who wants the people around him to assume the same attitude as the courtesy norm that he himself assumes – one who wishes to follow the categorical imperative. Polite, sensitive people are simple: impolite, sensitive ones are not. However, two simple persons are *not* necessarily compatible: a polite, sensitive person and an impolite person who wishes to associate only with others of his kind are both simple but incompatible. The categorical imperative, therefore, is not a sufficient precept to guarantee compatibility. Neither is it a necessary condition because some people not observing it – such as polite, insensitive ones – may well be able to enter into compatibility groups with others. The remaining sections of Menger's analysis are similar in form to the one just discussed, all involving the exploration of how sets of individuals may be partitioned according to their observance of various sets of norms to generate various equilibrated social structures. Proving several relatively simple theorems, Menger shows how, the more complicated the system of norms, the more complex the possible structures (see Figure 7.2).

Thus, Menger was right when he told Carnap that he had "half crossed over" to the philosophers, for his analysis was "philosophical" in name only. His reconfigured ethics was, above all, an analysis of the social order, a meditation on structural stability at a time when, in Austria, it was greatly threatened. Menger's reworked ethics, furthermore, echoed strongly his earlier redefinition of the boundaries of mathematics. To protect mathematics against arbitrary attacks like Brouwer's, and to make it as inclusive as possible, he had redefined it to concern only the structures built by mathematicians working with axioms and proofs, and not pronouncements about the value of such activities. Now faced with a political imbroglio of nationalism, anti-semitism, and state oppression, he did something similar with ethical theory. The study of ethics should concern only the social structures yielded by combining individuals with different ethical positions, and not pronouncements about the intrinsic value of their stances. He denaturalizes ethics, retreating to an abstract combinatoric analysis of the consequences of normative choices, with any discussion of their value or merit consigned, like the intuitionist claims in mathematics, to the separate realms of psychology and biography. Ethics had no more to say about the intrinsic value of German nationalist or anti-semitic morality than did mathematics about Brouwer's rejection of the law of the excluded middle for infinite sets.

Menger's was one troubled mathematician's response to social disorder and it served a meditative, therapeutic function for the man himself. His work was a thinly veiled plea for calm at a difficult time, for rational discussion in the irrational Vienna of 1934.[48] However, the ensuing period brought little respite. In 1935, he organised a second series of talks for the same audience of Viennese professionals and intellectuals he had addressed three years earlier. This time, he presented a synthesis of his reflections on the place of mathematics in social science, citing the latter's general lack of logical rigour and its deplorable reliance on various kinds of intuitive knowledge or "*von-innen-heraus-vestehen-können*" (possibility-of-understanding-from-the inside-out). He reviewed his earlier work on the Petersburg Paradox and discussed both the utility theory of value, dating from his father's contributions, and recent work by Abraham Wald in the Colloquium on the

---

[48] A letter from mathematician Eduard Çech looks forward to Menger's presenting a lecture on "Mathematical Ethics", probably in Prague in May 1934. (Çech to Menger, May 3, 1934, KMIT).

theory of general economic equilibrium.[49] Pride of place, however, was given to his treatment of ethics and compatibility groups, with theoretical examples of the coexistence of smokers and nonsmokers, the polite and the sensitive, once again serving as thinly disguised probings of Austria's political difficulties: the coexistence of dictator-loving and democratic groups, of Austrian nationalists and Pan-Germanists, of anti-semites and Jews. To an applauding audience, who, at the time, were nervously watching events in Germany and at home, Menger incantated that theoretical models did not offer advice or orders regarding political choices – they did, however, provide "general schemas and insights into numerous imaginable forms of which a few are found in social reality and many not at all, including perhaps those which have simply been overlooked..."[50]

By then, however, Menger's Vienna had begun to crumble. As part of Dollfuss's rout of the Social Democrats in 1934, the Social and Economic Museum had been repressed and Neurath, then in Moscow, fled for the Hague.[51] Hahn died unexpectedly in mid-1934 and Menger delivered a graveside oration. In July of that year, in an attempted *putsch*, the Nazis murdered Dollfuss, who was succeeded by his Minister for Justice, Kurt Schuschnigg. Schlick was murdered by a former student in 1936 and, one by one, most of the Circle and their associates headed westwards.[52] By the mid-1930s, it was becoming clear to Menger that his search for social equilibrium had been rendered academic. He started to inquire about employment in North America, and eventually left for the University of Notre Dame in 1937.[53]

---

[49] See Menger (1936a). The other lectures included Ernst Späth on physiology and Hans Thirring, Hermann Mark, and Werner Heisenberg on physics. Menger's talk closed the series. See also Menger (1934c), which dealt with the use and limits of mathematics in modelling the St. Petersburg paradox; and Wald (1934) and (1935).

[50] Menger (1936a), p. 128.

[51] In the Hague, the irrepressible Neurath continued the work of the Museum, developing the ISOTYPE method of pictorial statistics, with his 1939 book, *Modern Man in the Making* constituting something of a culmination in this respect. When the Germans occupied Holland, Neurath fled to England – by boat, penniless, without belongings – was quarantined on the Isle of Man while the British authorities ensured he was not a Nazi, and then settled at Oxford. For his remarkable story, see Neurath (1973).

[52] See Fleming and Bailyn (eds.) (1969), in particular the essay by Herbert Feigl, "The Wiener Kreis in America", pp. 630–73.

[53] See Letters, Veblen to Menger, Jan. 6, 1934; Dec. 24, 1934; Nov. 11, 1935; from Birkhoff to Menger, Oct. 6, 1935; and from Morse to Menger, Aug. 19, 1935, KMIT. How he wound up going to Notre Dame is unclear, but the Morse letter mentioned here seems to suggest Menger's particular interest in the possibility of working at a Catholic university. Up to 1941, he was involved with Carnap, philosopher Charles Morris, and linguist Leonard Bloomfield, in discussions of the "Chicago Circle". At the end of the war, he moved

Conclusion

Although powerless before the dynamics of Austrian political history, Menger's ethics provoked widespread response. His colleague and close friend Nöbeling hated it, objecting that formal methods were entirely inappropriate to ethical and social questions, and this reaction caused Menger to break with him for good.[54] Oswald Veblen, mathematician at Princeton and the Institute for Advanced Study, wrote to Menger in December 1934.

permanently to the Illinois Institute of Technology in Chicago, where he continued to organise a Mathematical Colloquium along the lines of the one in Vienna. By then, he had effectively abandoned social science, devoting himself to non-Euclidean and probabilistic geometries and, later, to pedagogy and teaching reform in mathematics.

54 Georg Nöbeling was born in 1907 in Lüdenscheid, Westphalia. Following studies in Göttingen, he came to Vienna in 1929 to study dimension and curve theory with Menger, where he participated in the Colloquium and edited some of the *Ergebnisse*. He and Menger seem to have been close, with their letters occasionally dealing with personal matters. But perhaps even those contain a hint of their future break. In one such letter in 1931, Nöbeling wondered about the values of art and science for their own sakes. What was education for, he wondered: "[T]hat one should be educated for something, that some kind of generally accepted norm is necessary, seems to be unquestionable. Now there are a large number of norms propagated by all kinds of groups of people. But that just proves that there are no norms, not even for a single people, our own. It is clear that such a norm cannot be constructed in any way whatever, but must be conceived and transfused with all the dark powers we have within us. The only thing that we can do, and this is hard enough in itself, is to lift them into our consciousness and live accordingly. The lack of a norm leads to the fear that our unconscious, the sole life-conserving powers that force us to communal living, to working beyond our daily bread, is not quite in order any more. If this is right, and a recovery is no longer possible, there is nothing to prevent our degeneration" (Nöbeling to Menger, Jan. 7, 1931, KMIT). Nöbeling continued that he had been wondering about such higher questions, but, in keeping with the dawning personal conviction that he was "not just Intellect", felt that a purely scientific attitude towards them was totally inadequate, that "Systematizing totally misses the essential".

Reacting to Menger's logical treatment of norms in *Moral, Wille* three years later, Nöbeling wrote: "I must confess that the book completely rubs me the wrong way.... What is decisive is that the whole formulation of the question (*Fragestellung*) fills me with loathing. Already, I don't especially like the formal in mathematics, and so don't like it at all in moral matters. That is seemingly only a question of taste, but only seemingly so. Your appreciation of this application of exact scientific thinking and my antipathy are based fundamentally on our divergent dispositions and intellectual structures, and these can be seen as the source not only of the differences of our tastes as far as the formal are concerned but also the differences of our life philosophies (*Lebensauffassungen*), our worldviews (*Weltanschauungen*), our concepts of science, etc." (Letter, Nöbeling to Menger, undated, 1934, KMIT). The next available letter from Nöbeling is completely formal ("Sehr geehrter Herr Professor"), with Nöbeling indicating that he will the "follow the wishes" expressed by Menger in his last letter. There follow only one or two similarly formal letters, containing veiled references to their rupture and with Nöbeling refusing to entertain any further criticisms of German mathematicians by Menger.

He praised the first chapter but, in an opinion then still common amongst mathematicians, doubted whether social science offered scope for a mathematician of Menger's prowess.[55] A Polish Catholic theologian and logician to whom Menger was close, one Monsignor Bochenski, wrote laudatory letters from the Vatican.[56]

Amongst the social scientists, Menger's recourse to mathematics resonated with the interests of a new minority. Ragnar Frisch, editor of *Econometrica*, then a new journal dedicated to the application of mathematical and statistical methods in economics, wrote to Menger, inquiring about the book and affirming, contrary to Veblen, that there was a great need for capable mathematicians to look towards economics and the social sciences.[57] At the time, Friedrich Hayek had not yet completely lost faith in the use of mathematics in social science, and in his important 1937 paper on "Economics and Knowledge", referred to Menger's work as a first attempt in the construction of a mathematics adequate to modelling social arrangements, but one that had yet to prove its worth.[58] Menger's key interpreter,

[55] Letter, Veblen to Menger, Dec. 24, 1934, KMIT.

[56] See Letter, Bochenski to Menger, Dec. 8, 1934. Bochenski distinguished between the philosophical and logico-mathematical aspects of the book, and had some doubts about the former. He cited the Thomists who, he said, believed in a psychological foundation of ethics, and thus in the possibility of a generally applicable ethics. However, he concurred with what he called the book's general "Weltanschauung" and wondered whether Menger realised how close he was to the Classical Thomist view, which held every religious belief to be an "actus intellectus assentientis veritati sub imperio voluntatis" or "a claim which is determined through the will". On the mathematical features, he greatly admired Menger's treatment and, although he wondered further about "the inner relationship between those of the same system", said that he drew great inspiration and joy from the book. "How did you get to the point of treating rather difficult problems so beautifully and clearly? I truly admire you" (*ibid*). See also Kecskeméti (1935), which interpreted Menger's account as a renunciation of ethical reasoning. Unlike logic and mathematics, he said, ethical theory had to respond to particular historical and social circumstances. Other reviews appeared in the *Economic Journal, Bulletin of the American Mathematical Society, Mind, Philosophical Review*, and *Blätter für Deutsche Philosophie*.

[57] Letter, Frisch to Menger, Sept. 18, 1935, KMIT. Frisch refers to his having met Menger at the Cleveland meetings at which the Econometric Society was founded. It was very probably through Frisch that Menger was invited two years later, in 1937, to participate in the third annual conference of the Cowles Commission for research in economics at Colorado Springs, where he presented his "social-logic" to an audience of mathematical economists and statisticians. See Menger (1937), the full version of which appeared as Menger (1938).

[58] Hayek (1937) p. 38, n. 1, a footnote that disappeared in the version of this essay republished in 1945. Hayek had left Vienna in 1931 for the Tooke Chair at the London School of Economics but remained close to the Austrian economics community.

however, was the person who, by 1931, had replaced Hayek as director of the Rockefeller-financed Austrian Institute for Business Cycle Research (*Österreiches Institut für Konjunkturforschung*). It was in Menger's meditations on ethics and social theory that Oskar Morgenstern saw a glimmer of response to his own theoretical preoccupations.

EIGHT

From Austroliberalism to *Anschluss*

Morgenstern and the Viennese Economists
in the 1930s

Introduction

One autumn evening in 1927, Karl Menger gave a talk to the collected members of the Vienna Economics Society, the group rehabilitated by Hayek. Amongst them, those close to Ludwig von Mises as leader of the Austro-liberals included Hayek, Fritz Machlup, Gottfried Haberler, and Richard Strigl. Somewhat apart from this nucleus, closer to Mayer, stood Paul Rosenstein-Rodan, Alexander Gerschenkron, and Morgenstern himself.[1] Other figures worth noting include Steffie Braun, one of the few women in this male-dominated group; Richard Schüller, the senior civil servant; economist Ewald Schams; and members of the business community, such as banker Karl Schlesinger and Menger's friend Felix Kaufmann, a philosopher of law and social science who worked for a petroleum company.[2]

As mathematician and scion of Vienna's most distinguished economics family, Menger was comfortable with the two communities.[3] That evening in 1927, he was not long back in Vienna, after his acrimonious sojourn in Amsterdam with Brouwer. As protégé of Hans Hahn, Menger the student was close to the mathematicians and philosophers of the *Schlick Kreis*, or Vienna Circle, which included, in addition to Hahn himself, Moritz Schlick,

---

[1] On the Austro-liberals, see Klausinger (2008).
[2] Schlesinger and Schams were probably the most mathematically minded of those present. Schlesinger had arrived in Vienna in 1919, when, like the von Neumann family, he fled the Communist Revolution of Bela Kun. In his 1914 *Theorie der Geld- und Kreditwirtschaft*, his development of Walras' monetary theory, he used simple mathematics extensively, something that distinguished him in the German-speaking economics literature. (See Morgenstern [1968], Weintraub [1985]). Ewald Schams is a neglected figure, competent in mathematical economics.
[3] As Menger later put it, he had "two souls reside within [his] breast" (1973, p. 38). On Menger's work in mathematical economics and ethics, see Leonard (1998).

Rudolf Carnap, and Otto Neurath. These figures were well known to the liberal economists such as Mises. In particular, Otto Neurath long had been an opponent of Mises, beginning in Böhm-Bawerk's famous seminar at the University, which also included socialist Otto Bauer, and continuing in the 1920s when Neurath was a key target in Mises' anti-socialist writings.

As we have seen, Menger had grown up with economics. Before beginning his studies in mathematics, he had written an Introduction to the posthumous revised edition of his father's *Grundsätze*, so that even if the revision was perceived to add little to the original, all the economists were familiar with the son's editorial contribution. It was also in 1923, after reading his father on the question of uncertainty, that he started to write on the Petersburg Paradox, the subject of his talk the evening in 1927 in question.[4]

At this time, Morgenstern was temporarily back in Vienna, between Rockefeller-funded stays in Harvard and Rome. Although he was still completing his *Habilitation* thesis, *Wirtschaftsprognose*,[5] with the benefit of hindsight, it could be said that Menger's talk that evening was something of a watershed for Morgenstern. It marked the beginning of a decade during which he gradually distinguished himself from his Austrian mentors, actively promoting the development of mathematical economics and learning to question the necessity of any connection between liberal politics and economic analysis. These two developments were related, with the insistence on "purity" and logical rigour being thought necessary by Morgenstern to preserve economics from political influence. In all of this, the presence of Karl Menger was decisive.

The Limits of Mathematical Representation

It is curious that Menger later remembered Mayer's finding his talk that evening to be too mathematical, for the paper was, as much as anything, about the limitations of the formal treatment of choice behavior.[6] It

---

[4] As pointed out by Borch (1973), in the first edition of the *Grundsätze* we find a paragraph in the first chapter dealing with "Time and Error" (4. Zeitirrtum), and in the revised edition of 1923 – the one introduced in detail by Menger junior – there appear in the second chapter two new parts, one dealing with the time element (5a. Das Zeitmoment), the other with uncertainty (5b. Das Moment der Unsicherheit). Menger the son was thus quite aware, at this point, of the issue of uncertainty in economics. The time element, in turn, was the subject of Morgenstern (1934b), which was published alongside Menger's in the *ZfN*.

[5] See Leonard (2004).

[6] See Menger (1979) p. 259; also Craver (1986a), p. 12.

concerned the Petersburg Paradox. Player A offers player B a bet, in which a coin is tossed and B takes 2^{n-1} when heads first occur at n (that is, if all the first $n-1$ throws show tails, and the nth, heads). Although B's mathematical expectation for such a game is infinite, it is usually observed that B will not accept such a bet, something which, Menger correctly notes, is not so much a logical "paradox" as a discrepancy. Menger then considers three of the existing theoretical resolutions of the problem – based on the perception of a discrepancy between changes in utility and changes in wealth, boundedness of the utility function, and ignoring small probabilities in the calculation of expected utility – and shows why they are insufficient to resolve the paradox.[7]

The "solution" Menger reaches is not really a solution at all, but rather a general, qualitative description of the behaviour of a person faced with the question: "how much am I willing to pay for the probability p of winning an amount D – that is, for the chance (p, D)?" Several features, he says, can be regularly observed in such kinds of evaluations. First, when the possible loss associated with a bet is large, even a large gain will be undervalued relative to its expected value. Second, an individual will generally be willing to risk only a *part* of his total wealth in games of chance of any kind. This proportion, w, varies from person to person, but generally is closer to 0 than to 1. Only in cases of extreme desperation is an individual willing to risk all his wealth on a particular bet. Finally, the behavior of individuals in buying a chance (p, D) depends on the probability p. When p is very small, it tends to be undervalued so that a divergence appears between observed behaviour and that conforming to expected values. In general, Menger concludes, chances are undervalued both when probabilities are very small and very large. Only in the middle range is behavior according to expected values likely to be observed. Even here, however, the existence of

---

[7] First, as suggested by Bernoulli, it is possible that B's utility of his winnings, which are additions to wealth, is not linear in their monetary value. B's subjective evaluation (utility) function may exhibit diminishing marginal utility of wealth so that his expectation, for a given game, becomes finite. Such a solution, however, says Menger, is *ad hoc*. Given any utility function, as long as it is unbounded, it is always possible to design a Petersburg game that yields an expectation of infinity (see Bassett [1987]). Secondly, therefore, one might impose boundedness on the utility function. However, he says the discrepancy remains for the reasons that it is clear by introspection that B's unwillingness to bet is also attributable to the fact that it is very unlikely that he shall win a very large amount, and that there exist games in which B would not even bet the amount corresponding to his *finite* subjective evaluation. Imposing boundedness, therefore, is not sufficient. Thirdly, B may disregard small probabilities entirely, taking account of only the first k terms in the calculation of expected earnings. This Menger dismisses as *ad hoc* and as an incomplete description of behavior.

roulette and other games shows that chances are often overvalued. Menger says that the probabilities at which the maximum overvaluation occurs for an individual depends on his wealth, the potential gain, and other personal circumstances.

To Menger, the value of a general analysis of this type was that it allowed one to speak with greater precision of personal characteristics. For example, given wealth and a possible gain, a person who is willing to bet at a minimum probability that is smaller than most other people could be called a *gambler*. One who estimates the likelihood of a gain higher than most others might be called *optimistic*, and so on. Menger says that to the extent this forms a general, qualitative, description of gambling behavior, it provides an explanation of the Petersburg Game, but, again and again, he insists on the limits of what mathematical representation can achieve.

It should be...stressed that in formulating general regularities in behavior with regard to games of chance we have confined ourselves to rather quali-tative statements and have refrained from defining specific functions possessing those qualitative properties. This has three reasons: 1) The observed qualitative properties are shared by many, and even by many simple, functions so that there may be no reason to prefer a particular one to the others. 2) An exact description of actual evaluations is not supplied by any of these functions. 3) [T]he numbers and functions involved in the description vary from person to person. A comprehensive function, therefore, would include numerous parameters. In fact, the number of parameters would be so large, and the dependence on these parameters so com-plicated, that the general function would lack any transparency. And what would be gained by setting up such formulae? Could more than general remarks corres-ponding to the qualitative observations (ascertained directly from the empirical material) be predicated about the functions? (Menger 1979, p. 271).

The reaction to his talk was mixed. Mises, as one of the few economists present who was familiar with Bernoulli, was quite taken by Menger's proof that the distinction between marginal utility and marginal wealth – which Bernouilli had constructed as a response to the Petersburg Paradox more than 200 years previously and which had been regarded ever since as provid-ing a solution – was, in fact, insufficient to resolve it. Mayer did not like the paper and, as editor of the *Zeitschrift für Nationalökonomie*, advised him against publishing it. In the version finally published seven years later, Menger alluded to this:

Those who believe that by writing $f(D,W)$ they introduce a mathematical formu-lation superior to verbal expressions, are under a misapprehension, just as are, on the other hand, those who when seeing the mere symbol $f(D,W)$ suspect the use of mathematics which, because of ignorance or on the ground of general preju-dices, they reject in the social sciences. Simple special functions, however, as the example of Bernoulli's logarithmic evaluation demonstrates, are not in the theory of value borne out by experience as they are in physics.... [W]e therefore have refrained from setting up such formulae for $f(D,W)$ and have followed rather the method of the Austrian school of economists – hopefully without being suspected of underestimating mathematics (*ibid*, p. 272).[8]

Morgenstern, on the other hand, was quite taken by Menger's nuanced discussion. It was an encounter with the scientific attitude he had seen in Whitehead and other recent reading. Furthermore, it was coming from the son of Carl Menger, whose views on the place of mathematics in economic theory had been quite different and had been perpetuated by his Austrian successors.[9] In Morgenstern's eyes, Menger cut an intriguing figure.

Morgenstern thus began to distinguish himself from his teachers, foraging in mathematics to the extent that his training allowed. During his stay in Rome in 1928, as we have seen, he wrote with enthusiasm in his

---

[8] The paper first appeared in published form as a note by Menger (1934a), "Bernoullische Wertlehre und Petersburger Spiel" ("Bernoullian Economics and the Petersburg Game") in the fifth volume of the proceedings of the Menger's Mathematical Colloquium. The full version was published, seven years after the original talk, as "*Das Unsicherheitsmoment in der Wertlehre*" (1934c) in the *Zeitschrift für Nationalökonomie*. This was done at the behest of Morgenstern, who had become managing editor of the *Zeitschrift* in 1930, a position he held till 1937. In the English translation published in 1979, "The Role of Uncertainty in Economics", Menger says that Morgenstern had shown great interest in the paper ever since the original seminar (1979, p. 259).

[9] Members of the Austrian School emphasise the Aristotelian dimension of Menger Senior's a priorism and his opposition to the use of mathematics. They point to his correspondence with Walras in 1884, in which he said that economists "do not simply study quantitative relationships but also the nature (das Wesen) of economic phenomena. How can we attain to the knowledge of the latter . . . by mathematical methods". (Quoted in Smith (1986), p. 3). See also Smith (1990), (1995) and Caldwell (2004). For later Austrian critiques of the employment of mathematics, see Mises (1933) and Mayer (1932). Menger the son saw the matter differently: "How is it that men of such high intelligence and such specific logical-analytic talent as the members of the old Austrian school did not have a better knowledge and understanding of mathematics, especially of calculus . . . even if, for whatever reasons, they decided to refrain from using it in economic theory?" (*op cit*, p. 44). He says part of the answer lies in the fact that the Austrians came to economics from jurisprudence and government activities, and had not received instruction in mathematics in the *Gymnasien*. "Still, they might have resorted to a self-study from textbooks in later life; and in the 1890's my father indeed started such a self-study, as is clear from a three-page introduction into the elements of differential calculus in his handwriting, which he had bound into his copy of the second edition of Walras's *Eléments*. . . . But I am afraid that he did not acquire an operative knowledge, let alone a critical insight into calculus. The psychological problem is thus reduced to explaining why such eminent minds as the founders (and perhaps also several younger members) of the Austrian School were, as mature men, unsuccessful in their self-study of analysis" (pp. 44–45). He goes on to explain this in terms of the inadequacy of the textbooks, something which gave him a lifelong interest in mathematical education.

diary about attending a conference that featured contributions by mathematicians David Hilbert, Hermann Weyl, Emile Borel, and Oswald Veblen, amongst others.[10] He also plunged into the writings of the Vienna Circle, writing to his friend Haberler the following year that Moritz Schlick's *Allgemeine Erkenntnislehre* had impressed him more than any reading since Kant, and that he was now reading Carnap's "*Der logische Aufbau der Welt*", which he found to be a "first-class effort". Nothing could be done, he now felt, without a thorough knowledge of mathematical logic and epistemology.[11] At the same time, if only in his diary, he became increasingly critical of his mentors, registering his frustration with Mises and complaining about Hans Mayer's political intrigues with Othmar Spann.[12]

In a 1931 overview of the field of mathematical economics for the *Encyclopaedia of the Social Sciences*, Morgenstern wrote that there was no reason why mathematics might not be applied to the social sciences and to economics in particular. The objections, he wrote, tended to identify mathematics with the use of the infinitesimal calculus and to involve the claim that, in economics, one dealt with discretely varying quantities and with relationships that were not "mechanical". He said this overlooked the existence of other branches of "discrete" mathematics and there was nothing inherently "mechanical" about mathematics of any kind: it was, in logical empiricist fashion, simply a machinery for drawing inferences. There were very few instances in which mathematics was absolutely necessary, he admitted, but it facilitated the prosecution of the argument and was most useful when the problems selected were "too complicated to be tackled by ordinary means" (p. 367). Walras, for example, had demonstrated the existence of an equilibrium set of prices and quantities by the construction of a set of simultaneous equations that represented the demand for each good and arbitrage conditions.[13] Morgenstern concluded: "Another mark of progress would be the achievement of a closer integration between the psychological and mathematical orientations, a development which would not be

---

[10] See Diary, May 27, Aug. 28, and Sept. 4, 1928, OMDU.

[11] OM to Haberler, Mar. 28, 1929, OMDU, Box 4, Folder Correspondence, 1930–1932, S – Z, translated by Cornelia Brandt-Gaudry.

[12] On Mayer, see Diary, Dec. 22, 1928 and Apr. 19, 1929. In March, between his reading of Schlick and Carnap, he wrote: "Friday was the Economics Association. Mises spoke about worn-out methodology, and his concluding talk especially was just impossible. Lots of Jews" (Diary, Mar. 22, 1929).

[13] "Since the number of unknowns is equal to the number of simultaneous equations it follows that the problem of general equilibrium is capable of a theoretical solution", Morgenstern wrote (*ibid*) – not quite correctly, as Abraham Wald would soon show.

hindered by any fundamental disagreements between the exponents of the two types of economic theory" (p. 368).

Thus emerged the main tension that was to characterize Morgenstern's work in the 1930s: on the one hand, upholding the Austrian conceptual orientation – the "psychological" approach to time, expectations, and equilibrium – on the other hand, in contrast to his teachers, promoting mathematics as the appropriate means of doing so. Just as he had earlier broken free of Othmar Spann's demagoguery, moving into Mayer's and Mises' orbits, so, in the 1930s, he moved on again, gradually shunning his mentors and embracing the mathematicians. There was more to this, however, than the simple choice of scientific method. Morgenstern's alliance with the mathematicians was a way for him to achieve authority and create a distinct professional identity, independent of his mentors.

Social Planning and Contemporary Civilization

In September 1930, Hayek sent a memo to the Rockefeller Foundation describing the activities of the *Osterreichisches Institut für Konjunkturforschung* or Austrian Trade Cycle Institute, of which he was director.[14] He described how the affair had been set up in 1927 by Mises, with the financial help of the Austrian Chambers of Commerce, Labor, and Agriculture; the Austrian National Bank; various industry and banking groups; and the Federal Railroads. In the intervening period, the Institute had produced a monthly bulletin of economic conditions, carried out some special investigations, and begun producing monographs, the first of which was Hayek's own 1929 *Monetary Theory and the Business Cycle*. In the near future, they wanted to pursue special studies, including on the history of business cycles in Austria, the relationship between credit and the business cycle, and the elimination of seasonal fluctuations from time series. With a staff of five, and two research workers, they were stretched and needed more funds, he said, especially to hire short-term researchers: $3,000 per year for five years would do.[15]

Hayek's memo was part of a campaign begun earlier that year by Mises to attract Rockefeller support. At the Foundation, the psychologist, Beardsley

[14] Hereafter, we shall use the now more commonly accepted "Austrian Institute for Business Cycle Research".
[15] Memo, Hayek to Rockefeller Institute, Sept. 23, 1930, Austrian Institute for Business Cycle Research Records, Rockefeller Foundation Archives, Rockefeller Archive Center, Pocantico Hills, New York (hereafter AIRAC), Record Group 1.1, Series 705, Sub-series S, Folder 36: "Austrian Institute for Trade Cycle Research, Vienna, 1930–1934".

Ruml, had been replaced in 1929 as head of the social science division by Edmund E. Day, the business-cycle economist.[16] Day had been trained at Harvard and had, up to then, been at the University of Michigan. His staff included Miss Sydnor Walker, who remained from the Ruml régime, and the newly appointed John Van Sickle, a Michigan colleague, who became assistant director of the Foundation's social sciences office in Paris before later moving back to New York in 1934 and being replaced by Tracy B. Kittredge. Whilst Ruml had been interested in promoting interdisciplinary work, Day preferred to support projects in specific fields. With the collapse of Wall St. in late 1929 and the ensuing Depression, he emphasised the urgent need for research on economic stabilization.

The costs imposed by serious business depression – of demoralization, broken health, disorganized families, neglected children, lowered living standards, permanent insecurity, impaired morale, as well as financial distress – are so appalling when viewed socially as well as individually that no problem of this generation calls more clearly for solution than this of economic stabilization. It is no exaggeration to say that unless the problem can be solved or at least measurably reduced the present social order is in serious jeopardy ... No more important contribution could be made by the Foundation to the wise development of that social planning and control which seems ultimately so necessary and inevitable if contemporary civilization is to survive.[17]

Thus, in the early years under Day, inspired by Charles Bullock's Harvard Economic Service, the Foundation made new grants to economic research institutes at the University of Oslo and in Rotterdam, Kiel, Bucharest, and Heidelberg.

In Vienna, one of the first to seize the Rockefeller opportunity was actually von Mises, who, in 1930, approached Van Sickle for support. This although he must have viewed social planning and control as quite antithetical to contemporary civilization. The latter sought second opinions elsewhere. At Harvard, Bullock expressed reservations about the excessive Austrian emphasis on theory and deductive methods as opposed to the empirical style favoured by the American economists, but he spoke highly of Hayek and Morgenstern and underlined the need to preserve Viennese economics in the face of economic decline and inadequate university salaries.[18] At the League of Nations, Hans Staehle regarded the Austrian institute as "the

---

[16] See Craver (1986b).
[17] Day, Edmund E. "Proposed Foundation Program in Economic Stabilization", September 1931, quoted in Craver (1986b), p. 212.
[18] See Letters, Bullock to Day, May 22, 1930, and Bullock to Van Sickle, May 22, 1930, AIRAC, Box 4, Folder 36.

best equipped institute in German-speaking countries," adding that it was superior to the Berlin Institute, where Director Ernst Wagemann was "not equal to his task," as most of his workers were mere youngsters.[19]

In his professional diary, Rockefeller's Van Sickle hesitated. He was concerned that it would apparently be only a matter of time before Hayek received a call from elsewhere. He also knew that Mises, who, because of his Jewishness, could never hope to be more than a *Privatdozent* in Vienna, was supposedly in negotiation with a German university. He also wondered about the wisdom of funding in light of "present dissension in the SS [social science] field, and the anti-Jewish feeling [which] would complicate future relations of the RF [Rockefeller Foundation] in Vienna".[20] However, he was by and large well disposed towards the "very good men in Vienna".[21] A September dinner with Mises seems to have sealed the affair, and in November 1930 the Foundation guaranteed the Institute a generous $20,000 for the period till 1935.[22]

From 1928, Hayek and Morgenstern had been co-directors of the Institute, with its tiny staff. Then, in 1931, Hayek did, indeed, receive his call – from Lionel Robbins at the L.S.E. – and departed, leaving Morgenstern in charge. After the collapse of the Creditanstalt Bank that year, Morgenstern became increasingly involved in public economic debate. As Klausinger (2008) reports, for the next three years, along with Fritz Machlup, Morgenstern wrote in the *Neues Wiener Tagblatt*, advancing Austroliberal arguments: criticising the inflationary effects of any credit injections to save banks, opposing exchange controls as a way of defending exchange parity, and favouring the austerity of domestic price adjustment over protectionism, as a way of dealing with the trade deficit. Instead of resorting to public works as a means of countering depression, the Austroliberals promoted *Auflockerung* – namely, price flexibility and the removal of market restrictions in the spheres of both production and employment. Elsewhere in the

---

[19] John Van Sickle, Diary, May 9, 1930, AIRAC, *ibid.*

[20] John Van Sickle, Diary, May 21, 1930, AIRAC, *ibid.* Kiel was the university with which Mises was in discussion. Note the irony of Austrian anti-semitism in 1930 sending him in the direction of Germany.

[21] *Ibid.*

[22] Although Mises was pessimistic as to the immediate future and believed that union with Germany would ultimately take place in one form or another, he was, according to Van Sickle, optimistic as to the long-run future of Vienna as a cultural and economic centre. He regarded as Vienna's first-rate minds, philosophers Schlick, Carnap, and Wittgenstein; economists Hayek, Morgenstern, Haberler, Machlup, Schütz, and Rosenstein-Rodan; and philosophers of law Kaufmann and Schreier. John Van Sickle, Diary, Sept. 18, 1930.

capital, public speeches by Mises and Hayek promoted the same economic philosophy.

Until 1934, when the new corporate state became hostile to liberalism, Morgenstern was involved in liberal circles at the intersection of academia and business. For example, his participation in the *Neues Wiener Tagblatt* grew out of discussion with Machlup, Victor Graetz (director of the *Steyermühl* company which owned the newspaper), and Julius Meinl, head of the coffee emporium. Others to whom Morgenstern was close included Victor Kienböck, President of the Austrian National Bank; Rost van Tonningen, of the Finance Committee of the League of Nations; and banker Karl Schlesinger. From mid-1932 to mid-1933, Morgenstern was instrumental in organizing economic policy conferences aimed at the promotion of liberal policy amongst industry leaders. He also became involved, to his intellectual discomfort, in a pump-priming project, advanced by certain industrialists and aimed at subsidizing the employment of new workers. Throughout the early 1930s, Morgenstern became quite prominent and was rumoured to be favoured for positions of influence, including General Secretary of the *Hauptverband der Industrie*. This brought him into conflict with Mises, as the post was already occupied by a friend of the latter.[23]

When the time came to knock again on the Rockefeller door, in 1935, Morgenstern was able to write a glistening report of the Institute's activities in the interim, mentioning the continued monthly Bulletin, and the consulting activities to government, where, especially in the light of recent political turmoil, its impartiality was greatly respected. The turmoil in question was, of course, the rise of an autocratic government, as of March 1933, a development that put the Austroliberals – that is, those of them who remained in the country – on the philosophical defensive. By the end of 1934, however, Morgenstern appears to have weathered the transition. After a one-year gap in his personal diary, he emerged as a key advisor to the Austrian state, being a member of the team that negotiated the treaty with foreign creditors of the Creditanstalt and an advisor to that bank and the Ministry of Commerce on matters of railroad regulation.

In his 1935 report, Morgenstern also put special emphasis on the Institute's "purely scientific work," mentioning the publication of several monographs, including Hayek's 1931 *Prices and Production* and his own 1934 *Die Grenzen der Wirtschaftspolitik* (trans. 1937 *Limits of Economics*); the establishment of a reading room; and the cultivation of links to the University

---

[23] On Morgenstern's activities as policy advisor, see Klausinger, *op cit.*

by means of lectures and seminars, including those by Karl Menger and Franz Alt.

On the basis of the experiences of the last years I have worked out a program for research which I beg to outline briefly. This program provides for purely theoretical work as well as for empirical studies. These assume even relatively more importance than before; they are necessitated in order to examine theories of the trade cycle and procure a basis for new abstract thinking. It is my particular desire to harmonize more than has been done before both ways of research. I am absolutely convinced that abstract theoretical work, even making use of mathematical analysis or of the modern methods of Logic that have not yet been applied to Economics, are just as necessary as the systematic collection of facts.[24]

"Economists have so far entirely neglected the progress of mathematics and notably of logic during the last 30 years, so that it seems indispensable to subject economic theories of various kinds to the more rigorous test of these new ways of thinking and research", he said.[25] He noted the availability of several excellent people from Menger's Colloquium who could work on questions in pure theory, mentioning Abraham Wald in particular.

"Betweenness" in a Cultural Space: The Case of Abraham Wald

Abraham Wald had first knocked on Karl Menger's door at the University's Mathematical Institute in the autumn of 1927. He was a twenty-five-year-old German-speaking, Hungarian-accented mathematician of orthodox Jewish family, from Cluj, Rumania. He came from a large family; his father was a baker. Because of the conflict between Saturday classes and the Sabbath, Wald had been educated mainly at home by his brother, Martin, an engineer.[26] He was particularly interested in geometry, he told Menger, and had been reading Hilbert's *Grundlagen der Geometrie*, in which he thought improvements could be made by dropping some postulates and relaxing others. Menger recalls that Wald registered at the University, but was not seen for more than two years, as he did not attend classes and had to serve in the Rumanian army. Early in 1930, he reappeared and Menger put him

---

[24] Morgenstern "Report on the Activities of the Austrian Institute for Trade Cycle Research 1931–1935", Feb. 13, 1935, AIRAC, Folder 37, Austrian Institute for Trade Cycle Research, Vienna 1935–1936, p. 11.

[25] *Ibid*, p. 14.

[26] See Menger (1952). Until World War I, after which the area fell to Rumania, Cluj had been Klausenburg, belonging to Hungary – hence Wald's accent – and part of the Austro-Hungarian empire. On Wald, see also Morgenstern (1951b), Weintraub (1985), and various review-type articles by Freeman (1968), Menger (1952), and Tintner (1952).

to work on the problem of "betweenness". Within a month, Wald had characterized "betweenness" in the ternary relations in a metric space, yielding four publishable papers.[27] Menger invited him to join his Mathematical Colloquium.

The Colloquium had been organised by Menger beginning in 1928, and over the course of the next few years, brought together a number of mathematicians, including Kurt Gödel, Franz Alt, Georg Nöbeling, Olga Taussky, G. Bergmann, and Otto Schreier. Amongst the foreign visitors were the Polish mathematicians, Knaster and Tarski; the Czech, Çech; and, on his annual trips between between Princeton and Budapest, John von Neumann. Papers were formally presented and discussed, and later published in the seminar's proceedings, *Ergebnisse eines Mathematischen Kolloquiums.* A glance at that journal reveals a wide range of mathematical topics, with logic, topology, and the theories of dimension, curve, and measure dominating. Indeed, for the first five years, the *Ergebnisse* is without reference to economics or social science.

We have mentioned that many of the Viennese mathematicians at this time were Jewish, at a time of rising anti-semitism. This traditional Viennese prejudice had been particularly strong just after the war, had declined somewhat in the latter part of the 1920s and had risen dramatically with the onset of economic depression after 1929. This time, it took the form of protests by Austrian Catholic and German nationalist student fraternities against the disproportionate number of non-Aryan professors and students at the University of Vienna, with frequent public demonstrations, class disruptions, violent outbursts, and beatings. Matters were not helped by the fact that Vienna's police had no authority in the self-policing university. In anti-semitic student diatribes, Menger himself was incorrectly labelled as Jewish on at least one occasion and Hahn, the only member of the Academy of Sciences who was both Jewish and socialist, was also targeted. To both the Austrian Catholic and pan-German nationalists, the Jewish socialists of Red Vienna were targets of opprobrium.

As one of the *Ostjuden,* or Eastern Jews, Wald stood at the lower end of the established hierarchy amongst the Jews of Vienna. Families like his had flooded into the Leopoldstadt, Vienna's quintessential poor Jewish ghetto. He therefore likely would have been conspicuously different in accent and appearance from his assimilated counterparts, such as Mises or Schlesinger,

---

[27] A point q is "between" the points p and r if, and only if, $p \neq q \neq r$ and the three points satisfy the equality $d(p, q) + d(q, r) = d(p, r)$, where $d(\bullet)$ is "the distance between". See Wald (1931a, b, c; 1933).

who were of Jewish origin but culturally integrated. Menger, as an outsider in the Brouwer circle in Amsterdam, a gentile amongst amongst Jews, a mathematician amongst economists, and in a minority at the University in his resolute opposition to German nationalism, was sensitive to difference, to marginality. He felt that Wald "had exactly the spirit which prevailed amongst the young mathematicians who gathered together about every other week" at the Mathematical Colloquium, and he took him under his wing (1952, p. 15).[28]

Because of straitened financial circumstances, Wald was often absent from Vienna and it appears that, at some point, he became responsible for his ageing parents, adding to his burdens. In late 1931, he wrote to Menger saying that he could not return to Vienna for financial reasons, but that he had been taking a university course in insurance methods and was continuing to work on the topology of the k-dimensional interval, on which he was enclosing results. Further letters follow in 1932 with results on axiomatics and the theory of convex spaces. Then, in 1933, Wald was back in Vienna, desperately seeking some position that would allow him to remain in the city, close to Menger and his Colloquium. Given his background, however, and in the middle of the Depression, which perhaps hit Austria harder than any other European country, Wald stood no chance whatsoever for any kind of university appointment. So Menger turned on his behalf to Schlesinger and Morgenstern.

A banker with the time and inclination to engage in intellectual pursuits, Schlesinger had published a 1914 book on the Walrasian system and was an active participant in the Viennese Economics Society. According to Menger, he was interested in improving his mathematical skills and therefore receptive to the offer of Wald's tuition. Out of this conjunction came Schlesinger's 1933 paper on the modified Cassel system, which introduced inequalities into the general equilibrium problem and thus dispensed with Walras' simple counting of equations and unknowns.[29] Wald, in turn, produced

---

[28] Elsewhere, Menger notes that amongst his university colleagues, his "friend Hahn was the only mathematician who knew Wald personally. No one else showed the slightest interest in his work" (*ibid*, p. 18).

[29] The Colloquium's developing interest in economics and social science appears in the fifth volume, which concerns the meetings of 1933–1934, with reference to two notes by Menger, on "Bernoullian economics and the Petersburg game" and on the relationship between finite sets and the formalization of ethics, and to the papers of March 1934 on general equilibrium by Schlesinger and Wald. The notes by Menger (1934a, b) were essentially short communications concerning what were published subsequently as (1934c) and (1934d) respectively. On the general equilibrium papers of Schlesinger (1935) and Wald (1935), see Weintraub (1985), pp. 59–107, and Ingrao and Israel (1991), pp. 175–210.

several papers dealing with systems of equations in mathematical economics, including the production and exchange variants of the Walrasian general equilibrium equation system and the Cournot duopoly model.[30]

As for Morgenstern, his relationship with Wald began in earnest in 1933, as indicated by a small grant from the Rockefeller Foundation to the Institute for the employment of Wald "to undertake a methodological study of the decomposition of statistical series".[31] For the next few years, Wald worked as researcher at the Institute. In early 1935, Morgenstern wrote to the Foundation, praising Wald's statistical and mathematical work, which, he said, was very reassuring and indicated that there was "still very much purifying to be done".[32] Amongst his accomplishments here, Wald constructed a procedure for seasonal decomposition, different from that of Persons' method of "link relatives", which Morgenstern presented at Louvain and Paris that year.[33] When applied to Austrian unemployment data for the period 1923–1934, Persons' method of deseasonalization yielded a residual series that "tracked" the original series reasonably well until 1932 – in the sense that the deseasonalized data were above the annual average when the seasonal oscillation was positive, and vice versa. However, the data then "contradicted" the original data in the remaining years, lying *below* the annual average when the oscillation was positive and vice versa. Wald's correction produced a better fit and Morgenstern's talk included a graphic display of the results. Wald's work here culminated in a 1936 book, *Berechnung und Ausschaltung von Saisonschwankungen.*[34]

---

[30] See Wald 1934, 1935, 1936. The duopoly paper showed how the model's equilibrium depended on the shape of the demand function for the commodity faced by the two firms. For example, the existence of a unique, stable, equilibrium point requires that the demand function cut the price and quantity axes and have a negative first derivative and nonpositive second derivative. See Tintner (1952), p. 22. A third general equilibrium paper by Wald was lost in the flurry in 1938.

[31] July 24, 1933, Research Aid Grants, Paris, Rockefeller Foundation, Box 4, Folder 36, AIRAC, Vienna, 1930–1934.

[32] Feb. 13, 1935, "Report on the Activities of the Austrian Institute for Trade Cycle Research 1931–1935", Box 4, Folder 37, AIRAC, Vienna, 1935–1936.

[33] See "La nature et le calcul des variations saisonnières", Memorandum per Dr. A. Wald, distribué à l'occasion de la conférence de Dr. Oskar Morgenstern, Wien, 6 mai, 1935, à l'Institut Scientifique des Recherches Économiques et Sociales, Paris, located in KMIT. A time series $\varphi(t)$ is assumed to be combined of trend, seasonal, and accidental components: in Wald's notation, $f(t)$, $S(t)$, and $Z(t)$. Persons' decomposition method assumed, amongst other things, that the seasonal component was multiplicative – that is, $S(t) = f(t) \cdot p(t)$, where $p(t)$ is a periodic (12-month) function of time. Wald's innovation was to assume that seasonal variation took the form $S(t) = \lambda(t) \cdot p(t)$, where $\lambda(t)$ is a nonnegative function that varies "slowly" with time.

[34] On Wald (1936) see Morgan (1990), p. 84, n. 10.

Wald also provided Morgenstern with instruction in basic mathematics – algebra and differential calculus – succeeding Franz Alt in that role.[35] Wald had a considerable impact on him and Morgenstern wrote frequently of him in his diary.[36] By the end of 1935, Wald was assuring him that he would soon understand nearly everything in mathematical economics. This, Morgenstern noted with delight in his diary, but these were private writings. Publicly, the imperious Morgenstern seems to have kept Wald at arm's length, with letters remaining formal even years later, and he never wrote publicly of taking lessons from him. Although this rift between private rumination and public decorum was quintessentially Viennese, it is also possible that there was a certain ambivalence in Morgenstern's relationship

[35] Franz Alt (b. 1910) entered the University of Vienna as a student of mathematics in 1928 and was a participant in the Menger Colloquium and Hahn's seminar. Upon graduation, he recalled in a 1997 interview, Menger felt guilty that he could not provide him with some employment and recommended him to Morgenstern, who appointed him as private tutor in mathematics at 20 Schillings an afternoon. "Morgenstern . . . very interesting, very intelligent. . . . He was convinced that mathematics was important . . . He told me once that he had wanted to study physics, but right after World War I all the interest was in the social sciences, and so he felt he should go into that . . . He had me help him read books on mathematical economics. It helped that I knew languages. We read English mostly. There was a man named Bowley who wrote a book here on mathematical economics. It was just as interesting for me as for him. I had to prepare each meeting, read a chapter in the book, and the we discussed it. He knew as much about it as I did, but perhaps once in a while I could explain something". (From a May 1997 interview with Alt, at his New York home, conducted by Seymour Kass, Bert Schweitzer, Abe Sklar, and Mrs. Annice Alt.). Through Morgenstern, Alt met various figures, including Oskar Lange, and Paul and Alan Sweezy, and was led to publish a 1936 article on utility theory in the *Zeitschrift für Nationalökonomie* (see Alt 1936). In 1938, Alt moved to the United States, where he was introduced by Morgenstern to Harold Hotelling. The latter, in turn, introduced him to Charles Roos, formerly of the Cowles Commission, whose 1934 book, *Dynamic Economics*, Alt had reviewed for the *Zeitschrift*, and who had by then left Cowles to set up a private economic forecasting consultancy, the Econometric Institute, in New York. Alt later left economics and made his career in computing. I am grateful to Professor Seymour Kass for permission to quote from this interview, the manuscript of which has been deposited in the Vienna Circle collections both at the University of Pittsburgh and in Vienna.

[36] "Yesterday more mathematics. I am beginning to see deeper and deeper, and through the ongoing implicit repetition it all seems to settle down" (Diary, Oct. 19, 1935, OMDU). And later: "Another mathematics lesson, very interesting. I feel as though I am making real progress. Wald told me of his new works. An amazing thing. It isn't enough, as Walras assumed, to consider only monotonically decreasing utility functions, because he [Wald] proved that with with many of them, simple exchanges never lead to an equilibrium! Similar paradoxes for the addition of demand curves, which were considered before to be totally harmless! That should have far-reaching consequences . . . Wald is really intelligent. I consider these works to be very important; they throw new light on the application of mathematics to economics. One will not be able to do without these at all" (Diary, Nov. 2, 1935, OMDU).

with the gifted *Ostjude.* After all, it had been only a decade since he wrote of the pollution of German culture and much less since he cast aspersions on the Jews of the Mises and Geist Kreise.

In his relationship with Menger, however, Morgenstern's exercise of power could not take the same form. Menger was a faculty member and a mathematician of international reputation, so their relationship was more equal. In many respects, also, Menger's work was more directly useful to Morgenstern than was Wald's. Although Morgenstern could trumpet Wald's successes to others, whether in seasonal analysis or general equilibrium theory, he could not incorporate them directly into his own personal, theoretical, work. On the other hand, drawing on Menger's papers, as we shall see, he was able to make suggestions regarding how the mathematician's ideas might be useful in resolving questions in economics. Thus, it is reasonable to suggest that he felt Menger's and Wald's influences in subtly different ways, each operating through the filter of power relations connecting him to them.

In Menger's Orbit

The deepening of the relationship between the economist and the mathematician coincided with the period of political tension and change in Vienna. In March 1934, the year after Hitler's rise to power and the Austrian government's suspension of the constitution, the Dollfuss regime cracked down on the municipal socialist government of Red Vienna, bringing cannon fire and upheaval to the city. In this context, as we have seen, Menger turned earnestly to the mathematics and logic of social science and became very concerned with respecting the demarcation between social analysis and politics. In the latter regard, Menger's list of sinners included Neurath and – even if Menger was perhaps politically closer to him than to the socialists – Mises. It was one thing to be of liberal inclination; it was quite another to say that it could be legitimized by science. As for Morgenstern, having spent several years promoting liberalism in the policy sphere and having now become, perhaps, the city's most prominent economist, he found himself adviser to a regime that had sympathy for neither liberalism or socialism. He was in a difficult position and one of his responses was to follow Menger in his insistence on the integrity of economic science. It was almost as if, in these critical years in the mid-1930s, Morgenstern and Menger together carved out a sanctuary, a sphere impervious to the surrounding political pressures. In the process, Morgenstern became at once more humble and more suspicious of Hayek and Mises.

In March 1933, Morgenstern wrote:

Saturday I was to dinner at Menger's. He gave, in a manner of speaking, a lesson on curve and dimension theory. We talked about a math. course that he wants to give, which will probably be excellent. We plan to meet again in August; until then, he is going to read lots of books and articles which I have lent him, and we are going to construct an axiomatics of economic theory. It could be of importance (OMDU, July 11, 1933).

Throughout 1934, that bond strengthened, with Morgenstern spending part of his holidays with Menger and fiancée in Ramsau and Strobl, and then with Menger alone in the Burgenland. Morgenstern read Menger's book on ethics, which had been prompted by the civil strife of that year, and, in a seminar taught with Richard Strigl, he used Menger's paper on the Petersburg Paradox.[37] Beyond that, he attended the International Congress of Philosophy in Prague with Menger and Schlick, and in Vienna, teas and social gatherings with Menger, Wald, and Institute economists Reinhard Kamitz, Ernst John, and Strigl. All of these influences worked their way into Morgenstern's writings of the mid-1930s, several of which we shall consider.

Like much of Morgenstern's work in this period, "The Time Moment in Economic Theory" ("*Das Zeitmoment in der Wertlehre*") is not always clear and offers little by way of constructive detail, the salient points being largely criticisms of existing theory.[38] Drawing on Schlick's 1930 *Questions of Ethics* and Menger's "Uncertainty" paper, he dismisses the criticism that utility theory is tautological because it can be applied to explain any observed economic choice behaviour. Individuals may well act "not according to the most dominating pleasure motive but . . . [according to] the strongest impulse of displeasure". Unlike the essentially analytical theorems of logic and mathematics, the contradiction of which would be absurd, utility theory can be contradicted by observation. This discussion is inspired directly by Menger, whose work, according to Morgenstern, has shown that "no

---

[37] "We have just started a mathematical course which is given by the famous mathematician Karl Menger" (OM to Eve Burns, Mar. 6, 1934, OMDU, Box 4, Corresp. 1928–1939, Burns, Eve M.). On Prague, see Morgenstern's Diary, Nov. 4, 1934, OMDU. Then: "Monday a very good seminar, very enlightening. We started to clear up risk theory. Menger's article on the Uncertainty Moment was much help" (*ibid*, Nov. 29, 1934).

[38] Introducing the paper to Frank Knight, Morgenstern wrote: "I put much thought and work into that article and, what's more, much breadth results, the limits of which I cannot see at the moment". (OM to Knight, Sept. 12, 1934, OMDU, Box 6 Corresp. 1928–1939, Knight, Frank H.).

one is really fully aware of the relationship between 'logic and economics'" (Morgenstern, [1934b], in Schotter (ed.), p. 152).[39]

The bulk of the paper is devoted to rather vague considerations regarding how time might be incorporated into utility theory, also probably inspired by Mayer's (1932) criticism of the Walrasian system. His main target here is H. L. Moore's (1925) "A Moving Equilibrium of Demand and Supply", which introduced time coefficients into the Walrasian system of equations to yield a moving equilibrium. Morgenstern criticises the Walrasian system for assuming infinitely fast reaction times, presumably of prices, and suggests having different reaction times for different prices. This, he says without further elaboration, "should give results". This leads to a natural emphasis on foresight and uncertainty, issues that "might, for reasons not to be given here, make even better material for the application of mathematics to economic theory than was evident until now in an unfortunately large number of cases of mathematical economics" (*ibid*, p. 158).[40]

Morgenstern continued this logical critique of the foundations of orthodox theory in his 1935 "Perfect Foresight and Economic Equilibrium". Here, he criticises Walras and Pareto for failing to make explicit their assumptions about what subjects can foresee, and Hicks (1933) for assuming that perfect foresight is a precondition for equilibrium.[41] We must ask, says Morgenstern, "the foresight of whom? of what kind of matters or events? for what local relationships? for what period of time?" (1935a, pp. 171–72). Without this, the entire concept of general equilibrium is jeopardised. As it stands, the assumption of perfect foresight implies that individuals have complete insight into all economic processes concerning prices, production, and income. Given the interdependence and complexity of the economic

---

[39] See Menger (1934c), p. 278, n. 17.

[40] Responding to the paper, Knight pulled no punches: "I read your article . . . and must say frankly that my reaction was not very enthusiastic. Not that I found anything to disagree with, but that it seemed to me the whole argument was rather in the domain of such a degree of refinement of conception and doctrine that I did not get the feeling of very great importance in the contribution" (Knight to OM, undated, OMDU, Box 6, Corresp.: 1928–1939, Knight).

[41] In the opening paragraph, Morgenstern refers to the discussion of Wald's work on general equilibrium in Vol. 6 of Menger's *Ergebnisse*, 1935, which revealed that: "The mathematical economists present an especially noteworthy example [of logical carelessness]. They, indifferent to whether it is a question of a general or of some particular equilibrium, have been content to assert that there are present as many equations as there are unknowns, rather than from the start proving in an exact mathematical fashion that there is a solution at all – and a unique solution – for these equations" (1935a, p. 169). Strident tones from one who, only four years previously, had made this very claim of the Walrasian system (see 1931, p. 367).

system, this implies "incredible powers on the part of the economic agent", who must not only know exactly the influence of his own transactions on prices but also the influence of every other individual, and of his own future behavior on that of the others".[42] Such persons of perfect foresight are not mortals, he says, but "demi-gods" (*ibid*, p. 173). Not only does it imply that economics has posited the existence of an economic subject that perfectly knows economic science already, but it leads to a paradoxical situation of the Holmes–Moriarty type, with perfectly endowed actors outguessing each other. What, in his *Wirtschaftsprognose* of 1928, had provided an analogy enabling Morgenstern to refute attempts at economic prediction was now serving to undermine the existence of a general equilibrium: "Unlimited foresight and economic equilibrium are thus irreconcilable with one another". The theoretical matters concerned "are so extremely complicated that only far-reaching employment of mathematics could help to suggest the reciprocal dependencies. The relationship between human behaviors dependent on one another, even without the assumption of foresight, is almost inconceivably complicated, and it requires cogent examination" (p. 174): "Up to the present time, the only examination of a *strictly* formal nature about social groups, even though it is carried out in another field and is limited to the co-existent individuals independent of one another, is a work by K. Menger [*Morality, Decision and Social Organization*, 1934d] which it is hoped, will become known to economists and to sociologists because of its importance in laying the foundation for further work" [pp. 174–75].[43]

Although he never submitted it to the *Journal of Political Economy*, as he considered doing,[44] the "Foresight" paper was a better success than

---

[42] Innocenti and Zappia (2005) point out that, in the discussion of perfect foresight, Morgenstern's target here is also Hayek, following his 1933 lecture in Copenhagen, "Price expectations, monetary disturbances and malinvestment" (reprinted in Hayek 1939, *Profit, Interest and Investment and Other Essays on the Theory of Industrial Fluctuations*, London: Routledge & Kegan Paul, pp. 135–56). Hayek noted that equilibrium theory was now taking account of the time factor by making assumptions about the attitude of persons towards the future – that is, "essentially that everybody foresees the future correctly and that this foresight includes not only the changes in objective data but also the behaviour of all other people with whom he expects to perform economic transactions" (1933, pp. 139–40, quoted in Innocenti and Zappia, p. 74).

[43] He also makes a vague suggestion concerning the application of Russell's Theory of Types to questions of economic knowledge, proposing that degrees of knowledge of the economic system might be grouped into different types, ranging from low to high. Even given the same economic resources, he seems to say, different levels of knowledge and different degrees of insight might yield different kinds of equilibrium (1935a, pp. 176–77).

[44] "I have a certain interest to have [sic] this article appear in English because Mr. Keynes is preparing a book on the theory of money largely based on the element of expectation

the "Time Moment" one, capturing the interest of Menger, Wald, and others.[45] Even Hayek liked it. From London, he wrote: "It will interest you that we recently had an interesting discussion about your essay in our seminar... the results were really valuable and enlightening".[46] He had asked someone to write down the proceedings of the discussion, which he

and anticipation" (OM to Knight, Dec. 18, 1935, OMDU). Morgenstern's resistance to Keynes' economics is a recurring theme in this period. In a letter to Eve Burns in 1934, he claimed to have proved that, in Keynes' theory of money, "his equations completely don't hold up" (OM to Burns, Mar. 6, 1934). Then in his diary in 1935, he wrote: "Wald finds my article on Keynes mathematically alright. Now I am going to prepare it for publication and I am going to send it to Chicago" (OMDU, Oct. 26, 1935). Writing to Haberler on the *General Theory*, he said: "Keynes' book is simply horrible, to the extent that I have read it" (OM to Haberler, Apr. 4, 1936, OMDU, Box 5). To Morgenstern, Knight wrote "What do you think of Keynes' book?... a couple of friends whom I consider pretty competent judges say outright that Keynes is losing his mind" (Knight to OM, May 1, 1936, OMDU Box 6). See also later discussion in connection with the 1937 translation of his 1934 book, *The Limits of Economics*, in which he attacks the *General Theory*'s lack of rigour. I have been unable to trace any of Morgenstern's other writings on Keynes.

[45] "Yesterday I had lunch with Karl Menger... we quickly discussed 2 1/2 hours. He had carefully read the article on Foresight, agrees, and wants me to deal more with these interesting questions. He is now busy with completing a large work on the calculus of variations, but then he wants to immediately return to social-scientific questions. It was, like always with him, a very stimulating meeting" (OMDU, Sept. 11, 1935). "I believe that everything is correct", wrote Wald to Morgenstern. "One can also understand by 'foresight' that the economic subject has a subjective conviction to foresee any kind of economic things, which however do not have to be congruent with reality. Foresight in this sense I want to call 'subjective foresight'. The complete subjective foresight of an individual then means the subjective conviction that the person has the capacity to form an overview of all future economic phenomena. The full subjective foresight of two individuals need not necessarily be in agreement. The assumption that every economic subject has full subjective foresight could be free of contradiction. There are functional connections between subjective foresight and different economic phenomena. The assumption that every economic subject has full foresight in the usual sense means that every economic subject has the same full subjective foresight, and that this is congruent with the future true turn of events. Such an assumption then leads to a contradiction when situations come to pass where the economic subject wants to adjust his actions so that they are in opposition to his evaluation of the foresight of other economic subjects. This is probably the case in economics. But there are also conceivable areas where human actions and foresight play an essential role, and nevertheless full foresight in the objective sense would be free of contradiction" (*ibid*; Letter, Wald to Morgenstern, Aug. 2, 1935, KMIT). In the same letter, Wald mentions having begun reading "the book by Weber", suggested by Morgenstern, which, he found, gave a good orientation of many problems in economics, but treated them "rather superficially and not strictly". This was probably Weber's essays on *Economy and Society*.

[46] Letter, Hayek to OM, Feb. 9, 1936, OMDU. This time, Knight, too, was enthusiastic: "It seems to me that in your article on perfect anticipation you have done a major piece of work". He went on to add that "the market for high grade economists in this country seems to be quite 'bullish' at the moment. Are you interested?" (Knight to OM, Mar. 12, 1936, OMDU, Box 6, Corresp.: Knight). Haberler, too, wrote expressing his admiration (see Haberler to OM, July 30, 1935, OMDU, Box 5, Corresp. 1928–1939, Haberler, Gottfried).

hoped to submit for publication in the *ZfN*. No such paper appeared in the *Zeitschrift*, but when one rereads the now well-known meditation on equilibrium theory that appeared two years later, Hayek's "Economics and Knowledge" paper in *Economica*, many of the themes broached by Morgenstern surface again, and even Menger's work on ethics is cited for its promise.

Like Morgenstern, Hayek emphasises the relationship between equilibrium and foresight, but whereas the former saw perfect foresight as a form of omniscience that became self-defeating once placed in a social setting, Hayek took a different tack:

[T]he concept of equilibrium merely means that the foresight of the different members of society is in a special sense correct. It must be correct in the sense that every person's plan is based on the expectation of just those actions of other people which those other people intend to perform, and that all these plans are based on the expectation of the same set of external facts, so that under certain conditions nobody will have any reason to change his plans. Correct foresight is then not, as it has sometimes been understood, a precondition which must exist in order that equilibrium may be arrived at. It is rather the defining characteristic of a state of equilibrium. Nor need foresight for this purpose be perfect in the sense that it need extend into the indefinite future, or that everybody must foresee everything correctly. We should rather say that equilibrium will last so long as the anticipations prove correct, and that they they need to be correct only on those points which are relevant for the decisions of the individuals (1937, pp. 41–42).

Thus, for Hayek, the salient questions concern knowledge – who knows what, or believes what, about the economy, and about others? These are empirical questions, ignored by mathematical equilibrium theory, the Pure Logic of Choice, and, as Hayek says allusively (*ibid*, p. 54), the a priorism of older economists. But having highlighted this empirical question, Hayek subtly shifts gear. It is clear, he says, that economics, poor specification of knowledge or not, has come closer than any other social science to answering "that central question of all social sciences, how the combination of fragments of knowledge existing in different minds can bring about results which, if they were to be brought about deliberately, would require a knowledge on the part of the directing mind which no single person can possess" (p. 52). This allows Hayek to downplay the extent to which the knowledge question opens up any fruitful line of empirical research. He says it is not clear that such research would teach us anything: it is more important to be clear about the principles and, he says opaquely, about when the argument becomes subject to verification.

Here, we touch on the central, subtle, issues separating Morgenstern and Hayek. Although Hayek, emphasising the importance of knowledge, was willing to skirt dangerously close to the suggestion that inferior knowledge placed the economic equilibrium in jeopardy, in the final analysis he retreated to the safety of oblique, invisible-hand type references to the superiority of the coordinative market order. By the mid-1930s, Morgenstern was no longer willing to take such a step. As long as the logical underpinnings of the theory remained inadequate, he was ready to emphasise the possibility of disorder and disequilibrium rather than the coordinating powers of the market. As we shall see, he was even ready to muse about changes in people's mentality and the desirability of governmental intervention, all of which was a far cry from Austroliberalism. At this point, Hayek and Morgenstern, who had shared so much by way of formative influences, were on diverging paths, and the distance separating them would ultimately widen to that which separated the *Road to Serfdom* from the *Theory of Games and Economic Behaviour*.

Morgenstern's "Logistics and the Social Sciences" of 1936 was essentially a rebroadcast – to an economics audience – of Menger's 1932 public lecture on "The New Logic". Morgenstern says that only by adopting a formal language can one remove the "traps and contradictions of a sentential language" (p. 398). Once again, Menger's book on ethics is appropriated: "It would obviously be important if one could formalize the achievement of a concrete economic policy measure in the same manner as this has been done for normative systems . . . Although space forbids [entering] upon the relation which Menger's book has to economic theory [it] may suffice to mention that his logic of wants or wishes would provide a most useful model for a logically satisfactory economic theory of wants" (p. 404).

The aforementioned papers were intended by Morgenstern as chapters in a book to be called *Time, Profit and Economic Equilibrium*.[47] With the exception, perhaps, of the "Perfect Foresight" paper, they accomplished little theoretically, but they illustrate perfectly how Morgenstern grappled with theoretical emphases inherited from the Austrian economists yet turned to formal mathematical work for clarification. That encounter with Menger and with logic served to draw him steadily away from his mentors.

The Limits of Liberalism

Yesterday in the Economics Society, Menger gave an excellent presentation about the law of diminishing returns. It was an exemplary piece of work for the proof of the necessity of exact thinking in economics. It was interesting that Haberler failed

[47] See OM, Diary, Dec. 9, 1934, OMDU.

totally in the discussion . . . Of all these exact things he, by far, doesn't understand the most essential. Mises talks pure nonsense.

Morgenstern, Diary, Dec. 31, 1935

Ludwig von Mises gave stimulating lectures without, however, clearly separating the ideas of economic theory (which he presented with an idiosyncratic opposition to the use of even simple mathematics) from his idea of complete *laissez-faire*.

Menger (1994), p. 11

In Morgenstern's early years, as we have seen, Mises was a silent presence, never featuring explicitly in his work, but exerting an influence nonetheless on his writings and career. Morgenstern's critique of institutionalism and prediction was very much in the spirit of Mises, and his assuming the helm at the Institute could hardly have been done without the active encouragement of its founder. Throughout the early 1930s, Morgenstern's public economic commentary and policy advice were very much in the spirit of the Austrian liberal economists.

As time went on, however, Morgenstern expressed increasing resistance towards Mises, explicitly in his diary, more allusively in his writings. All of this was under the influence of Menger, who took umbrage at the way in which the elder scholar both opposed the use of mathematics and referred to logic when reinforcing his "scientific" liberalism. Following his experience with Brouwer, and even Hahn and Neurath, Menger was exceedingly sensitive to any intrusion of normative or political preference into scientific work. In Menger's company, Morgenstern learned to reject much of Mises' philosophy of economics, including his a priorism, his views on mathematics, and the way he used the discipline to justify laissez-faire.

A sense of Mises' priorities at the time may be found in his 1933 *Grundprobleme der Nationalökonomie*, later translated as *Epistemological Problems of Economics*. Here, Mises continued his onslaught against German historicism, arguing that the study of the unique and unrepeatable events of history could never lead to theoretical insight, that theoretical understanding was *a priori*, being rooted in the nature of human action and constituting the prior analytical scheme by which one selected amongst the confusing mass of data presented by historical reality. Mises argued the insistence by Sombart and the *Kathedersocialisten* on empirical methods and their arguments against the possibility of a universally applicable theory rooted in their political bias towards interventionism. He said that were they to concede that humans, throughout known time and space, were purposeful in their behaviour, directing it towards improvement of their situation, entering increasingly into economic exchange, and generating the economic phenomena of markets and prices, they would be forced to admit the universality of economic theory. They would also concede that the liberal order was the system of political organisation that best facilitated the unhindered pursuit of economic ends by individuals. "[T]he science of economics proves with cold, irrefutable logic that the ideals of those who condemn making a living on the market are quite vain, that the socialist organization of society is unrealizable, that the interventionist social order is nonsensical and contrary to the ends at which it aims, and that therefore the market economy is the only feasible system of social cooperation" (p. 196). This was Mises' message, repeated throughout the various essays of *Epistemological Problems* and later expanded in his 1949 magnum opus *Human Action.*

Mises persisted in his opposition to the use of mathematical formalism in economics, his main argument being that it not only was unnecessary, being merely an embellishment of insights gained independently of mathematical reasoning, but harmful, in that it induced a simplistic, mechanical, perception of the social domain. The problems faced in the social sciences are so complex that "even the most perplexing mathematical problems" appear simpler, says Mises. Those who wish to resort to mathematical methods are welcome to it, he says, but "[t]hose theorists who are usually designated as the great masters of mathematical economics accomplished what they did without mathematics. Only afterwards did they seek to present their ideas in mathematical form. Thus far, the use of mathematical formulations in economics has done more harm than good" (*ibid*, pp. 116–17). He goes on to condemn the Trojan horse of "mechanism" smuggled in with mathematics, as discussed earlier. He points to the natural sciences, in which the role of mathematics is different from that in the social sciences, insofar as the discovery of empirically constant relationships is possible; they are similar insofar as "even the mathematical sciences of nature owe their theories not to mathematical, but to nonmathematical reasoning" (p. 117).

To Menger, statements like this were naive. They suggested that Mises was unaware of the distinction between quantification and the use of mathematical symbolism, was unfamiliar with the generative role of mathematics in the development and refinement of concepts in physics, and viewed mathematics as some sort of uncontested, homogeneous, and neutral tool, to be "applied" in the natural sciences when the occasion demanded. It is little wonder that Menger, with characteristic restraint, described Mises' opposition as "idiosyncratic".

Nor could Menger go with Mises' contention that economic theory, the best developed branch of the science of human action, was *a priori*, not empirical. According to Mises,

Like logic and mathematics, [economics] is not derived from experience; it is prior to experience. It is, as it were, the logic of action and deed . . . logic and the universally valid science of human action are one and the same . . . What we know about the fundamental categories of action – action, economizing, preferring, the relationship of means and ends, and everything else that, together with these, constitutes the system of human action – is not derived from experience. We conceive all this from within just as we conceive logical and mathematical truths, a priori, without any reference to experience (pp. 13–14).

After his experience with intuitionism, Menger was all too familiar with justifications of mathematical and logical truths "from within". Looking "within", Brouwer had found grounds to reject the axiom of choice, the law of the excluded middle, and nonconstructive existence proofs. Menger was highly suspicious of appeals to intuition, the authority "within", as the basis for any kind of mathematics, as they usually translated into attitudes of intolerance. His counterattack against Brouwer was a gesture against monotheism in mathematics.[48] In that counterattack, Menger emphasised the possibility of multiple logics, so he was especially sensitive to Mises' cavalierly appealing to "the" logic in order to undergird his conception of human action. Also, his own work on the Petersburg Paradox had emphasised the precisely *empirical* nature of the question: some people accepted very favourable bets, others did not. Recourse to *a priori* reasoning here did not carry one very far in determining how individuals behaved: one would have to know much more about their particular circumstances. One can understand why Menger learned to regard Mises with suspicion, viewing his a priorism as scientifically inadequate and rejecting the way in which he incessantly sought to put economic theory to political use.

Certainly, there was much in Mises' writing with which Menger could agree: the rejection of Spann's universalism, the Austrian emphasis on individualism as the appropriate methodological approach in social science, the distinction between "cold, hard" science and the consolations of metaphysics. Menger also would have completely endorsed Mises' nominal separation of the irrefutable "facts" of economic science from the domain of political or ethical choice, but, again and again, Mises himself blurred the very boundaries he proclaimed to maintain. Notwithstanding his claim

---

[48] Menger's rejection of Neurath's programme for Unified Science was motivated by the same attitude.

that ethical choice and economic science occupied different realms – that even if economic theory pointed to the efficiency of classical liberalism, one was always free to reject it on political grounds – Mises' entire rhetoric in *Epistemological Problems* is intended to promote the politics of laissez-faire. This is presented as the reasonable political stance issuing from economic science, that collection of *a priori* truths evident to all clear-thinking observers. This Mises reinforced with frequent reference to the natural sciences, logic, and mathematics, much of which would have appeared foreign and therefore inassailable to his economist audience, but not to Menger, who knew more than Mises about actual practice in all of these fields.

Thus, the week following Christmas 1935, Menger explicitly challenged Mises on a question of logic and proof, presenting a paper on the law of diminishing returns to the Economics Society. He was responding to a claim by Mises, in his *Grundprobleme*, that certain propositions of economics could be proved, an example being the law of diminishing returns.[49] In the paper, later described by Schumpeter as a reading of "the logician's riot act" to economists (1954, p. 587), and by the author himself as the first instance in economics of a clear separation between the question of logical interrelations amongst propositions and that of empirical validity, Menger examined the existing proofs of the law of diminishing returns. He took the analyses of Wicksell, Böhm-Bawerk, and von Mises apart with a fine-tooth comb, showing how they failed "to meet the requirements which logic places on a sequence of inferences intended to constitute a proof". The talk created something of a stir, and, as indicated by the diary quotation with which we opened this section, it impressed Morgenstern, adding to a rift with Mises that was well in the making by then.[50]

In mid-1933, Morgenstern had written to Hayek that he was completing a book – "mainly a summary of discussions ... with practitioners", "for a wider audience", that would not "go too much into methodological details". Sending a copy to Knight in early 1934, he confirmed that his "methodological line [was] rather different from the one followed by Robbins,

---

[49] This was published as (1936b). For Menger's recollection of the time, see (1979, p. 279).

[50] From London, Hayek wrote to Morgenstern: "I have heard from Schütz about a discussion between Mises and Menger about the law of diminishing returns, and according to this report there seemed to be confusion on both sides. I would very much like to talk about this *if* I could get more detailed information about the content of this discussion beforehand. Supposedly a longer correspondence took place and a manuscript by Menger is available. Could one see some of this?" (Hayek to OM, Feb. 15, 1936, OMDU, Box 6, Corresp.: Hayek).

Mises and Hayek".[51] The book in question was his 1934 *Die Grenzen der Wirtschaftspolitik*, translated in 1937 as *The Limits of Economics*. A rambling book, it is critical, rarely constructive, and targets a range of established economists, including Robbins, Mitchell, Keynes, and von Mises. Indeed, it is for the latter that Morgenstern reserves his sharpest barbs.[52]

"[T]here are but few sciences", writes Morgenstern, "which are in such an objectively unsatisfactory condition as economics" (1934, p. 19). The discipline is riddled with "value judgments". In the *Foreword*, he reiterates Robbins' emphasis on the requirement of rationality of economic policy, and the need for "absolute precision of thought . . . when we are forced to be the unhappy witnesses of an almost unprecedented decay of intellectual life in so many countries" (p. vi). Widespread economic depression throughout the early 1930s pointed to the importance of directing theory towards "the mastering of practical life", without which it was but "an intellectual plaything, similar to chess, and [serving] to satisfy only a perverted desire for purely mental exercise" (p. 4). However, he says this is not an argument for empirical studies because Whitehead (1925) demonstrated that it is "impossible to grasp reality without the construction of theoretical formulae", that the "utmost abstractions are the true weapons with which to control our thought of concrete facts" (p. 6). Thus Morgenstern rails against the redundant doctrines of the historical school and their disguised successors, the institutionalists, criticising both throughout the book. Mises would, of course, have been in agreement with all of that.

Elsewhere, however, Morgenstern challenges Mises directly:

[T]he thoroughly empirical character of economic theory cannot be stressed too strongly. *A priori* theory would be very easy if it were possible to dispense with the necessity of dealing with reality and with the flux of economic events and if it were sufficient to lock oneself in a room and invent the world of facts, adopting

---

[51] Letter OM to Hayek, July 11, 1933, OMDU. Letter, OM to Knight, Feb. 9, 1934, OMDU.
[52] Of the original German version, a reviewer, Henry Laufenburger, wrote in the French *Revue d'Economie Politique*: "[Morgenstern] believes in the autocratic State which, according to him, can resist the demagogic demands of the parliamentarians, form long-term economic (five-year) plans and assure a better distribution of wealth. Without doubt, Mr. Morgenstern would like to have dictatorial power subject to certain control, but this would be organized by the controlled themselves. Why, given this, did Mr. Morgenstern not choose a title which would allow the reader to guess the content of his book? By this means, he would have avoided wasting the time of those actually interested in 'economics'" (1934, p. 1085, my translation). Another brief review by E. Phelps Brown in the *Economic Journal* alluded to similar frustrations (1934). The 1937 English version, *The Limits of Economics*, on which this account is based, is claimed by Morgenstern to be considerably revised and, therefore, "not . . . simply a translation" (1937, p. vi). The attack on a priorism and liberalism was present in both editions.

the attitude that if theory and reality did not agree, so much the worse for reality. 'Theory' of that kind can neither be confirmed nor refuted: nothing easier could be wished for. But, unfortunately, it has nothing to do with the real world (*ibid*, p. 10).[53]

Then, citing Neville Keynes, Cairnes, Weber, and Robbins on the separation of the positive and normative realms, Morgenstern lumps Mises in with the socialists in that both he and they allow political values to enter their theorising and seek support for their politics in economic analysis. He writes that liberalism is paradoxical in that it argues against government intervention without acknowledging that it may be necessary to maintain free competition in an age of rising monopoly power. Rigid systems, in general – be it liberalism or socialism – also ignore changes in the 'economic mentality', such as the appearance of a general desire in people to have the State systematically attend to their welfare.

It is difficult to know whether Morgenstern really believed this or whether it was a gesture of deference towards the Austrian State. Between 1934 and 1937, many of his economist colleagues had left Vienna and Morgenstern found himself as policy advisor to the corporate regime. In his attempt to walk a fine line between providing "neutral" advice and toeing the governmental line, he appears to have been successful. Many of the allusions in this book speak to the difficulties Morgenstern faced throughout this period.

He says the exclusive task of economics is to determine the effects of policy and, alluding to the Austrian situation, he proceeds with lengthy dissections of exogenous shifts or policy changes, of primary and secondary effects, of economic and psychological consequences. The book's two guiding metaphors are those of physical and spiritual health, with abundant references to medicine, psychological stability, and pathology. Menger, Gödel, and Wald are all harnessed in attacks on the imprecision of Keynes

---

[53] Mises is named only in the Appendix, where his *Grundprobleme* is described as "an attempt to find an *a priori* basis for economics... one of the points where he diverges fundamentally from the view point put forward [here]" (1937, p. 154). In the same passage, Morgenstern castigates Robbins' *Nature and Significance* for presenting the Austrian economists as being more uniform in their views than was actually the case. In another oblique reference to Mises, he continues: "It is, moreover, worth noting that in practice the difference is one of method only for the few surviving apriorists are obliged in practice to make so many concessions that in the actual theorems themselves they abandon their original position, so that in the end both they and the empiricists are speaking the same language. What is really the most unfortunate result of their methodological position is their tendency to identify economic theory with a particular system of economic policy" (*ibid*, p. 10).

and, especially, the political biases of Mises. As we shall see subsequently, blows are struck in the context of the power struggles amongst the Viennese economists. With this volume, Morgenstern distanced himself definitively from Hayek and Mises, rejecting not so much a liberal style of economic policy *per se* as the idea that it was the *only* policy conclusion to which economic analysis could lead.

On reading the book, Hayek became testy:

If one is supposed to be grateful for being sent a book, and one does not agree with it at all, and one knows the author too well to handle the matter in one phrase, the only way is to make the letter a counter conclusion. But for that I haven't had enough time. And you make the discussion very hard for me. To be honest, your book is a collection of, often brilliant, aphorisms, but it lacks the consistent argumentation with which one can start a discussion. Furthermore, that you were rude to some of my friends makes it even more difficult.... [We] can only hope that, through the years, with many applications of the principles to specific problems, we can convince each other.[54]

Hayek and Morgenstern, however, never did convince each other. Allying himself with the mathematicians and the "positivists", Morgenstern drew further and further away from Hayek. Although they had emerged from the same milieu, with shared Austrian theoretical concerns, by the mid-1930s they had grown apart. Hayek was in London, campaigning against socialism and planning, condemning "scientism" in economics, pointing to the natural economic order (see Hayek 1942–44). Morgenstern was in Vienna, criticising liberalism and a priorism, and sceptical of the very concept of equilibrium in the absence of a logical mathematical explanation of the interaction of beliefs and opinions. It is not difficult to see why the extant Morgenstern–Hayek correspondence not only thins in the late 1930s but is devoid of serious intellectual engagement.[55]

---

[54] Hayek to OM, April 2, 1934, OMDU. Neither did Knight like the book. On reading the 1937 translation, he wrote: "Frankly, I hardly know how to comment on your book. I have not read the English version in its entirety. It seems better than the German edition, but I have not made any detailed comparison. I hope it will not give offense if I say frankly that it did not seem to me, or to some colleagues whom I have heard comment, that the book represented a terribly serious effort on your part to penetrate to the more fundamental issues. We have been inclined to infer that it was written rather for a semi-popular audience than with a view to making some real contributions to the discussion, which you are certainly capable of making" (Knight to OM, July 31, 1939, OMDU).

[55] Morgenstern's diary is replete with criticisms of Hayek, most of which are impressionistic. For example: "Hayek... has written to Knight that he should give up economics and rather plant potatoes. He is totally crazy. Now my view is confirmed that Hayek is never going to become anything" (Jan. 9, 1935). See also Sept. 15, 1933; Sept. 14, Oct. 26, and Nov. 2, 1935, OMDU.

Flight

Perhaps Morgenstern's repeated references to disequilibrium and psychological unease were omens of what was to come, for, in a few short years, beginning in 1934, the intellectual community of which he was part collapsed entirely. A unique perspective on that slide into oblivion is provided by the surviving records of the activities of the Rockefeller Foundation.

Recall that, in 1930, the Foundation made a five-year grant to the Institute. In July of the following year, no doubt encouraged by the institute's success, president of the Austrian National Bank, Richard Reisch; Hayek; Pribram; and – testament to the power of lucre – enemies Mayer and Mises, sent to the Foundation a jointly signed "Memorandum on the Situation of Research in Social Sciences in Austria".[56] They requested money for the support of politically-independent research in the poorly funded social sciences. They wrote,

After the war, these difficulties have become immense because of the general impoverishment and because the influence of party-politics, which is so particularly dangerous to social sciences, has become overwhelming. The small means which are available are mostly under the administration of more or less political organizations which, quite naturally, use it for purposes which seem most important from their respective partisan point of view and which are not in the first place guided by scientific considerations. . . . There is, therefore, at present no body or organization whatever in Austria which could assist independent and unbiased research in social sciences.[57]

They distinguished their project from the newly funded business cycle Institute, which covered only a small section of economics, leaving many young men and women without support and compelled to earn a living by uncongenial means. They had in mind work in social history (the transition from monarchy); sociology (the problems arising from the "racial and national mixture of population in Central Europe"); economics (problems of changes in economic structure and others needing quantitative measurement, which did not fall into the Institute's ambit); and political science (the transition from autocracy to democracy). Without naming a duration, they requested $15,000 per year. There is no evidence of a reply from the Foundation.

---

[56] Memorandum on the Situation of Research in Social Sciences in Austria, July 27, 1931, AIRAC.
[57] *Ibid.*

In March 1933, Rockefeller's Van Sickle met Mises in Paris and they spoke of the effect of Hitler's accession to power on the development of economics in Germany and Austria.

[Mises] was inclined to take a very pessimistic view, and in his opinion we had probably seen the end, for at least a generation of any intelligent economic research in the German-speaking countries. He felt that the dictatorial regime in Germany and the extension of nationalistic tendencies in Austria will destroy any intellectual freedom in the field of economic studies, or will make it impossible for any properly qualified economists to obtain academic positions. He felt that the National Socialists would attempt to develop their own economic theories based on false premises with disastrous results for Germany and the almost complete suspension of the development of economic science.[58]

The 1931 social science proposal was then brought up again by Pribram in October 1933, when he called to see Van Sickle in Paris. The allusion to racial issues in the original memo now came to the surface directly. Pribram suggested that the directorship of such a social science institute might consist of:

Aryan	Prof. Richard Reisch, representing Economics
Jew-Aryan	Prof. Mises or Prof. Hans Mayer, representing Economics
Aryan	Prof. Karl Bühler, representing Psychology
Aryan	Prof. Verdross, representing Law and Political Science
Jew	Prof. Pribram, representing Modern Social and Political History[59]

Pribram emphasised the importance of the proposed institute's being independent of the University, where the majority of the professors were frankly Nazi: "The directors of the proposed Institute would all be members of the university, but they are all Liberals and independent. There would be only one or at most two Jews in the Direction" (*ibid*).

Van Sickle indicated to Day that he supported the proposal, saying that the situation in Vienna was now so serious that the Foundation might be justified in "backing frankly the minority liberal element" (*ibid*). The group should be financed for the next two years, until 1935, he said, at which point the grants to Morgenstern's and the Psychological institutes would have expired and the matter could be reviewed. He added, significantly: "We must reckon, of course, with the fact that institutes now receiving direct

[58]   TBK, Internal Rockefeller Foundation Memo, re Conversation with Prof. Mises, Paris, Mar. 3, 1933. Mises also forecast Jewish professors having to leave Germany and the use of income tax laws to seize Jewish property in both countries. There were already cases in Austria, he said, where the entire personal capital had been confiscated through bloated tax claims.

[59]   Letter, JVS to Edmund E. "Rufus" Day, Oct. 10, 1933, AIRAC.

aid from the Foundation will no longer be so keen for a general institution whose Board of Directors might not treat them so generously as we have" (*ibid*). Indeed, although there is no "smoking gun", Morgenstern's actions and writings with regard to Mises, including *Limits* and the other attacks discussed earlier, are entirely coherent with his having felt stung by this bid to usurp the role of his "own" institute. His extension of the Institute's activities beyond business cycle work to mathematical economics and to the study of the Danube Basin, may also be seen as an attempt to thwart the funding manoeuvres of the larger group. By December 1934, writing in his diary of his plans for reading rooms at the Institute, Morgenstern could say that Mises and Mayer were not going to be asked any more.[60]

Van Sickle pursued the matter. He visited the economic institutes in Heidelberg, where political interference suggested that Foundation support should be reduced, and then Vienna, where more was justified: "The general opinion is that Austria will survive as an independent state with an authoritarian government, and that social science will be reasonably free. I was impressed on this visit, as on every former one, with the genuine interest in research and the suprising vitality of scholarship. There are warring factions, but there are good scholars who stand between them and who can be trusted to administer any funds we might place at their disposition".[61] He had lunched with Pribram, Verdross, and Degenfeld, who agreed that, in order to administer a grant, a Committee for Promotion of Social Science Research should be formed, independent of the University, and minus "any of the prima-donnas – notably Mayer, Mises and Spann" (*ibid*). He had later explained to the latter why they were being excluded. He continued:

I have suggested to Pribram that in the letter of request the Social Sciences should be so defined as to exclude support of the pure Romanticism and the vituperative propaganda of Spann, yet permit support of precisely defined problems by younger scholars of the Spann School.

... There are distinct hazards in this proposal, which arise out of deep personal animosities. It is my hope, however, that these animosities can be reduced by a tactful and impartial committee. I am particularly desirous of drawing Spann into the circle of beneficiaries because I believe that he will then find it more difficult to continue his present destructive opposition to all objective liberal research.

---

[60] Diary, Dec. 9, 1934, OMDU. In May, Van Sickle had written to Day of Morgenstern's intention to expand the field of Institute activities beyond business cycle work. Letter, JVS to EED, May 1, 1933, AIRAC. Funds for Wald followed in July, and for another price study in August.

[61] Letter, JVS to EED, Oct. 28, 1933, re "Social Sciences in Vienna", Box 4, Folder 35.

Thus, if one of his men receives Committee support, it would be harder for him to characterize as 'stuff and nonsense' another piece of work accomplished under committee auspices by a man of the rival marginal utility school, and to oppose his 'habilitation' at the university. To do so would be an affront to the whole committee (*ibid*).

In an immediate reply, Day quashed Van Sickle's proposal, citing Austrian political instability. Van Sickle fought back, with letters travelling back and forth between him and head office into early 1934.[62] In January of that year, in a telegram to Day, he announced Dollfuss's suspension of the constitution, but insisted that the situation was not so bad as to endanger scientific work.[63] A few days later, having spoken to Professor Charles Rist in Paris, he elaborated further:

Authoritarian government bids fair to spread in Europe. We shall doubtless have to learn to distinguish between good authoritarianism and bad authoritarianism. Even such an old Liberal as Professor Rist appears to be swinging around to a belief that some modified form of dictatorship may be the only way out of the present mess. The democracies seem paralyzed by the conflicting aims, aspirations and appetites of their constituents. Freedom appears to be a luxury that we cannot afford after our triumphant war to make the world safe for Democracy. Unless Nazism sweeps Austria, and I don't think it will, the type of authoritarianism will be one compatible with reasonable freedom of research and expression. I hope that my proposal of October last is only postponed, not discarded.[64]

Following the bloody events of February, Van Sickle said that he understood the international public outrage, but that the whole affair was "very human". The public had seen only the visible and best aspects of Social Democratic domination of Vienna; the model tenements, progressive schools, improved hygiene, and so on. What they did not see was the "slow, steady expropriation of the middle classes by a variety of class taxes. Only one who has lived in Vienna can realize the bitterness and despair provoked by this policy.... Then too the anti-religious attitude of the party... deeply offended the provinces with their large catholic populations. Total result: the provinces and the entire middle class against the Socialists and only waiting their chance to destroy them".[65] He admired the Socialists, but was not surprised that it ended the way it did.[66]

---

[62] See Letter, EED to JVS, Nov. 6, 1933; and JVS to EED, Nov. 20, 1933, and JVS to Sydnor Walker, Dec. 1, 1933.

[63] Cable, JVS to EED, Jan. 19, 1934, Box 4, Folder 35.

[64] Letter JVS to EED, Jan. 24, 1934, Box 4, Folder 35.

[65] Letter JVS to EED, Mar. 10, 1934, Box 4, Folder 35.

[66] Concerning the events of March 1934, Morgenstern wrote somewhat cryptically to Eve Burns in the United States: "The time of the shootings was really bad, since it is really

Van Sickle felt that the new regime could either swing towards the German Nazis or towards Italian fascism, but that a compromise between dictatorship and liberalism was likely. If this obtained, then social science research in Vienna could continue. He hoped to encourage the Viennese Committee to submit one or two modest proposals, and added a postscript:

A word is perhaps in order regarding the Jewish situation in Vienna. If Nazism triumphs there will be a Jewish exodus even greater relatively than from Germany. If one or the other solutions prevails, the Jews will officially enjoy protection, but there will be little or no chance in academic life for younger men not yet in secure positions. These men will try to get out as fast as they can find openings abroad (*ibid*).

A month later, he said that all those he had talked to were of the opinion that the Dollfuss regime was growing stronger and could hold out indefinitely against any domestic Nazi pressures. On the other hand, he wrote, "Pribram was the most pessimistic, but his attitude is probably a function of his age, his poor health and his race. Naturally the Jews are the most uneasy".[67] That year, Pribram left for the United States and Mises left for Geneva.

Abraham Wald was another such uneasy Jew. At this point, he had been scraping along for three years, thanks to Schlesinger and to Morgenstern's Institute. Like many others, he began to consider leaving Austria. Morgenstern, like Pribram before him, became a key person in the attribution of Rockefeller student grants and fellowships in Vienna. Throughout the 1930s, some of his underlings at the Institute were awarded travel grants to study abroad, with several of them going to Harvard, as he had. In mid-1935, the Foundation's Tracy Kittredge interviewed Wald in Vienna, on Morgenstern's suggestion that he would benefit from some time in the United States or England to work on time-series problems.[68] Nothing

no pleasure to shoot cannons in the middle of the city, and what's more to be shot at by them. One will have to wait to see what else will happen since the great task in such events is not to surmount them but rather how to liquidate them, and this process has only just started and no-one can say where it is going." Switching to his book, *Grenzen*, which Burns had read, "I am very sorry to have disappointed you with my book with its negativism, but I have the feeling that what is really necessary today is pitiless criticism, and I can tell you in confidence that I have just started with it now. My second book will also be overwhelmingly critical because only through that can the rubble of tradition be removed", (OM to E. Burns, Mar. 6, 1934, OMDU, Box 4, Folder Corresp. 1930–1932, S – Z).

[67] Rockefeller Foundation Internal Memo, JVS, re The Status of SS [Social Science] in Vienna, JVS Visit to Vienna, Apr. 12, 1934, Box 4, Folder 35.

[68] See Note, undated, concerning Kittredge interview with Wald on July 9, 1935, AIRAC.

came of it, and Wald continued his search for stable employment.[69] With Menger's recommendation, Morgenstern secured more money to employ Wald, and continued to press the Rockefeller Foundation on the question of a fellowship. As of 1936, however, the question of Wald's background arose increasingly often in the Rockefeller correspondence. In February of that year, Kittredge interviewed Wald yet again, in Morgenstern's presence, and wrote supportively to Van Sickle, who was, by this time, in New York.[70] The latter, who had spent several years in Vienna and was close to Morgenstern, replied:

Although Wald's work is too mathematical for me to have any opinion based upon direct examination of his publications, I have no doubt that he is one of the very ablest of the men working upon problems of statistical technique as applied to business cycle analysis. It is a pity that his nationality and race combined make his future so precarious . . .
[However, we] have given so many fellowships to Morgenstern's group that I think we should lay our emphasis elsewhere for a while after we have made an award to Dr. John. Wald should be kept under observation, but I am not inclined to recommend any early award.[71]

A few months later, in July, Kittredge interviewed Wald yet again. Morgenstern was still pushing to have Wald visit Princeton, at either the University's mathematics department or the Institute for Advanced Study. In his notes, Kittredge wrote that, because of his Jewishness, Wald would be very unlikely to secure a university appointment in Vienna, or to "ever become a permanent member of the staff of the Institute". The Foundation had no provision for funding someone in Wald's position, he said, but, at least, Wald had recently invented some new device for improving radio apparatus and so was assured of at least a minimal income.[72]

In September, Van Sickle was still holding off on Wald, who was "obviously a man of exceptional ability but, unfortunately, a man without a country . . . is impossible to foresee what the future holds in store for him. His development should be kept under observation as he may prove in

---

[69] Later in 1935, a possibility arose in Palestine, through Jacob Fraenkel at Jerusalem, but it, too, fell through. Wald wrote to Menger of his intention to go to Palestine anyway, if he could get the entry permit and the money. He had been working on geometry and metrical geometry, he wrote, but it was difficult as he had to work with his brother and did not have the necessary peace. As always, he looked forward to getting back to Vienna and the Colloquium. Letter, undated, Wald to Menger, KMIT.
[70] Letter, Kittredge to Van Sickle, Feb. 23, 1936, AIRAC.
[71] Memo, Van Sickle to Kittredge, 27 Mar. 1936, AIRAC. Ernst John was one of Morgenstern's economic researchers at the Institute.
[72] Note on Interview TBK with Wald, July 11, 1936, AIRAC.

time to be one of those rare individuals whom we are justified in aiding regardless of immediate prospects. It is hard on him, but I am satisfied that we should not recommend him for a fellowship in the near future".[73]

Given the political situation, and with Pribram and Mises now gone from Vienna, the social science project had been shelved. The Foundation also began to worry about the Trade Cycle Institute, which seemed to be drawing too close to the government, but expressed confidence that if anyone was capable of "maintaining standards", it was Morgenstern.[74]

In early 1937, Wald continued to worry. Morgenstern continued to press his case. Van Sickle continued to resist: "In spite of Morgenstern's guarantee of employment in the Institute on his return to Vienna, I doubt whether there is any real future for [Wald] there. Growing anti-semitism has closed the doors to such men throughout most of central Europe. It is a tragic situation but I don't see how we can use our fellowships to combat the trend. If an award were made to Wald to study in this country I am convinced that he would use the sojourn here to seek permanent employment".[75] He suggested that they contact other scholars, just to be sure that Wald was "really gifted". In the meantime, Morgenstern had Wald send a reprint of his *Zeitschrift* general equilibrium paper to Van Sickle. Finding it impenetrable, Van Sickle sent it onto Warren Weaver at the Rockefeller offices in New York, explaining Wald's case: "He is one of those homeless Jews whom it is very difficult to place".[76] Weaver sent it onto Harold Davis at the Cowles Commission, saying the same thing.[77]

---

[73] Letter, Van Sickle to Kittredge, Sept. 16, 1936, AIRAC.

[74] See Memos, JVS to TBK, Sept. 25, 1936 and TBK to JVS, Oct. 13, 1936, AIRAC. A month later, Gerhard Tintner, one of the young associates of the Institute, fled Vienna. Meeting with Van Sickle, on his way to the Cowles Commission in Colorado, he said that the Italo–German agreement augured poorly for Vienna's Jews, whose lot would be serious. Freedom had already disappeared, he said, and the Institute's Monthly Bulletins, which he had been writing, "no longer reflect the views of the staff. Interpretations are consistently colored to suit the government, though the statistical data . . . have not been tampered with. Morgenstern meantime plays a larger role in Austrian public life, has secured reasonably adequate public support and . . . appears to have consoled himself for the loss of freedom by the thought that he can work freely within the government. Tintner thinks that Morgenstern's role there is thoroughly salutary. If Tintner's interpretation of the situation is correct it would seem that our relations with the Institute will have to be carefully reviewed at the time our present grant terminates" (Memo, JVS to TBK, Nov. 16, 1936, AIRAC). Morgenstern quickly intervened, dismissing Tintner's pessmism as excessively gloomy. (See Letter, OM to JVS, Nov. 23, 1936.)

[75] Memo, JVS to TBK, Feb. 9, 1937, AIRAC.

[76] Memo, JVS to Weaver, June 16, 1937, AIRAC.

[77] Letter, Weaver to H. T. Davis, June 18, 1937, AIRAC.

In his letters to Menger, Wald appeared increasingly anxious. He worried about the renewal of his contract at the Institute, sent reprints to Hotelling and Schultz, and waited. Then, thanks to Morgenstern, he was invited by Hans Staehle, director of economic research at the League of Nations, to Geneva for September and October to work on price indices as part of cost-of-living analyses being conducted by the International Labour Office. Here, building on earlier work by Haberler, Leontief, and Staehle, he showed how an improved approximation to the true cost-of-living index could be constructed, under the assumption that the utility function could be approximated by a second-degree polynomial, and given certain other restrictions on the indifference mapping. By the same means, he showed how statistical data could be used to numerically estimate the underlying utility function and hence the demand functions.[78] Staehle was moved to write to Kittredge at the Foundation, singing Wald's praises, explicitly recommending a fellowship, and suggesting that Frisch, Menger, Tinbergen, and Haberler be consulted.[79] Kittredge remained recalcitrant, reiterating Van Sickle's argument about the risk of having Wald enter the American labour force.[80] Then, that same day, he wrote privately to Van Sickle, reporting a turn taken in the conversation with Morgenstern re Wald:

OM of course shares Staehle's views as to AW's quite unusual abilities... [but] Morgenstern still feels however that if only one appointment from Vienna can be envisaged in 1938, he personally would give preference to the candidacy of Kamitz. K. has become Morgenstern's chief of staff and has been sharing increasing responsibility for the theoretical as well as for the practical investigations of the Institute. If an exceptional appointment could be made to Wald in addition to the ordinary fellowship appointment requested for Kamitz, Morgenstern would be delighted.[81]

Why did Morgenstern, at this moment, choose to hold back in promoting Wald? Was it because he had information about other possibilities for him, and knew that Kamitz would not face the same opposition? We shall likely never know. Why did the Rockefeller Foundation continue to create obstacles for Wald, yet look favourably upon Kamitz, who was not of the same intellectual standing? Might it have been because Kamitz was not

[78] The results were published in Wald (1937, 1939, and 1940). See Tintner (1952).
[79] Letter, Staehle to TBK, Nov. 26, 1937, AIRAC.
[80] Letter, TBK to Staehle, Dec. 1, 1937, AIRAC.
[81] Memo, TBK to JVS, Dec. 1, 1937, AIRAC.

Jewish, Rockefeller Foundation's committment to scientific detachment notwithstanding?[82]

The Foundation sought opinions on the relative merits of Wald and Kamitz. Both Haberler and Tintner rated Wald "head and shoulders" above Kamitz, whom they also rated below Ernst John, the previous Rockefeller Fellow.[83] On the other hand, Howard Ellis, at Berkeley, endorsed Kamitz, who, he said, was of "convincing and businesslike appearance and address [ensuring] no lost motion in awkwardness or vagueness concerning objectives".[84] Van Sickle spoke to Morgenstern, who was by now in the United States on a Carnegie fellowship for the first few months of 1938, visiting Vanderbilt, Princeton, and elsewhere. After the conversation, the Foundation officer stuck to his guns:

> I am quite ready to believe that Wald is quite unusually gifted. I still do not see how we can give him a fellowship, in view of the fact that he would be almost certain to use the fellowship to secure a permanent position in this country . . . Morgenstern yesterday . . . said that Wald had been offered a Cowles Commission fellowship. This offers $1,000, but nothing for travel. As Wald is responsible for his parents in Rumania, he has not been able to save anything and cannot, therefore, finance the trip to Colorado. Morgenstern expressed the hope that we might be able to make a grant-in-aid to get him over here. I told him that I did not see how we could possibly do so, much as I should like to help Wald. I suggested that he attempt to interest some well-disposed American Jew in Wald with a view to getting the slight assistance that was needed.[85]

Head-and-shoulders notwithstanding, the fellowship went to Kamitz, and Wald was refused travel money. It was late January 1938. During this time, Nazi activity in Vienna rose visibly, with groups of youths roaming the streets molesting people of Jewish appearance, graffiti appearing on the walls, and petrol bombs being thrown into synagogues. Early in February, Hitler dismissed his senior generals, making himself supreme commander of the German armed forces. On February 12, he summoned Chancellor

---

[82] In late 1937, precisely when it was rebuffing Wald, the Foundation was ruminating over whether to continue funding the Institute in the light of the political climate in Austria which had "greatly abridged the freedom of the individual". However, they were reassured by a prominent League of Nations official – probably Staehle – that the Institute was "a bright spot in a part of Europe not conspicuous for scientific detachment . . . The work is not subjected to political pressure and . . . the Director's views are scientific and detached as far as any human being's can be". See Comment on Foundation appropriation to Institute of 30,000 Austrian Schillings per year for 1938–1940, no date, probably late 1937, AIRAC.

[83] See memo JVS to TBK, Jan. 6, 1938, AIRAC.

[84] Letter Ellis to JVS, Feb. 21, 1938, Series 705E Austria, Folder 1214.

[85] Memo JVS to TBK, Jan. 21, 1938, AIRAC.

Schuschnigg to a now-famous meeting at Berchtesgaden, his mountain retreat, where the Austrian capitulated to Hitler's demand that the Nazi von Seyss-Inquart be admitted to the Austrian cabinet as Minister of the Interior, with control of the police. On Thursday, February 24, Schuschnigg made a radio broadcast, pleading for a unified Austria, but without defiantly challenging Hitler. Then, in early March, he threw down the gauntlet, declaring that a plebiscite would be held in which Austrians could vote for or against a free, German, independent, social, Christian, and united Austria. Two days later, on March 11, to the dismay of Austria's Jews, he announced in another broadcast speech that the plebiscite had been cancelled and that Hitler had demanded that the Federal President Miklas appoint a cabinet of his – Hitler's – choosing. Otherwise, German troops would be sent into Austria. With this, Schuschnigg stepped down as Chancellor, making the way for Hitler's Seyss-Inquart. That night, as George Clare (1982) recalls, crowds of Nazis on the backs of lorries were again roaming through the streets of Vienna under the swastika.

On March 15, Ernst Wagemann, the director of the Berlin Institute, arrived in Vienna with instructions to liquidate the Trade Cycle Institute. He spent a week there, dismissing most of the staff, including Wald and the absent Morgenstern, and retaining only the politically acceptable Kamitz and John. The former was made acting director, and instructed not to communicate with Morgenstern or any foreign institutions, including the Foundation. However, early in May, in an out-of-the-way café on the outskirts of Vienna, Kamitz met secretly with Kittredge.[86] He told Kittredge that he had suggested to Wagemann that the Foundation might be willing to continue offering financial support if the independence of the Vienna Institute could be assured, with it reporting independently on Austrian conditions and doing basic theoretical research. Wagemann had insisted, however, that economic reports and analysis would have to conform to instructions from Berlin and that he was personally opposed to the theoretical investigations, so the monograph series would be scrapped.[87]

---

[86] See Memo, TBK to Sydnor Walker, May 19, 1938, AIRAC.

[87] How the Institute could have even attempted to maintain its previous program, given its virtual dismantling by Wagemann, is not clear. Like many Austrians, Kamitz seems to have played his cards pragmatically. At the same time as he went to the trouble, and ran the risk, of meeting Kittredge, telling him about the plight of the Institute, he was able to inform him that he had "no personal difficulties" having been asked to take over lectures at the *Hochschule fur Welthandel* to replace professors who had recently been discharged. It even looked likely that he would be appointed to a dozentship so that his "prospects for an academic career . . . seemed good" (Memo, TBK to Sydnor Walker, May 19, 1938, AIRAC).

On March 19, as president of the Vienna Economics Society, Hans Mayer wrote to all members: "In consideration of the changed situation in the German Austria I am informing you that under the respective laws now applicable also to this state, all non-Aryan members are leaving the Economics Society".[88] In actuality, by then many of those members – Christian, Jewish, and the "mixed group" alike – had already left or were, in one manner or another, leaving Vienna. Mises was in Geneva and Hayek had long been in London. Menger was now at Notre Dame, Tintner in Iowa City, Haberler in Harvard, Machlup in Buffalo. Morgenstern was in the United States, searching for a new university. In late 1937, he had broken with the Austrian regime over its unwillingness to face up to agrarian special interests in the matter of necessary downward price adjustments. When the Nazis took over the Institute in March 1938, Morgenstern, in the United States, was deemed *persona non grata.*

In the streets of Vienna, Jews were forced into demeaning acts and coerced into acts of sacrilege. Shops were defaced and looted, property destroyed, and apartments plundered. By April 3, Morgenstern, in Wisconsin, wrote to Van Sickle that Schlesinger and Kunwald, another economist, had committed suicide.[89] On April 11, the Institute's Monthly Bulletin appeared with a foreword by Wagemann:

The vast historical development of these days, which has inspired and widened the life of the German people in all its aspects, emphasises also new ways for this publication. Out of the union of Austria with the Reich there has developed on the economic side two important issues. It will now be necessary, in general, to provide for the fusion of the economic and constitutional life of these two different State economies and, in particular, to overcome the economic distress of Austria. This has to be accomplished by the powerful and quickly-effective means and methods which National Socialism has developed and which were completely lacking in the former Austrian government with its remarkable lack of understanding.... The

---

[88] Quoted in Mises (1978), p. 99. In these *Recollections,* written in 1940 when he had just arrived in the United States and was bitterly upset at the turn of events, Mises condemns Mayer as a Nazi collaborator and dismisses him as an economist. I suspect that the lack of historical interest in Mayer's economics from the outbreak of the War onward was shaped by his *Anschluss* actions, and by Mises' 1940 condemnation. Not until 1994 was some of Mayer's work translated into English, in a volume of Austrian readings, edited by Israel Kirzner (1994).

[89] On the treatment of Jews following the Anschluss, see Oxaal, Pollack, and Botz (eds.) (1987), Wistrich (ed.) (1992), and Pauley (1992). Botz (1987) reports that despair amongst the Jewish upper middle classes dramatically increased the number of suicides in the months following the *Anschluss,* with 220 reported in March alone; "The Jews of Vienna from the Anschluss to the Holocaust" in Oxaal, Pollack, and Botz (eds.) (1987), pp. 185–204.

close collaboration of both [the Berlin and Vienna] research organizations will make possible our fruitful collaboration in the great tasks which lie before us.[90]

Excluded from this project and fearing the power and effect of National Socialism, Abraham Wald was still in Vienna. He wrote to Menger about the bureaucratic difficulties being created by the Rumanian government, who would only issue a three-month passport, whereas the Cowles position was for one year. He hoped Cowles would not make any difficulties for him: "It would be a great misfortune for me were I to lose this position. I would then be facing the abyss and would not even have the financial means to travel anywhere.[91] He could not even leave Austria to go home to Cluj because the Rumanian government had forbidden reentry without the special permission of the Ministry of the Interior. Then, at the eleventh hour, Wald got out, making it to Colorado via Cuba. With the exception of one brother, who also made it to the United States, all of Wald's immediate family disappeared in a concentration camp.

In Vienna, for all the expression of venom and hate in early 1938, the city was struck silent. Mises, Menger, Morgenstern, Machlup, Tintner, Haberler, Wald – all were gone.[92] Schlesinger and Kunwald were dead. The Institute was but a shell. Of those present at the young Menger's talk on the Petersburg Paradox a decade ago, there remained only Mayer, now presiding over a spectral Economics Society.

Conclusion

During the decade leading to 1938, Morgenstern achieved considerable power in the Viennese economic community. At the helm of the Institute and with the confidence of the Rockefeller Foundation, he wielded influence over the type of research done. He also provided support for researchers in a difficult environment. Throughout the decade, he played the role of largely liberal policy advisor, but found himself in an increasingly difficult position, growing skeptical of dogmatic Austroliberalism, ever more attentive to clarity in argument, and relentless in his insistence on the separation of economics and politics.

His theoretical writings, themselves a curious *mélange* of vague suggestion and harsh critique, reveal several concerns, from the emphasis on

---

[90] Austrian Institute for Trade Cycle Research, *Austrian Monthly Monthly Bulletin*, Apr. 11, 1938, p. 12 (translated from German).
[91] Letter, Wald to Menger, April 18, 1938, KMIT.
[92] On Mises' flight from Geneva to the United States, see his (1978) *Notes and Recollections*.

expectations, beliefs, and other psychological factors as the most important manifestations of time in economics; to the need to examine the logic and consistency of the field in a manner similar to that in the branches of mathematics; to the need to rid the discipline of all element of political apology.

His passage from Vienna was easier than that of Wald. Not only did he arrive in the United States on a fellowship from the Carnegie Endowment for International Peace, but the Rockefeller Foundation helped him settle at Princeton by paying half of his salary for a while. Although Princeton was then still a sleepy gentleman's college and worlds away from Vienna, Morgenstern knew some members of the faculty, including Fetter, and it was close to the cultural benefits of New York city. In 1939, Morgenstern was still thinking about beliefs, interaction, and the limitations of the economic viewpoint when he wrote once more to Frank Knight: "[T]here are only a few people, if any, interested in methodological questions. Those with whom to discuss such problems are principally the mathematicians, of which we have some excellent ones in town. I have now been stimulated by these talks and proceeded to jot down notes on a further paper of what I called maxims of behavior. In this paper I shall endeavor to investigate a very curious relationship between the quantitative limits which maxims may have. I hope to be able to show you something of this in the not too distant future".[93] Of course, the unnamed stimulus here was John von Neumann who, like all the Austro-Hungarians, was affected by the upheaval of the times.

[93] Letter, OM to Knight, Nov. 8, 1939, OMDU, Box 6, Corresp.: 1928–1939, Knight. In this connection, see Morgenstern (1941b).

Plate 6. Morgenstern Family Portrait: left, sister Hannichen; seated, parents; standing at back, Morgenstern. *Credit:* Courtesy of Special Collections, Perkins Library, Duke University.

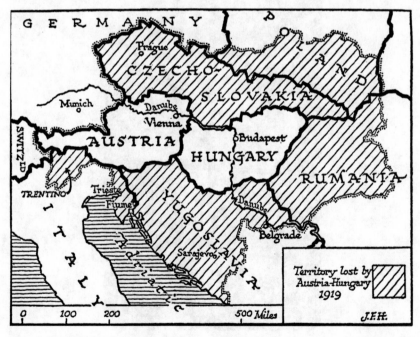

Plate 7. Territory lost by Austria-Hungary, 1919. *Credit:* From J. F. Horrabin (1934), *An Atlas of Current Affairs*, New York: Alfred Knopf.

Plate 8. Othmar Spann. *Credit:* Courtesy of the Österreichische Nationalbibliothek.

Plate 9. Hans Mayer. *Credit:* From Grass, N. (ed.) 1952, *Österreichische Rechts- und Staatswissenschaften in Selbstdarstellungen.* Innsbruck, Austria.

Plate 10. Ludwig von Mises, 1935. *Credit:* Courtesy of the Österreichische Nationalbibliothek.

Plate 11. Oskar Morgenstern. *Credit:* Courtesy of the Österreichische Nationalbibliothek.

Plate 12. Hans Hahn. *Credit:* Courtesy of the Österreichische Nationalbibliothek.

Plate 13. Karl Menger. *Credit:* Courtesy of Rosemary Menger-Gilmore.

Plate 14. Burning of the Justizpalast, July 15, 1927. *Credit:* Courtesy of the Österreichische Nationalbibliothek.

Plate 15. Demonstration in Lastenstrasse, July 15, 1927. *Credit:* Courtesy of the Österreichische Nationalbibliothek.

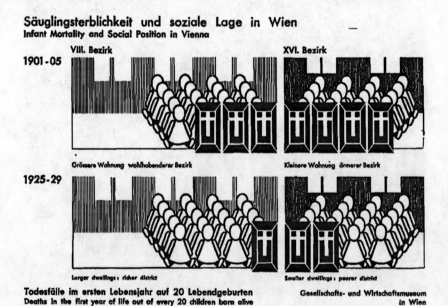

Plate 16. Infant Mortality and Social Position in Vienna. In this graphic, designed by Gerd Arntz, each small coffin signifies a death within the first year of life for every twenty children born alive. *Credit:* From *Bildstatistik nach Wiener Methode in der Schule. Mit 24 zum teil farbigen Tafeln* (Deutscher Verlag für Jugend und Volk: Wien/Leipzig 1933).

Plate 17. Karl Marx Hof after the Siege, February 18, 1934. *Credit:* Courtesy of the
Österreichische Nationalbibliothek.

Plate 18. Zerschossenes Arbeiterheim, February 1934. *Credit:* Courtesy of the Österreichische Nationalbibliothek.

PART THREE

FROM WAR TO COLD WAR

NINE

Mathematics and the Social Order

Von Neumann's Return to Game Theory

Dear, my dislike for Europe has nothing to do with Mariette. I feel the opposite of a nostalgia for Europe, because every corner *I knew* reminds me of the world, of the society, of the excitingly nebulous expectations of my childhood . . . of a world which is gone, and the ruins of which are no solace . . . My second reason for disliking Europe is the memory of my total disillusionment in human decency between 1933 and September 1938, the advent of Nazism and the reaction of humanity to it – in that period I suffered my life's greatest emotional shock . . .

John von Neumann to Klari von Neumann, Thursday (no date), 1949[1]

From Berlin to Princeton

Notwithstanding the fact that his family had nominally converted to Christianity upon the death of the father in 1923, in the climate of the late 1920s, von Neumann knew that his chances of obtaining a chair in mathematics in Germany or Hungary were negligible. Socially, he was still perceived as Jewish, and that in a Germany where many dozents were competing for promotion. Stan Ulam also remembered him speaking of the worsening political situation, which made him doubt that intellectual life could be pursued comfortably. Thus, von Neumann readily accepted when, at the beginning of the 1930s, Oswald Veblen, Princeton mathematician and occasional visitor to Göttingen, arranged to have him spend six months per year at Princeton. For the next two years, von Neumann commuted from Berlin to Princeton, by cruise-liner, first-class as always, to a professorship in the Mathematics Department, shared with his fellow Hungarian, mathematical physicist Eugene Wigner.

---

[1] Klari von Neumann-Eckhart Papers, in the possession of Professor Marina von Neumann Whitman of Ann Arbor, Michigan (hereafter KEMNW).

Princeton's strength in mathematics in the 1930s resulted from its having two centres of gravity: the university's Department of Mathematics and the nearby, but independent, Institute for Advanced Study. The latter had been officially incorporated in 1930 through a large endowment by supermarket millionaires, Louis Bamberger and his sister Mrs. Caroline Bamberger Fuld. Also involved in the inception was the Institute's first director, Abraham Flexner.[2] It was decided to locate the Institute at Princeton University because of the latter's excellent library and the quality of its mathematics department, in the building of which, Fine Hall, the Institute was first set up. The first full faculty member, secured by Flexner in 1932, was Albert Einstein, then keen to leave Germany and being courted by universities the world over. By the time the Institute opened its doors in the fall of 1932, Oswald Veblen, von Neumann, and James Alexander were on the faculty, having transferred from Princeton's mathematics department. Einstein physically arrived in 1933. The Institute, which at the outset had only a School of Mathematics, paid lavish salaries, averaging twice those of Princeton professors. In 1933, it moved to a new building, constructed on a site bought for the purpose, just south of Princeton campus, where it has remained to this day.

Princeton mathematics was known for its strength in topology and algebra, two areas important in the growth of American mathematics. Veblen was a leader in combinatorial topology, a field in which Alexander and Solomon Lefschetz also worked. Other mathematicians included Bob Robertson, who would become a friend of von Neumann, Luther Eisenhart, the logician Alonzo Church, Marston Morse, Carl Siegel, Albert Tucker, and statistician Sam Wilks. During the 1930s, it was one of the finest departments in the United States. In addition to the quality of faculty, the social occupation of space seems to have been important: all those in reminiscence about Princeton mathematics point to the importance of the Fine Hall Common Room, where afternoon tea and the playing of games made for a certain *esprit de corps*, quite unlike the mathematics departments at

---

[2] Flexner had been author of the famous "Flexner Report", a slamming indictment of the medical college system and, in particular, of its then-large number of "quack" colleges, where well-paying students could quickly become medical doctors with less than a minimal amount of medical training. Following the report, a number of these colleges were forced to close down, and Flexner became quite famous. He was writing another book, comparing the American, English, and German university systems, when he was approached by Bamberger and Fuld. On the Institute, see Regis (1989) and, especially, Batterson (2006).

Columbia or Harvard, where nothing comparable existed.[3] Von Neumann was a stalwart at Princeton teas.

From the beginning, von Neumann took greatly to life in the United States. Temperamentally, the country suited him, and, although he would always dress formally – including when on horseback and on the beach – he seems to have appreciated the freshness of life in the States. Arriving with his first wife, Mariette Kovèsi, herself from a prominent family in Budapest, where her father was head of the Jewish Hospital, they soon moved into a large house in Princeton with domestic staff. When not working hard, von Neumann "played hard". Their home became known for its famous parties, complete with caviar imported from Russia. Von Neumann drove a powerful car, dangerously, some say. Indeed, he crashed cars several times, on both continents, so that, at Princeton, a particular curve in the road became known to acquaintances as "von Neumann Corner". Although he embraced American life generously, he returned to Hungary virtually every summer, fleeing the heat and humidity of summertime Princeton, just as his teachers such as Fejér continued to flee those of Budapest.

A Time of Instability

Von Neumann had no reason to regret his decision to leave Germany. In mid-March 1933, a few weeks after the Reichstag fire and days after the sweeping Nazi election victory under Hitler, he wrote from Budapest to Flexner back at the Institute. His summer plans were not yet fixed, he wrote in his still imperfect English. He had hoped to spend the summer lecturing in Berlin, but the "newer german developments" (sic) had thrown this into question. He didn't think conditions would improve.[4] A week or so later, in April, the German "Restoration of Civil Service Act" was passed, effectively allowing the Nazi government to dismiss academics for reasons relating to politics or race. It marked the beginning of the systematic release of Jewish faculty members from the German universities. Flexner wrote

---

[3] With tea at 4.30 P.M. every afternoon – an institution begun by Anglophile Veblen – the Common Room allowed professors and graduate students to mingle. Games were played: cards, chess, *Kriegspiel* (a form of blind chess), and invented challenges involving stacking the chess pieces on top of each other. Some students, such as Merrill Flood, relied on all-night poker sessions to supplement their income.

[4] John von Neumann to Abraham Flexner, Mar. 18, 1933, Faculty Files, John von Neumann, Folder "1933–1935", Von Neumann Papers, Institute for Advanced Study (hereafter VNIAS).

from Princeton, condemning the German government's madness, and its destruction of the Göttingen faculty in particular.[5]

From Budapest, von Neumann wrote to Veblen about American economic affairs in detail, and about European politics with irony: "There is not much happening here, excepted that people begin to be extremely proud in Hungary, about the ability of this country, to run into revolutions and counter-revolutions in a much smoother and more civilized way, than Germany. The news from Germany are bad: heaven knows what the summer term 1933 will look like. The next programme-number of Hitler will probably be the annihilation of the conservative-monarchistic – ("Deutsch National" = Hügenberg) – party".

You have probably heard that Courant, Born, Bernstein have lost their chairs, and J. Franck gave it up voluntarily. From a letter from Courant I learned 6 weeks ago (which is a very long time-interval now in Germany), that Weyl had a nervous break-down in January, went to Berlin to a sanatorium, but that he will lecture in Summer.

I did not hear anything about changes or expulsions in Berlin, but it seems that the "purification" of universities has only reached till now – Frankfurt, Göttingen, Marbürg, Jena, Halle, Kiel, Königsburg – and the other 20 will certainly follow.

I am glad to learn from your letter that these things received the full attention and appretiation [sic] in America which they deserve. It is really a shame that something like that could happen in the 20th century.[6]

He chose not to go back to Berlin. After a leisurely summer, with weekends spent on Lake Balaton and in the Hungarian countryside, he returned permanently to the Institute at Princeton and never set foot in Germany again.[7]

Amongst von Neumann's Hungarian correspondents was the previously mentioned Rudolf Ortvay, a physicist and an important mentor eighteen

---

[5] Flexner to von Neumann, May 6, 1933, Faculty Files, John von Neumann, Folder "1933–1935", VNIAS. On Göttingen after 1933, see Segal (2003).

[6] VN (Budapest) to Veblen, Apr. 30, 1933, Veblen Papers, Library of Congress (hereafter VLC), Box 15, Folder 1.

[7] Aspray (1990) indicates that von Neumann visited Germany, Austria, Hungary, Italy, and France in 1930 and 1931; Germany and Hungary in 1932; Hungary and Italy in 1933; Hungary alone in 1934; England and Hungary in 1935; France in 1936; and Hungary alone in 1937 and 1938 (p. 256, n. 35). We might also mention that chessplayer Emanuel Lasker, too, left Germany in 1933, his property confiscated by the Nazis. He drifted for a number of years, staying in the Netherlands and then England, and then accepting a research post at Moscow's Institute of Mathematics. There, Gerald Abrahams says, the great man was somewhat inert, being content to throw out mathematical ideas but not pursue them seriously. Abrahams fails to emphasize that Lasker was by then more than 65 years of age.

years his senior. Born in Miskolc, in the northeast of the country, Ortvay also studied at Göttingen and, following a period at Koloszvár and Szeged, moved to Budapest in 1928, where he ran the Institute for Theoretical Physics. Like Lipót Fejér, he was a family friend of the von Neumanns and a frequent guest at their dinner table in the 1920s. He had followed the young von Neumann's career from the beginning and they maintained a revealing correspondence all through the 1930s.[8]

In October 1933, von Neumann wrote to Ortvay describing Einstein's arrival the previous day at the Institute, the German physicist having been slipped off the boat and spirited away so as to avoid the reception committee awaiting him on the dock. "What is new in Budapest?" von Neumann continued. "How is the mood in general, and especially with regard to the German situation?"[9]

By then, Ortvay had become quite pessimistic about European politics and his resonant letters from here on were at once a lament for cultural decline, an inquiry into the place of the scientist in society, and a meditation on the vagaries of the human spirit. Von Neumann's replies, in turn, not only shed light on his character, but gradually reveal the extent to which his social scientific reflections and political preoccupations were intermeshed.

From 1934 onward, while continuing his mathematical work on the spectral theory of Hilbert space, ergodic theory, rings of operators, and Haar measure, von Neumann was increasingly preoccupied by politics, entering into the finest detail in his letters. Faced with a relatively emotional Ortvay, he tended to maintain a certain detachment: "What you write about the uncertainty of the future of European civilization is regrettably plausible. There is one consolation in it, but even this isn't an excessively certain conclusion: the war demoralised principally the countries that lost, and in history after a lost war experimentation with a state structure of tyranny or dictatorship, and the rise of a romantic, irrational nationalism, is neither a new nor rare phenomenon. Naturally references to historical analysis are

---

[8] The von Neumann–Ortvay letters, written in Hungarian, are located in the Library of the Hungarian Academy of Sciences in Budapest and in the von Neumann papers at the Library of Congress, with copies in the Stan Ulam papers at the American Philosophical Society in Philadelphia. There are sixty of them, running from May 9, 1928 to Feb. 16, 1941. Most of them have been reproduced in Hungarian in Nagy (1987) and a few of them in English in Rédéi (ed.) (2005). With a few exceptions, which are indicated, all the following quotations in this chapter are based on translations of the letters found in Nagy, *op cit*, conducted for the author by Mr. Andrew Szirti.

[9] Von Neumann to Ortvay, October, no exact date, 1934.

especially arid and hopeless, since if these could be trusted, then new wars could not be avoided".[10]

In November 1934, von Neumann wrote of the many new doctoral students arriving at the Institute, amongst whom there were two German Rockefeller Fellows: "Whether they are genuine Nazis, I don't know, they are fairly discreet . . . How do you now judge Central Europe?" he continued, where "the situation . . . seems to be so tense that in the end there will be trouble! There are so many uncertain and easily misunderstood circumstances in the European 'balance' that there may exist a government that jumps into an adventure".[11] England and Italy were equally indecisive, or, rather, hypocritical, he said, and the weight of Russia was just as incalculable as it had been in 1914. Later that month, he grew sombre: "The European political situation appears to be quite dark even from here . . . ; to wit, here the people have already accepted that the lesson was for nought, and that in Europe there shall be a war in the next decade".[12]

Von Neumann's rupture with the German mathematical community was made final early the following year, when he responded to the infamous affair involving Ludwig Bieberbach, the *Deutsche Mathematiker Verein*, and Danish mathematician Harald Bohr. In an attempt to demonstrate the relevance of mathematics to the new regime and to justify the ouster of many Jewish mathematicians, prominent German mathematician and Nazi sympathiser Bieberbach had taken a novel position in the longstanding philosophical debate between the Intuitionists and the Formalists.[13]

Occupying a chair in Berlin as of 1921, Bieberbach was a successful mathematician of high standing in Germany. He wrote on numerical analysis and produced successful textbooks on differential and integral calculus, the theory of complex functions, and differential equations.

In his early philosophical writings, Bieberbach was sympathetic to Formalism, insisting upon a sharp distinction between pure and applied mathematics. Formalism was the best way to mathematical knowledge: "The

---

[10] He then went on to discuss Roosevelt's foreign exchange policy and negotiations with Congress over his spending plans. Von Neumann to Ortvay, Jan. 26, 1934.

[11] Von Neumann to Ortvay, Nov. 2, 1934.

[12] Von Neumann to Ortvay, Nov. 28, 1934.

[13] Bieberbach (1886–1982) studied under Paul Koebe and Felix Klein at Göttingen, completing a doctorate in 1910. Having taught at Zurich and Königsberg, he became full professor at Basel in 1913, and at Frankfurt two years later. In 1921, he was appointed successor to C. Carathéodory at Berlin, which was then Germany's second mathematics centre, after Göttingen. My account of Bieberbach relies primarily on Herbert Mehrtens' (1987) excellent article on the subject.

truth of mathematics rests solely in its logical correctness and consistency", he declared in 1914 (Bieberbach, quoted in Mehrtens 1987, p. 201).

In his textbooks of 1917 and 1918, however, he emphasised the importance to mathematics of outside, empirical stimuli, saying that they were an essential part of mathematical creativity. In 1919, he was becoming increasingly anti-Formalist in his outlook: "The tendency towards formalism must not cause us to forget the flesh and blood over the logical skeleton ... The formalism of university education in mathematics has forgotten about applications and the cultural meaning of mathematics" (quoted in Mehrtens, *op cit*, p. 202). Mehrtens explains how the shift towards the study of logic and foundations soon provoked a crisis as some German mathematicians embraced the emphasis on *Anschauung* (which variously connoted visual perception, geometrical intuition or visualization, or Kantian intuition) promoted by Brouwer, in contrast to the formal style associated with Hilbert and Göttingen. Mehrtens suggests that this shift of values owed something to the German defeat in the Great War. When that defeat was compounded by the conflict between Brouwer and the International Union of Mathematicians over the latter's boycott of German mathematics, epistemological positions took on a political connotation.

By 1926, when von Neumann went to Göttingen, Bieberbach had openly sided with Brouwer and was attacking Hilbert's "ideal of science" (*ibid,* p. 205). Bieberbach then drew on a 1927 book by French mathematician Pierre Boutroux, *L'idéal scientifique des mathématiciens,* which posited the existence of two conflicting styles of mathematical thought: intuitive vs. discursive; synthetic vs. analytic; order of invention vs. order of proof. Adapting Boutroux's taxonomy, Bieberbach sought to promote *anschauliche Gegenständlichkeit,* or intuitive, visual, object-related thinking, in opposition to modern formalist style associated with Hilbert and then epitomized by van der Waerden's much-discussed (1930) *Moderne Algebra.* Mehrtens writes:

What was called 'modern' in those years was the kind of mathematics which, after the turn of the century took the lead in the field of mathematical research and proved most productive ... If the advent of this 'modern' program may be interpreted as a 'scientific revolution' in mathematics, then Bieberbach was on the side of the losers in this revolution. His description of formalism may well reflect his fears of being left behind by the establishment of new standards in assessing research – standards to which he either could not nor did not wish to conform (p. 207).

Bieberbach's denunciation of the formalist ideal grew steadily from here on. In a 1926 lecture, he accused the proponents of Formalism of wishing to impose a prototype for all mathematics, one that overestimated the importance of purely logical form and removed the "sense and meaning" from mathematics. The intuitionism of Brouwer and Weyl, on the other hand, provided an escape from this "nightmare", and allowed the restoration of meaning.

The 'loss of meaning' deplored by Bieberbach (and others) during the early twenties was very much a result of a gradual but basic change in the forms of scientific communication, namely in the language of mathematics. With modern axiomatics and logical symbolism, the formal justification of proofs and theorems had become the primary purpose of the language of mathematics. Imprecise elements – like imagery or motivational arguments that communicate some 'meaning' other than scientific validity – were increasingly excluded. The official 'modern' communication of a piece of mathematics presents a statement or a series of statements and, by the mode and means of expression, its rigorous justification, nothing more (*ibid*, p. 208).

Bieberbach condemned this reduction of mathematics to the examination of "pale skeletons in the sand of the desert, of which nobody knows what they mean or what purpose they serve" (quoted in Mehrtens, *ibid*, p. 209). Mehrtens notes that Bieberbach saw the value of Brouwer's Intuitionism not so much in the technical stance it adopted on foundational issues, such as the acceptability of the law of the excluded middle in the case of infinite sets, as in its promotion of a politically congenial mathematical *style*. Because it rooted mathematics in the simple human practice of doing one activity after another, Bieberbach felt Intuitionism showed the proper relationship of mathematics to human life. Hilbert's Formalism, by contrast, was isolated from "matters of education, application, *Volk*, or nation" (p. 210). In 1926, when Bieberbach was making these claims, the debate on foundations had begun to diminish in importance, but Göttingen remained what Mehrtens calls the "stronghold of mathematical modernism" and "the center of social power within the discipline" (p. 212). Bieberbach's critique from Berlin was thus an attack on the power structure within the discipline.

By 1930, Brouwer had withdrawn from the foundations debate and in 1934 he refused an offer of a chair from Göttingen. At a 1932 celebration of Hilbert's seventieth birthday, Bieberbach was quite isolated. It was on the foot of this experience that he embraced the Nazi cause. In 1933, he quoted his teacher Felix Klein's celebrated remark of 1893, made at Northwestern University: "It would seem as if a strong naïve space-intuition were an attribute pre-eminently of the Teutonic race, while the critical purely logical sense is more developed in the Hebrew races" (quoted in Mehrtens, p. 219). Whereas this remark was neither intended to be, nor was perceived as, anti-semitic in 1893, in Bieberbach's hands forty years later, it became so.

In various public statements, beginning with his Fördverein lecture of 1934, Bieberbach promoted intuitionist mathematics as being more truly German, insofar as it emphasized the concrete and the empirically relevant. Formalist mathematics, on the other hand, was too given to purely abstract manipulations and tended to be favoured by Jewish mathematicians. Bieberbach went very far in this matter. He justified the student boycott of Göttingen Jewish professor, Edmund Landau, citing it as proof that representatives of different races could not successfully mix as students and teacher. In his Fördverein lecture, Bieberbach sought to connect different mathematical styles to the race psychology of Erich Jaensch who, in his 1931 *Über die Grundlagen der menschlichen Erkenntnis,* had postulated the existence of two psychological types. The S-type demonstrated "unstable psychic functions, internally generated synaesthetic perceptions, and a tendency towards disintegration", whereas the J-type, in contrast, showed "stable psychic functions, in which perceptual imagery and conceptual thinking were strongly integrated" (quoted in Mehrtens, p. 228). Jaensch attributed the crisis of modern culture, with its separation of intellect and soul, to the dominance of S-type thinkers. The Nazi movement provided an opportunity for a cultural renaissance that would change this.

With similar political intentions, Bieberbach applied Jaensch's psychology to the domain of mathematical creativity. J-type mathematicians, he said, were in the tradition of his teacher, Klein, whereas Hilbert's Formalism belonged to the S-type. The fact that Hilbert was East Prussian and not Jewish led to some juggling by Bieberbach, with Hilbert becoming a form of J-type open to influences of the S-type. On the foot of this race-psychological taxonomy, Bieberbach proposed to establish *Deutsche Mathematik*, a new journal that would promote a truly German, *Anschauuliche* mathematics.

In May 1934, when Harald Bohr published an article condemning Bieberbach, the latter responded with an open letter excoriating Bohr and his "hatred of the new Germany".[14] At the meeting of the Deutsche Mathematiker Vereinigung (DMV) in Bad Pyrmont in September of that year, Bieberbach's followers succeeded in having a resolution passed expressing regret for his behaviour but also criticizing Bohr's.

Bieberbach's actions met with consternation amongst mathematicians abroad, and one of the first to respond was Oswald Veblen at Princeton.

---

[14] Quoted in Mehrtens (1987), p. 221. On the Bieberbach affair, see also Segal, *op cit,* pp. 263–88 and chapter 7, *passim.*

In a May 1934 letter to Bieberbach, he wrote of the "various degrees of sorrow, derision and contempt" with which the latter's ideas were being received abroad. Von Neumann would thus have been very much aware of Bieberbach, from his Göttingen days in 1926 at the inner sanctum of formalist modernism, through his time at Bieberbach's Berlin itself, to his present position at Princeton. Indeed, it is probably reasonable to believe that Bieberbach represented, in a particularly extreme form, the kind of mindset that contributed to von Neumann's decision to leave Germany for Veblen's Princeton in the first place. In January 1935, triggered by the Bad Pyrmont affair, von Neumann followed Veblen's suit and wrote to William Blaschke at the DMV: "Although not a German, . . . I had received my scientific education in the German speaking part of the World and have spent part of my scientific career in German Universities – a part, which remains for me unforgettable for ever . . . Nevertheless I cannot reconcile it with my conscience to remain a member of the German Mathematical Society any longer . . . It is my hope that my paths and those of the D.M.V., whose true interests I still believe to be serving, are not separating for ever".[15]

A sign that Nazism was now reaching into the discourse of mathematics itself, the Bieberbach episode contributed to the growing gulf between von Neumann and the Germany in which he had been trained. It was a contributing element to what would become by the end of the decade, by his own admission, a defining trauma of separation and loss, his "greatest emotional shock". In light of Bieberbach's extremism and the subsequent events of the 1930s, it is possible to see von Neumann's commitment to Göttingen-style "mathematical modernism" as no longer merely an intellectual or scientific preference, but one that was becoming laden with political import. His creation of game theory must be viewed in this particular context, for it was a declaration of commitment to modern formalist mathematics, this time in the social domain. The fact that it was a new mathematics of politics and social organisation gave it a particularly charged and reflexive quality.

At this time, Polish mathematician Stan Ulam entered von Neumann's life. They met in Warsaw in 1934, when the latter was returning from a Moscow conference with Birkhoff and Marshall Stone. Years later, Ulam would remember meeting him on the platform: "The first thing that struck me about him were his eyes – brown, large, vivacious, and full of expression. His head was impressively large. He had a sort of waddling walk . . . At once

---

[15] Von Neumann to W. J. E. Blaschke, Jan. 28, 1935, in Rédéi (ed.) (2005). Original in German; translation by M. Rédei.

I found him congenial. His habit of intermingling funny remarks, jokes, and paradoxical anecdotes or observations of people into his conversation, made him far from remote or forbidding".[16] The two hit it off immediately. Both were what Ulam describes as third- or fourth-generation wealthy Jews, comfortable with each other, and they were also linked through mutual acquaintances, for Ulam's widowed aunt had married Árpád Plesch, one of the richest men in Budapest, who was, of course, well known to the von Neumanns. When Ulam moved to the States in 1936, he was von Neumann's assistant at the Institute, before being appointed Junior Fellow at Harvard, through the intervention of George Birkhoff. The two emigrés relied on each other as the decade wore on.

Throughout the mid-1930s, von Neumann devoted his efforts to pure mathematics, including continuous geometry. He worked incessantly, even when on holiday in the Canadian woods with Abraham Flexner and his family. However, political concerns became disruptive. By January 1936, he was writing about the effect of Mussolini's Italy on the European situation and, within a few months, was predicting a cataclysm: "Here Europe is judged darkly, as with every affair that is distant and complicated. But even I cannot bring myself to tranquility. The danger of war appears to be truly great, even if the catastrophe does not take place this year. I hope that from near by, the picture is not this desolate. How do you judge it?"[17]

Crisis

Implausible as it may initially sound, there is some truth to the claim that von Neumann's expansion of game theory into the social realm was triggered by his divorce from Mariette Kovèsi in late 1937. It is true that a year previously, in late 1936 or early 1937, he gave a popular talk at Princeton on what, according to the *Science News Letter*, was for him "a mere recreation", his analysis of games and gambling. All of it appears to have referred, however, to the work he had done at Göttingen a decade previously. There was no mention of anything other than two-person parlour games. He spoke about "stone-paper-scissors", showing that by "making each play the same number of times, but at random, . . . your opponent will lose in the long run".[18] Also briefly reported are his comments on the probabilities of

---

[16] Ulam (1976) p. 67.
[17] Von Neumann to Ortvay, Apr. 1, 1936.
[18] *Science Letter News*, Apr. 3, 1937, "Princeton Scientist Analyzes Gambling: You Can't Win", p. 216.

making particular plays in both dice and a simplified poker. Amongst those attending the Princeton talk was Merrill Flood, then a graduate student and later a mathematician at the RAND Corporation: "I don't know what the title of the lecture was, but I went because of von Neumann, whom I'd come to know well. He lectured on the minimax theorem, although he didn't call it that. In fact, he didn't tell us that there was such a theorem. He gave us examples of how mixed strategies could be used in games".[19]

At the end of 1937, however, von Neumann's divorce triggered a period of veritable crisis, one effect of which was to focus his attention as mathematician on the social structure. It was the divorce that prompted his remarriage, and it was the protracted preparations for the latter that saw him leave Princeton and spend the summer of 1938 in Europe. There, he was forced to confront the political situation directly and witness its contamination of Hungarian and von Neumann family life. This confrontation appears to have had a stimulative effect in his turning to a new social mathematics, which brought together considerations of rationality, social norms, and equilibrium.

Towards Christmas 1937, Mariette Kovèsi left von Neumann for Desmond Horner-Kuper, Princeton graduate student in physics and a regular guest at the von Neumann parties on Westcott Road. In some of his correspondence at least, von Neumann appears to have borne it all with equanimity: "Many thanks for your letter", he wrote to Ulam, " . . . and particularly for what it contained about my 'domestic' complications. I am really sorry that things went this way – but at least I am not particularly responsible for it. I hope that your optimism is well founded – but since happiness is an eminently empyrical (sic) proposition, the only thing I can do is to wait and see . . . ".[20] However, Ulam would later say that the rupture greatly shook his

---

[19] Interview with Merrill Flood by Albert Tucker, San Francisco, May 14, 1984, Transcript No. 11 (PMC11) of oral history project *The Princeton Mathematics Community in the 1930's*, deposited in the Seeley Mudd Library, Princeton University. Flood was impressed by the mixed strategy idea: "I remember going to Kleene and Einstein and half a dozen other people to find out if they had ever heard of that . . . Nobody came up with the idea of mixed strategy among all these bright people. That convinced me that that's a subtle thing". This was Flood's introduction to game theory and, a year later, he returned to it when asked to give a popular lecture to Princeton undergraduates in a bid to recruit mathematics majors. It was around that time that he approached von Neumann, who handed him "a 20-page manuscript in his handwriting in Hungarian, which was all he then knew about game theory . . . I had that paper for a couple of years . . . I was never able to read the darn thing. I was too reticent to go and persuade von Neumann to give me the time, which he would have done" (*ibid*). This was probably von Neumann's notes on a simple two-person poker, which he had also completed in Göttingen.

[20] Von Neumann to Ulam, Oct. 4, 1937.

friend, and if von Neumann's productivity at this time is any indication, the combination of marital difficulty and Hungarian upheaval truly affected him. His output of papers, normally volcanic, collapsed to one in 1938 and none the year after.[21]

The main reason why von Neumann returned to Hungary in mid-1938 was so that he could marry his second wife-to-be, Klára "Klari" Dán, and bring her back to the United States. Part of the same privileged Budapest circle, they had met properly only a few years previously, on one of von Neumann's summer trips home. It was in Monte Carlo, where Klari was holidaying with her first husband, Francis:

When we walked into the Casino, the first person we saw was Johnny; he was seated at one of the more modestly priced roulette tables with a large piece of paper and a not-too-large mound of chips before him. He had a "system" and was delighted to explain it to us: this "system" was, of course, not foolproof, but it did involve a lengthy and complicated probability calculation which even made allowance for the wheel not being "true" (which means in simple terms that it might be rigged). Johnny was a little bit bashful about his "system" and insisted that he really did not believe in it; nevertheless he was determined to test it thoroughly.

Francis went on to another table. For a while I wandered around watching the lunatic pleasure of people destroying themselves, then I went to the bar and sat down, wishing I had company with my drink. As I was sipping my cocktail, Johnny appeared. I shall never forget the meek and apologetic way he sidled up to by table and asked if he might join me. "Of course", I said, "pull up a chair; I hate to be a lonely drinker". Johnny, a little embarrassed, but with the cute cunning of a child who wants his ice cream but will not ask for it directly, exclaimed: "A drink – what a splendid idea – I would love to have one with you, but are you sure that you can afford it? You see, you will have to pay for mine. My system did not quite work and I am completely cleaned out".[22]

Von Neumann's charm evidently worked, for by February 1938, with him newly divorced, Klari had left not Francis but a second husband and had begun sitting out the six-month waiting period before her own divorce proceedings could begin.

Klari was to wait in an increasingly fraught environment. Early in March, in a well-known speech at Győr, close to the Austrian border, the Hungarian

---

[21] I think it reasonable to suspect that, in the straightlaced Princeton of the 1930s, in which the couple's social integration had been dependent upon the Veblens, the impact of the divorce upon von Neumann was dramatic. The tacit social pressure upon him to regain respectability through remarriage was probably considerable.

[22] Von Neumann-Eckhart, unpublished draft autobiography, pp. 10–11, KEMNW.

prime minister, Kálmán Darányi, outlined his plans for concrete legal measures designed to cope with the "zsidókérdés", the "Jewish Question". "I see the essence of the question in the fact that the Jews living within Hungary play a disproportionately large role in certain branches of the economic life, partly owing to their particular propensities and positions and partly owing to the indifference of the Hungarian race. Their position is also disproportionate in the sense that they live to an overwhelming extent in the cities, and above all in the capital. . . . The planned and legal solution of the question is the basic condition for the establishment of a just situation – a just situation that will either correct or eliminate the aforementioned social disproportions and will diminish Jewry's influence . . . to its proper level".[23] With the *Anschluss* of Austria a fortnight later, the Győr Programme became something of a national obsession in Hungary, giving rise to a three-month parliamentary debate on Bill No. 616, designed to ensure the "more effective protection of the social and economic balance".

From Princeton, von Neumann watched developments. To Ortvay he admitted being even more pessimistic than him, and felt that the catastrophe could not be avoided.[24] It was not a case of proving why it would happen, he said, but why it would not. He was certain that, if there were no other means to ensure an English victory, the United States would intervene on England's behalf, the latter being essential to U.S. security in the Far East. He was also very interested in how domestic politics in Hungary would be affected by Austria's demise.

Ortvay's pessimism deepened in turn. Even putting aside the danger of a catastrophic war, he said, he judged the whole development of culture very darkly. The "advance of the masses" was a negative feature of early twentieth-century modernity: the development of the popular press, the "adoration of the automobile and machinery", the excesses of propaganda, mass travel – this was "modern barbarianism, with all its technical superlatives as described so nicely by A. Huxley", and it prevented the emergence of a higher form of life. Ortvay said the problem was not how to further satisfy the masses but, rather, how to keep them under control. The obvious need for a strong moral stance in scientists, given that it could not be expected in politicians, served to underline the importance of emotions and spiritual qualities. Yet never before, wrote Ortvay, had there been so great a gulf

---

[23] Quoted in Braham (1981) p. 121.
[24] Von Neumann to Ortvay, Mar. 17, 1938.

between the scientist's technical capacities and his level of culture or moral state. It caused him great anguish daily, he said.[25]

In April, von Neumann fled Princeton, returning to Hungary to be close to Klari. It was the beginning of a short *Wanderjahr*, spent hovering around Budapest and travelling around Europe, waiting until Klari was free, watching as his own country changed by the month.

In May, Hungarian Bill No. 616 became Law No. XV, the famous "Balance Law", the purpose of which was to reduce to twenty percent the proportion of Jews in the professions and in financial, commercial, and industrial enterprises of ten employees or more. Those to be exempted included war invalids and those who had converted before August 1919 and their descendants. The aims of the law were expected to be achieved within five years, through the dismissal of 1,500 Jewish professionals every six months.

This legislative anti-semitism in Hungary of 1938 was indicative of a change of the rules of Hungarian life, the instantiation of a new norm, with the moral authority for same being provided by the Church and the Hungarian liberal-conservative leaders of the gentry and the old feudal order. It was a response to the popular perception of injustice as regards Jewish privilege, undertaken in such a way as to dampen the claims of the Hungarian radical Right, the *Nyilas*. The latter, under Szálasi, were clamoring not only for much harsher measures against the Jews, pointing to Germany, but also for significant reforms in the area of land ownership and the franchise. They represented a genuine threat to the traditional semi-feudal order and were duly feared. Therefore when Darányi appeared to be too close to the popular right, he was ousted and replaced as Prime Minister by Béla Imrédy. Other attempts to stall the far right included Horthy's forbidding civil servants to join extremist political parties, in April 1938, and the Interior Minister's banning the *Nyilas* Party less than a year later.

It was the underlying ethical shift that struck von Neumann: "I am familiarized by now with the state of mind, the bellyaches and the illusions of this part of the world – such as they are since the annexation of Austria. The last item (illusions) is rather rare, the preceding one not at all . . . Hungary was

---

[25] Ortvay to von Neumann, Apr. 4, 1938. In the next letter, he reported visiting Germany, for a celebration for the physicist Sommerfeld, his teacher, at which Planck, Heisenberg, and others were present. He had visited some of the Nazi architectural sites, *Haus der Deutschen Kunst*, the *Führerhaus*, which, despite their Spenglerian striving for gigantic dimensions, with their huge columns and stone cubes, he said, he preferred to the retrograde modern buildings.

well under way of being Nazified by an *internal* process – which surprised me greatly – in March/April. The new government, which was formed in May stopped this process, or slowed it down, but for how long, is not at all clear".[26]

In June, he was in Warsaw for a conference organised by the League of Nations' International Institute for Intellectual Cooperation, in which several physicists, including Bohr and Heisenberg, took part. He also gave a talk to Ulam's former teachers and colleagues, including the logicians Knaster, Kuratowski, and Tarski. Ulam later travelled down to Hungary to join von Neumann, visiting Budapest and travelling with him through the countryside. They visited von Neumann's teachers Lipót Fejér and Frigyes Riesz at Lillafüred, near Miskolc, where the mathematicians were spending part of their summer.[27] Lying in an attractive forested area in the mountains about a hundred miles from Budapest, Lillafüred, with its luxurious castle-like hotels, was then a favourite resort of the Hungarian elite.

Ulam would later remember their walking through those woods with Fejér and Riesz, talking about the possibility of war, after which he returned northwards to Poland, by train, through the Carpathian foothills: "The whole region on both sides of the Carpathian Mountains, which was part of Hungary, Czechoslovakia, and Poland, was the home of many Jews. Johnny used to say that all the famous Jewish scientists, artists, and writers who emigrated from Hungary around the time of the first World War came, either directly or indirectly, from these little Carpathian communities, moving up to Budapest as their material conditions improved". When later asked why these Jews were so creative, von Neumann felt that it was "a coincidence of some cultural factors which he could not make precise: an external pressure on the whole society of this part of Central Europe, a feeling of extreme insecurity in the individuals, and the necessity to produce the unusual or else face extinction".[28]

In the summer of 1938, when Ulam and von Neumann were there, the pressure on the area was real. For several months, Hitler had been dangling before Hungary the promise of the return of Subcarpathian Ruthenia and Slovakia, should the Hungarians cooperate with his plans for the rest of Czechoslovakia. Hungary held back, keen to involve Great Britain along with the Third Reich and Italy in settling these East Central European disputes.

[26] Von Neumann to Veblen, June 8, 1938, VLC, Box 15, Folder 1, emphasis in original.
[27] See Ulam (1976), p. 111.
[28] *Ibid*, p. 114.

By late summer, Klari had sent von Neumann away from Budapest, claiming that his meddling in these matters only made things worse. The tension surrounding the divorce, exacerbated by what Klari declared to be von Neumann's childish manner, left her sounding quite desperate at times. Her almost daily letters, many of which were written from the finest hotels and spa resorts in Lucerne, Venice, and Montecatini, are pervaded by signs of depression and even hints at suicide. Von Neumann's letters, written in locations ranging from Lund to Abbazia, are filled with attempts to placate her and reassure her, apologizing for earlier tantrums and promising that the future will be better. Although it proved to be a durable epistolary pattern for the course of their lives together, in the short term, it was political anxieties that prevailed. By late August, Klari wrote von Neumann that it had been decided that her sister, Böske, and children should absolutely leave the country, in a matter of days: "I don't know what fate will bring us, things look very dangerous at present and maybe in a few days we shall have such worries we won't have time to think of this (sic) [divorce] questions anymore".[29]

Isolated in various hotel rooms, Klari seems to have had few friends in whom she could confide, and she soon began to write about wishing to see the "Fellners". This was Vilmos "Willy" Fellner and his wife Valerie "Vally" Koralek. Like the von Neumanns and the Dáns, the Fellners were a prominent assimilated family, their fortune going back to the 1860s and the beginning of the liberal period during which Hungarian Jewry flourished. Von Neumann and Fellner had attended the same *gymnasium* in Budapest and were students at the same time in chemical engineering at Zurich. Fellner later said that it was von Neumann and a mutual friend, another Hungarian named Imré Revesz (later Emery Reves, confidante of Churchill), who were responsible for sparking his interest in economics. Switching to that field and accompanied by von Neumann, Fellner transferred to Berlin, where he completed a doctorate in economics in 1929. He then returned to Budapest to run the family manufacturing business (sugar, alcohol, and paper). Like a number of cultivated nonacademics, such as Schlesinger and Kaufmann in Vienna, Fellner pursued an active interest in economics, although without publishing anything of note during that period. He and his wife visited the United States in 1928 and 1934.[30]

---

[29] Klari (Grand Hotel & La Pace, Montecatini Terme) to vN, Aug. 28, 1938, VNLC, Box 1, Folder 7.

[30] On Fellner, see Marshall (1988), Haberler (1984), and Adelman (1987).

On September 10, having written the previous day about Budapest being in a "frantic state" with the tension "getting worse every day", Klari met the Fellners.[31] That evening, throughout a film and then a late-night circus cabaret, Klari and Willy Fellner talked politics till three in the morning. "[E]ven a huge snake fully alive", she wrote to von Neumann, "could not disturb our happy projecting of who is now going to be killed. Well I suppose this is what happens if two full-blooded pessimists meet. Poor Vally again tried to persuade us to watch the show or at least not to use certain names too often as the place was terribly crowded and people seemed rather interested in our opinion . . . I'm so worried that I don't talk of this (sic) matters with my family anymore. I don't want to know them (sic) how terribly scared I am. I don't know what's awaiting us in the future, but never as long I may live will I forget 1938 . . . If you go to England I should very much like an objective report from you whether the Jewish question is really getting so bad there as I heard".[32]

Their petty bickering was now beginning to look silly, Klári admitted. They should really save their strength and nerves for the times ahead. Yet she continued to alternate between chiding him for screaming over the telephone and reassuring him that she was calm. "Johnny dear you must understand this. I shall never ask you to come back . . . I shall never ask you to risk your life or anything for my sake. But don't expect me to ask you to come back in this dangerous corner [of the world]. I refuse to take over this responsibility. We had a long argument with Will [Fellner] over this question. He thinks that it might be easier for me to get out when you are here. I don't think so, unless we are already married, but that's still a long way of (sic) . . . I intentionally don't speak of politics, it's no use as it changes every second. Father keeps a wonderfully cold (sic) head and we all try to be cheerful for his sake. But I do believe that this is the age of heroes".[33]

Whereas von Neumann deliberately chose to appear optimstic with Klari, he was less sanguine in his letters to Veblen. From Lund, where he was visiting Marcel Riesz, he wrote: "I agree with you, that war at this moment is improbable, since neither side seems to want it just now – but the Sudeten-german (sic)- population seems to be very nearly out of control, so you can never tell. It also seems, as if Messrs. H[itler] and M[ussolini] were a little more emotional lately than rational, so you really cannot tell. So we

[31] Klari to vN, Sept. 9, 1938, VNLC, Box 1, Folder 7.
[32] Klari to vN, Sept. 11, 1938, VNLC, Box 1, Folder 7.
[33] Klari to vN, Sept. 14, 1936, VNLC, Box 1, Folder 7.

may be much nearer liquidation than it seemed 2 weeks ago. God knows what will happen . . . ".[34] He was on his way back to Copenhagen, he said, where he wanted to see Bohr and talk especially about the latter's ideas on biology. Yet, the following day, he wrote Veblen that his wandering around Scandinavia was beginning to seem futile, that he felt like getting back to Budapest straightaway. "The trouble is, that I don't see what I can do when I get there: My brother cannot leave now, my mother won't leave without him, and Klári cannot leave either during the next weeks. I hope that things won't be quite as bad when you get my letter, as they look now".[35] By late September, Klári was warning von Neumann against discussing politics over the telephone, and the Fellners had announced to her "their future plans".

Back in Budapest in early October, von Neumann wrote to Veblen that the Munich Non-Aggression Treaty between Chamberlain and Hitler had provided welcome breathing space. Following that agreement, Imrédy had visited Hitler at Berchtesgaden, with no satisfactory conclusion as regards the Czech territories. A month later, however, Hungary's claims were submitted to German–Italian arbitration, resulting in the First Vienna Award of November which granted the *Felvidék* in southern Czechoslovakia to Hungary. Imrédy sought re-election that month and formed a new government. All of this provided respite for von Neumann. In late November, with Klari's divorce at last secured, they were married and, a fortnight later, sailed together for the United States. The Fellners had already left for California. Like the vast majority of Hungarian Jews, however, von Neumann's and Klari's families clung to Hungary, soothed for the time being by the Munich outcome.

Rationality and Pathology

From Princeton, the von Neumanns watched events unfold in Hungary. If the retrieval of the *Felvidék* had been welcomed by all Hungarians, for whom Trianon had been a injustice, it also brought with it a population of one million, including several orthdox Jewish centres, and, into Imrédy's government, an anti-semitic minister, Andor Jaross. Thus emerged a contradictory feature of Hungarian politics during this period: territory was regained, satisfying a need shared by all Hungarians, but with it came pockets of orthodox Jews, the effect of which was to inflame anti-semitism. By

---

[34] VN (Grand Hotel, Lund) to Veblen, Sept. 15, 1938, VLC, Box 15, Folder 1.
[35] VN (Gand Hotel, Lund) to Veblen, Sept. 16, 1938, VLC, Box 15, Folder 1.

December, Imrédy was promoting a second anti-Jewish bill "Concerning the Restriction of the Participation of the Jews in Public and Economic Life". Then, in a strange twist, Imrédy himself was unable to refute an accusation by the radical right that there was Jewish blood in his own ancestry, which compelled him to resign in February 1939. Horthy swore in Pál Téléki a few days later.

A renowned academic and cartographer, the aristocratic Teleki was tolerant of the Magyarized Jews, but less so of the "Ostjuden". This issue became more significant with Hungary's acquisition of Subcarpathian Ruthenia in March 1939, which brought with it a substantial Jewish Orthodox population, whose urban politicised intellectuals were left-leaning.[36] This development stimulated the parliamentary debates on the second Jewish law, which took place in the first half of 1939. Compared with the law of the previous year, the new bill was more "Nazi" in content, referring not only to the Jewish threat to economy and culture but also to the racial, psychological, and spiritual difference of the Jews. Throughout, anxiety grew in Budapest. On New Year's day, 1939, Ortvay wrote to von Neumann in Princeton that Leo Libermann, an opthalmologist and university professor known by both of them, had just committed suicide. "In the state of the world, one cannot find great joy, I see it as slipping downward . . . ".[37]

If Ortvay persisted in his psychological probing throughout this time, von Neumann resisted it, and this was a theme in their correspondence. Yet, at the same time, he was ready to speak of what he called the "pathology" of the general situation. It was difficult to write about politics, he admitted, and especially difficult to be sure that his diagnosis was not simply the expression of his own desires – "*Wunschbestimmt*" – but he felt reasonably objective about the matter: the war was inevitable, he said, and the arguments that it was not necessary, or that it would not resolve the problems, were beside the point. "The whole affair", he wrote, "is a pathological process and, viewed clinically, is a plausible stage of further development. It is 'necessary' even emotionally – if it is permissible to use the word 'necessary' in this connection. It will bring the acute problems to a resolution insofar as it will diminish the moral and intellectual weight of the European continent and its vicinity, which, considering the world's structure, is justifed. May God grant that I am mistaken".[38]

---

[36] On the Jews of Subcarpathian Ruthenia, see Rothkirchen (1978).
[37] Ortvay to von Neumann, Jan. 1, 1939.
[38] Von Neumann to Ortvay, Jan. 26, 1939.

In his letters, the emphasis shifted subtly from the inevitability of catastrophe to the question of what would follow. Apologising to Ortvay for
not delving into the mathematics of the "spirit" – that is, emotions and attitudes – he dwelt persistently on politics, with the vocabulary of structure
and equilibrium creeping into his prose. Point by point, he went through
the issues. It was naive to hope that any outcome would be useful to
the Jewry stranded in Europe. One possibility was an outcome similar
to "the Turkish–Armenian affair during the World War" – the genocide of
Armenians by the Turkish government – to which Hitler had referred in a
recent speech, an outcome which, von Neumann said, was "superfluous to
analyse".[39] Even if this did not occur, there would be social chaos and lasting
division between the various sides in the vanquished countries, making it
impossible that a "state of equilibrium could take place", he said. A victory
to the Western powers would be, in many respects, Pyrrhic, with the rap-
prochement of the dissatisfied, in the form of a German–Russian coalition,
posing a future threat at least as worrying as the present one, he said. The
position of the Western powers vis-à-vis their allies and dominions outside
Europe would be at least as weak as after 1914–1918.

Economically, the United States stood to gain little from a war. The
wartime boom would be only temporary, with debts incurred never being
repaid and the American social structure dangerously loosened. Speculating
on the possibility of American imperialism emerging in the event of their
victory in war, von Neumann felt that this would be possible "only if the
war liquidates Japan too", which wasn't completely out of the question.
However, he felt that there was currently little popular support for such
a development in the United States: it was quite foreign to the ordinary
American, he said, and the terminology and symbols of politicians and
big business suggested that they looked in other directions to satisfy their
ambitions. It all depended on what happened to the British, as the Great
War had shown. American support for British power in order to maintain
world stability was "a very negative motivation, the avoidance of damage
rather than achieving a gain". Even the Roman Empire became imperialistic

---

[39] The Armenian catastrophe of 1915, in which the Turks razed Armenian villages and
walked more than a million people to their deaths in the eastern deserts, occupied an
important place in the imagination of many German and Central European intellectuals
during the interwar period. This was largely thanks to Franz Werfel's novel, *The Forty Days
of Musa-Dagh* (1934), which described the affair. In *Principles of Topological Psychology*, a
1936 book that Ortvay was reading at the time of his correspondence with von Neumann,
Berlin social psychologist Kurt Lewin draws on Werfel's account of the Armenians under
siege.

only in the second century A.D., said von Neumann, when it agreed that the permanent annexation of the Balkans was 'unavoidable' from the viewpoint of their security. The United States might go in such a direction but, for the moment, all instincts were isolationist. The war, he agreed with Ortvay, would indeed be a terrible cultural loss in Europe – indeed, such a loss was already being incurred – but neither should one exaggerate: when the Romans took over Greek culture, the ancient civilization remained essentially intact for another 300 years. "After all this", he concluded, "I believe the war is plausible in spite of all, and with the relatively early participation of the U.S.A. Because it is a pathological procedure, which does not take place because anyone considered it intelligently, that it is in his interest, but because certain abnormal spiritual tensions – which no doubt exist today in the world – search for 'resolution' in this direction. And because from a rational point of view, England and France cannot let one another perish, nor can the U.S.A. let England".[40]

During this time, von Neumann periodically apologized for writing so much about these themes of "war and peace", but persisted in it nonetheless. The pessimistic diagnosis, on which they now seemed to agree, he said, was much closer to reality than had been the illusions of last October. Replying at length, Ortvay felt that Western Europe was in decay, as evidenced by its excesses of capitalism and mechanisation, its shallow rationalism, "which consists in the fact that a few easily comprehensible viewpoints are fulfilled to the extreme", and its "excessive cult of the will", which conferred power upon "a very aggressive, half-cultured mass". America, although hampered by the absence of an aristocracy, still showed signs of cultural health and force, and thus bore a responsibility for regeneration – indeed, for the future of humanity. He implored von Neumann: if only a minority there could substitute for the absent aristocratic class and set an example for the rest of the population. He realised how non-modern his thinking was – it was, he admitted, as if he were living in Herder's time . . .

Just as von Neumann resisted what he perceived as Ortvay's excessive psychologising, neither was he drawn on another theme pushed by Ortvay, modelling "spiritual processes" or thoughts and feelings. Ortvay wrote that the brain must be considered as a complete system, to which individual spiritual processes must be subordinated, just as the quanta were ordered

---

40 Von Neumann to Ortvay, Feb. 26, 1939. In his letters, von Neumann makes several references to the history of antiquity, in which he was well-read. One of his favourite books was Thucydides' *The Peloponnesian Wars*, in which he was particularly fond of the Melian dialogues, a model of rationalist, *realpolitik*, discourse.

to the system's vibrations. Although the knowledge of the physical system of the brain was very imperfect, perhaps someone who was knowledgeable in mathematics and physics, Ortvay wrote suggestively, could construct an independent axiomatics of such a system: "Approximately that which mathematics gave to a great extent concerning the natural, perceptible materials . . . must be accomplished in the field of spiritual acts, social relations".[41] Perhaps it was the case that sensations were, indeed, located in the brain, but spiritual acts such as judgements and the taking of positions were determined independently: "the so-called material world would be but a region of a complete world that includes the spiritual as well . . .".[42]

As part of this quest, in May 1939, Ortvay turned to several books on set theory and logic, by Fraenkel, Heyting, Carnap, and others. In his *The Logical Structure of the World* and *The Logical Syntax of Language*, Carnap had attempted to analyse the logical structure of language and to construct axiomatics of other branches of knowledge.[43] Ortvay felt that an examination of the logic of living languages would lead to the establishment of interesting types of structures and eventually to a "fertile interaction of the mathematical and humanities' spheres of thoughts". On the other hand, he found Carnap's sketches of the axiomatics of the "sensory realms" to be extremely primitive. Much more empirical research was necessary before the field "becomes ripe for serious axiomatics". The excessive emphasis on sensations did not conform to the psychology of the time, and physics had shown that precipitate reductionism only led to insipidity. "Much more important is the elaboration of characteristic structures and perhaps their provisional axiomatisation as well".[44] He felt that Heidegger, in *Being and Time*, had shown an intuitive grasp of many characteristic structures but was unable to formulate them at all exactly. Neither did he agree with Carnap's aim of replacing the summary statements of all fields with sensation statements. The question was far from settled, he felt.

Von Neumann reacted angrily to Ortvay's mention of Carnap, making it clear that he had little respect for him.[45] Neither did he think much

---

[41] Ortvay to von Neumann, Apr. 10, 1939.
[42] Ortvay to von Neumann, *ibid.*
[43] Carnap (1928) and (1934) respectively.
[44] Ortvay to von Neumann, May 28, 1939.
[45] Yes, von Neumann confirmed, Gödel's results meant that there wasn't a complete axiom system, even in mathematics. However, from the mathematical perspective, Carnap's insights were very feeble and naive. Von Neumann said he simply did not have the objective knowledge minimally required to say something in this area. Carnap may well

of another book of which Ortvay's letter had reminded him, namely Karl Menger's *Moral, Wille und Gestaltung* (*Morality, Decision and Social Organisation*): "A few years ago, Menger wrote an axiomatic treatment of the field of human relations (ethics?). I found it to be completely 'flat' and say little. It is my feeling that most promising today would be the discussion of some specific psychological fields and maybe that of economics . . .".[46] Turning to politics, it was clear that, in the present state of transition, no further concessions should be made by the Western powers. Perhaps Chamberlain would do so, but he was no longer in a position to shape his own politics. The experience of the past few months had shown that concessions only evoked further demands.

For more than a year, at that point, emboldened by Hitler's advances, Hungary's German-speaking Swabians had been growing vocal in their demands for increased economic and cultural autonomy. By 1939, they had become an important political presence, providing a direct link with the Third Reich. As previously with Czechoslovakia, Germany wanted to have Hungary's support for its designs on Poland. Hungary resisted, Poland being an old ally, but it was also keen to placate Germany, the support of which it would need in its own claims on Transylvania, which it wanted to retrieve from Rumania.

If the law of 1938 had met with surprise but not opposition on the part of the Jewish population, the second anti-semitic bill brought protest. The Hungarian Jews proclaimed their patriotism, pointing to their sacrifices during the Great War and to their contribution to the economic, cultural, and scientific life of the country. They turned to the British Jews

---

have persuaded the school philosophers of the value of the philosophy of science, but that was about the extent of it. On the completeness of mathematical axiomatics, for example, he expressed "completely naive, simplistic views with a terribly important 'air'". If the affair were as simple as Carnap imagined it to be, von Neumann said, then "there would be no need for fundamental mathematical research – at least from a mathematical point of view!" Von Neumann said it was a pity that we had to rely on such a "turbid source" to learn about such solid topics! He was especially annoyed that Carnap, although always with Gödel's name on the tip of his tongue, "obviously has absolutely no understanding of the meaning of Gödel's results". Did Ortvay really believe that Carnap was saying something new about the structure of language? That what he was doing would be useful in the preparation of "ultimately serious efforts"? Von Neumann to Ortvay, July 18, 1939.

[46] Von Neumann to Ortvay, July 18 1939. Not until several years later, when stimulated by McCulloch and Pitts's 1943 paper, did von Neumann take up Ortvay's suggestions and seriously turn to the operation of the brain. See Aspray (1990), p. 180. Von Neumann discussed this in his Hixon Lectures of 1948 and in his 1950 talk to the Cybernetics Group in Atlantic City, which repeats verbatim many of the lines of argument raised by Ortvay a decade previously. See Nagy et al (1989).

for assistance, while Hungary's ecclesiastical leaders spoke in favour of the reforms. In February 1939, Szálasi's followers launched a grenade attack on people leaving Budapest's Dohàny St. Synagogue. In May, the second law was enacted, prohibiting Jews from obtaining citizenship (something aimed at recent refugees and those residents in the recently acquired territories), and ordering retirement of all Jewish court and prosecution staff by 1940, and primary and secondary teachers by 1943. Reintroducing the 1920 *Numerus Clausus* of a six percent limit on admission to universities, it also prohibited Jews from holding positions as editors or publishers of periodicals or producers or directors of plays or films. Licenses held by Jews for various kinds of businesses were to be withdrawn. Firms of five or fewer employees could have one Jew; those of nine or more employees could have two.[47]

Jewish historian Ralph Patai has written about the devastating cumulative psychological effect of the laws of 1938 and 1939: "Even if the laws did not immediately endanger their lives ... the new situation demanded a total rethinking of their own position in Hungary, something of which most Hungarian Jews were simply incapable".[48] Patai goes on to describe how that attachment to Hungary left many Jews somewhat paralysed. Many of them shunned Zionism, regarding themselves as patriotic Hungarians, so that even though the 1939 law made express provision for the emigration of Jews from Hungary, subject, of course, to financial restrictions, relatively few resorted to it.

Von Neumann's family and his in-laws were amongst those reluctant to leave. Thus, that summer, in July 1939, Klari returned to Budapest from Princeton to try to persuade them to do so. While she was gone, Ulam and von Neumann slipped away for a few days to visit Veblen at his summer home in Maine. On the way, they "discussed some mathematics as usual, but mostly talked about what was going to happen in Europe. We were both nervous and worried; we examined all possible courses which a war could take, how it could start, when".[49] When Hitler overran Poland the next month, Ulam felt as if a curtain had fallen on his past life, cutting it off from his future: "This was the period of my life when I was perhaps in the worst state, mentally, nervously, and materially. My world had collapsed ... There was a terrible anxiety about the fate of all those whom

---

[47] On the economic impact of the Jewish laws on the Hungarian economy, see Kádár and Vági (2004a, b).

[48] Patai (1996), p. 541.

[49] Ulam (1976), p. 115.

we had left behind – family and friends".[50] With Klari away in Europe for several weeks, von Neumann's anxiety, too, reached a new pitch. "What are your further plans in Europe?" he wrote to her. "Your father should not hesitate any longer . . . Can he not make up his mind? Don't be untimely (sic) sentimental, you might be the one who saves them by insisting on talking rationally!"[51] That August, after some delays, von Neumann's mother and brother arrived in New York. After further delays, the Dán family, too, left Budapest for the States. Ulam's family did not escape from Poland.

Amidst the great tension, Ortvay's letters now ran to several pages, ranging on subjects from axiomatics to God to Freud. He hoped that the European nations would wake up before European culture collapsed entirely. A desirable solution would see, not one side crushed by the other, but an entente, in which each recognised the other's virtues, their right to exist, as well as their faults and sins. With Freud's death in London that year, Ortvay wrote of his long-standing interest in Freudianism and of having been in contact with several of the psychoanalyst's followers, with sometimes unpleasant experiences. Freud, he said, had provided the first systematic exploration of the psychology of the subconscious and of repression. Ortvay was prepared to acknowledge the importance of sexuality, but not to the extent suggested by Freud. Drives such as aggression, the will to power, revenge, and envy were important, as were, in a few people, higher spiritual emotions. The Freudian view was quite unbalanced, Ortvay felt, and its success lay in its being drilled into his followers. Through effective propaganda, it took on a political or religious dimension.[52] Yet, looking at the war and the events leading to it, Ortvay felt that he could not deny the great importance of repression and of sharply distinguishing between the

---

[50] *Ibid*, p. 118.
[51] Von Neumann to Klari, Aug. 12, 1939 (year unmarked), KEMNW. His letters of this interlude, which also allude to Klari's father's depression, culminated in an Aug. 24 telegram: "PLEASE TAKE FIRST AVAILABLE SHIP PLEASE WIRE LOVE".
[52] It is interesting to speculate whether Ortvay had contact with Sandor Ferenczi, Freud's principal interpreter in Hungary. According to von Neumann's brother, Nicholas, Ferenczi was actually a close relative of the von Neumann family, and, like Ortvay and Fejér, a dinner guest at the von Neumann household during John's youth. Psychoanalysis was a frequent topic of dinner-time conversation. Ferenczi was initially one of Freud's closest disciples, accompanying Freud on his 1909 trip to America, with Ernest Jones and Carl Jung and, unlike most of Freud's intimates, actually undergoing an analysis with him, in 1914 and 1916. By 1920, however, their relationship was strained, with Ferenczi rejecting Freud's authoritarianism and favouring a more equal, and even emotional, relationship between the analyst and analysand. See Vonneuman (1987), p. 36. On Ferenzci's life see Rachman (1997). For his impact on the neo-Freudians and humanistic psychologists, see Hoffman (2000).

causes superficially believed to be important and the underlying mechanisms: "I believe that these are economic forces only to a very slight degree; rather they are enormously primitive and brutal passions, and the 'economic' reasons are in many cases only suitable for the purpose of letting modern man hide the real reasons from himself... Nietzsche already saw a great deal here".[53]

These extensive discussions of rationality and pathology, of politics and social relations, coupled, as we shall see, with the arrival of Oskar Morgenstern in the picture, brought von Neumann back to the mathematics of games. In November, he was planning to spend part of the following summer at the University of Washington at the invitation of Abraham Taub. In a letter to the department suggesting possible topics for some additional popular lectures, he included the theory of games: "I wrote a paper on this subject in the Mathematische Annalen 1928, and I have a lot of unpublished material on poker in particular. These lectures would give a general idea of the problem of defining a rational way of playing. I think that even stating the problem is not at all trivial and leads to a number of quite amazing considerations on the nature of games like chess on the one hand, and of another kind on the other hand, of which – I think – poker is the prototype. The discussion of games played by more than two persons leads to further questions which can also be discussed in a manner which I think will interest the intelligent but non-technical audience".[54] A week later, he returned to Ortvay: "Unproductive as it is to meditate upon political problems, it is hard to resist doing so. Maybe from Hungary the meaning of the European, and particularly East-European, situation's elements are clearer. But from here it makes a fairly complicated and confused impression. In particular, it appears in all likelihood that not 2, but 3 or 4, enemies are facing one another".[55] One might say the European situation was not a two-person game.

A week before Christmas 1939, the psychological difficulties of forced emigration were brought home to the von Neumanns in the starkest manner, when Klari's father and reluctant exile, Károly "Charles", Dán, committed suicide at a train station near Princeton. The Weyls and others rallied

---

[53] Ortvay to von Neumann, Sept. 26, 1939. He then continues with several further pages, most of the content of which is devoted to speculations about the possible relationship between the general axiomatic approach and the study of philology.

[54] Von Neumann to Prof. Carpenter, Nov. 29, 1939, VNLC, Container 4, File 3, Personal Correspondence 1939–40.

[55] Von Neumann to Ortvay, Dec. 8, 1939.

round the von Neumanns in their difficulty. From the Institute, Veblen's secretary kept him up with the news:

Mrs. von Neumann came to call on me yesterday afternoon! I hope she did not feel under any kind of compulsion . . . But it seemed to some satisfaction to her to talk. She looked shrunken, but did become natural in talking of general conditions – in England now for instance. She said she has now no courage to try to dissuade her mother from returning as soon as possible to Hungary; that she had insisted on her parents' coming here as the only best course she could then see. Now she questions whether alternative courses might not have been better. I told her it seems to me we must in such cases rest on the assurance that we did what seemed best at the time (which we should probably do again in the same conditions, with the same experience). Professor Weyl also has been conscious of this special cause of her depression.

She would like, if her mother were willing, to take her away somewhere for a little change; but unless her mother's re-entry permit into Hungary can be extended, she must be back there – January 22 I think was the date. She also would like Professor von Neumann to get at least a few days rest away from Princeton, "even 20 miles away". But she herself apparently needs it as much as anyone. She says she has been closely confined by her father all fall, conscious of his abnormality, trying to help him, and not wanting to expose his condition to other people.[56]

This tragedy would prompt von Neumann to take his wife with him on his visit to the West coast. In March, he confirmed with Seattle that he would give three evening lectures on games, covering "The case of chess; The notion of the "best strategy"; Problems in games of three or more players". He would leave in May and drive across the country.[57]

Ortvay continued to write about mass movements and war, insisting that rational, utilitarian, considerations played only a secondary role: the fundamental reasons were "primitive passions". This conformed to the Freudian mode of thinking, he said, but the passions were different from Freud's. Anything that challenged our self-worth evoked hate, which, in the case of mass movements, was directed towards destroying the object of the animosity. Even in business, where utilitarian considerations were perhaps strongest, a fundamental force was often the suppression of a competitor, who simply could not be tolerated, and not just for reasons of profit. Passions of this kind, he felt, were at the root of the last war, the present one, and the anti-semitic movements as well.[58]

---

[56] Dec. 27, 1939, Mrs. Blake to Veblen, VLC, Box 15, Folder 1.

[57] Von Neumann to Prof. Carpenter, Mar. 29, 1940, Container 4, File 3, Personal Correspondence 1939–40, VNLC.

[58] Ortvay to von Neumann, Mar. 30, 1940. In this connection, Ortvay asked for von Neumann's help for Rozsá Péter, who had lost her position because of the Jewish law.

It was around this time that Oskar Morgenstern entered von Neumann's life. He got to know von Neumann and his colleagues beginning in 1939, and first met the Neumann couple in early 1940 at the home of Hermann Weyl, at what appears to have been one of their first outings after their bereavement.[59] By March of that year, von Neumann had become "Johnny" and he was showing an interest in Morgenstern's concerns for the problems of foresight and decision. They drew closer in April, when the relationship between Morgenstern's concerns and von Neumann's work on games began to become clear. Before heading out West, von Neumann read and praised some of Morgenstern's earlier work. Under the mathematician's influence, the economist turned to Richard von Mises' (1939) *Kleines Lehrbuch des Positivismus*, and began to regret not having abandoned even sooner the universalism and idealistic philosophy he had encountered in Vienna.

The Trip Out West

By mid-May, von Neumann and Klari were driving across the United States on the way to Seattle, where he was to lecture from mid-June to end-July. Once again, von Neumann was on the move, this time into the heart of North America, driving westwards, his back turned to Europe, as if to flee the source of their troubles. From a hotel in Winslow, Arizona, he wrote a long, rich letter to Ortvay: "The travel is quite dreary until the middle of Kansas, but from then on the land is incredibly beautiful and varied – I am really ashamed that for 10 years I have always put it off till "next year". Furthermore, the most beautiful parts, the Grand Canyon, Northern California and Oregon are still ahead of us".[60] However, there was no getting away from politics: "Naturally, from the perspective of bringing Europeans over here, all that can be said is that the bottom has fallen out of the world – I don't even dare to think what the disintegration of the Scandinavian countries, the Netherlands, Belgium (and tomorrow and the day after, who knows what else?) will bring. But even if these – and other evident political possibilities – make even the slightest degree of success doubtful, I will do everything I can".

Turning to science, von Neumann agreed with Ortvay that theories that were unduly complicated could not be right. For these reasons, he was

[59] Although the highly cultivated Weyl was a proponent of Brouwer's Intuitionist philosophy of mathematics, a lover of German Romantic poetry, and a Heidegger scholar, on all of which counts he lay far away from von Neumann, he had been his teacher and now, abroad, the émigré couples were close.

[60] Von Neumann to Ortvay, May 13, 1940.

especially "horrified" by biochemistry. "I cannot accept", he said, "that a theory of prime importance, which describes processes which everybody believes to be elementary, can be right if it is too complicated, i.e., if it describes these elementary processes as being horribly complex and sophisticated ones".[61] He could not substantiate this with any detailed knowledge, but he felt it intuitively. There was a need for new terminology and new models in several fields.

One area in which von Neumann sought simplicity was in "politics and psychology". Thus, here, although he agreed with much of what Ortvay had written, he could not go with him entirely on the importance of primitive passions. We quote von Neumann at length:

I too believe that the psychological variable described by you, where resentment is the primary attitude, and the "egotistic-", "profit motive" only a secondary and (often not even quite plausible) rationalization – is an oft-occurring and important psychological mechanism. But neither is it permissible to forget entirely the other variable either: selfishness, in a wrapping of principles and ethics . . . In the present conflict, particularly given the antecedents, I would still find it difficult to believe that the enemies of the Germans are moved mainly by the first mechanism.

Concerning the practical chances, and the future, and what would be desirable . . . It is difficult to write about this, since the letter will travel for 3–4 weeks, and this time interval is not "negligible". You know that I do not believe 'compromise' to be either desirable or possible. The survival of the German power in any form signifies, among other things, the rapid liquidation of the European "Vielstaatlerei" [federation]. I don't believe that this would be a factor ensuring equilibrium from a small European nation's point of view. If the allies are victorious, then without doubt they will orient Europe to the "Vielstaatlerei". Viewed from afar, this is a retroactive development, but from the viewpoint of small European nations e.g. the Hungarian nation, it is the only chance at all. To speak of a German counterweight against Russia, I believe, is an impractical daydream.

That the war, in the case of the Western Allies also, even if they are victorious, will result in the extension of state power and the impoverishment of today's economically leading classes is very plausible to me too. But I believe that this has to be interpreted as follows:

If, in physics, it can be shown of a procedure that it is accelerated by all disturbances and entirely independent of the disturbance's nature, and clearly accelerated more the greater the disturbance – then it is usual to assume that the procedure leads to a state of equilibrium. This is most likely true in politics as well. Further, in politics, even more complicated is the fact that if such a procedure is carried out by means of a given political movement, then it soon becomes clear that efforts directed towards combatting this movement serve as at least a good mechanism in the same direction . . .

---

[61]   *Ibid.*

I don't believe that cultural wealth would be less in a centralised society than in the old, free economy. Although such a thesis could be defended dialectically, its opposite, I believe, could be defended just as well. Empirically, all that is clear is that the transition is harmful, but this, naturally, is no miracle.

Returning to the purely political theme: I don't see how both sides could acknowledge the other's raison d'être: If the German nation's frame of mind, which evolved during the last ten years, does not end with a very obvious cataclysm, then no one else on this earth has a raison d'être.[62]

Did all of this – the insistence on simplicity, the refusal to psychologise, the emphasis on political equilibrium – did all of this speak to the "*Wunschbestimmt*" that von Neumann had written of previously? Were they projections of his own desires, signs of his hopes for order beyond the European cataclysm, in the same spirit as his earlier reminders that Greek civilisation had remained intact long after the Roman conquest? And was it for confirmation of such hopes that he would turn to mathematics?

Klari later remembered this journey out West as being filled with drama. It was May 1940 and the Allies were suffering one setback after another. Rather than spend two days as intended in Denver, they stayed a week, in order to avail of newspaper extra editions and continuous broadcasts: "Holland was being invaded the day after we arrived in Denver ... By the time the negotiations for the surrender of Belgium had started, we had made it to Nevada".[63] She remembered von Neumann spending hours beside the car radio, or insisting, even during social gatherings, that the continuous chain of news be switched on. "Then, as soon as the news was being told, Johnny would start a running comment of his own, giving his interpretations of the day's events".[64]

Continuing northwards from Pasadena, where they presumably saw Theodor von Karman, to the Fellners at Berkeley, they heard the news of France's fall.[65] In Metford, Oregon, their car broken down, they spent the night in a tiny hotel, where the radio "was blasting President Roosevelt's famous 'stab in the back' speech. The Italians were coming in on the German side and all the little Central European nations were jumping on the Nazi bandwagon": "We must have played at least a hundred games of Chinese checkers that night (of course, Johnny won all of them), but all throughout he kept talking, going over and over the same arguments, like the broken

[62] *Ibid.*

[63] Von Neumann-Eckhart, unpublished draft autobiography, p. 21, KEMNW.

[64] *Ibid*, p. 24.

[65] Fellner had managed to secure a lecturing position in economics at Berkeley when he and his wife fled Hungary in 1938.

216 Von Neumann, Morgenstern, and the Creation of Game Theory

record running in the same groove; he was repeating the details of the last weeks' tragic events and then proving that, in spite of all the adversities, the Germans were going to lose in the end. He talked with the obsession of a maniac who, however, had clear logical arguments to prove his case".[66]

A Mathematics of Social Stability

Returning to Princeton in August, von Neumann was soon to be drawn into war work. What began as a consultantship on mathematical statistics and aerodynamics to the Ballistics Research Laboratory of the Army Ordnance Department at the Aberdeen Proving Ground in Maryland, was to become a complete immersion, marked by the same peripatetic frenzy that had previously carried him around Europe or across the States. Ahead of that, however, he threw himself into the development of a new social mathematics.[67]

One characteristic of his working practice as a mathematician was his apparent need for an interlocutor, even a passive one, in certain phases. That summer, the person was Israel Halperin, his former student, by then a young faculty member at Queen's University in Canada.[68] Having driven down to Princeton to be close to him for a few months, Halperin actually followed the von Neumann couple out to the University of Washington, where he attended von Neumann's game theory lectures. Once back at Princeton, he visited the house on Westcott Road several times a week, where von Neumann would "go over ideas or create them, and fill my head full of this stuff for an hour and a half. Then he would tell me to come back

---

[66] Von Neumann-Eckhart, unpublished draft autobiography, pp. 25–26, KEMNW.

[67] Discussions with Fellner at Berkeley were clearly important, for no sooner was von Neumann back on the East Coast than he wrote to him, clearly in the light of earlier conversations. The letter thanks him for reminding von Neumann of a paper by Gerhard Tintner, which he was reinterpreting in the light of game theory, and shows that von Neumann had begun working out a concept of "solution" to the three-person game. Von Neumann to Willi Fellner, Aug. 15, 1940, von Neumann Papers, National Technical Information Centre and Library, Budapest, original and translation kindly provided by Mr. Ferenc Nagy.

[68] Halperin had completed a doctorate in mathematics at Princeton a few years previously, as von Neumann's only doctoral student. Unaware that Institute professors were under no obligation to supervise theses, Halperin had approached him, and ended up working on continuous geometries. Of the Washington game theory lectures, he later remembered them as being on minimax theory and poker applications only. See Halperin Interview with Albert Tucker, May 25, 1984, Princeton University, *Princeton Mathematics Community in the 1930's*, Transcript Number 18 (PMC18). See also Halperin (1990).

the next morning ... It was my impression that he wasn't just talking about it, he was doing the work, and that the reason he sent me home after each morning was that he wanted to think alone for a while ... I realized I was right at the beginning of something very hot, but it wasn't the sort of thing I felt comfortable with".[69]

As of this time, Morgenstern gradually became a more active interlocutor. He was independently pursuing ideas that had grown out of his 1935 paper on the difficulties of assuming perfect foresight in economic theory. Unlike Halperin or the absent Fellner, Morgenstern could engage von Neumann on the economic aspects of the new theory.

Nonetheless, von Neumann worked away independently, exploring games of three and more players and, in a decisive move beyond his paper of 1928, creating the new concepts of coalitional equilibrium and stability. By October, he had produced an unpublished typed draft "Theory of Games I (General Foundations)".[70] Following a presentation of the two-person, zero-sum case, he presents the set function $v(S)$ for the n-person game. It shows the value (that is, gains) available to a coalition, S, who, by complete internal cooperation, play minimax against their complement. He conjectures that this set function, $v(S)$, will be sufficient to determine the strategies to be adopted for the entire game by each of the n players. "We now study the special case n = 3 for a clue as to what we should mean by a solution to our problem. Assuming a fully normalised game, $v(S)$ is here uniquely determined by ... :

$$
\begin{array}{cc}
0 & 0 \\
-1 & -1 \\
v(S) = 1 \quad \text{for a } (S) = 2 \\
0 & 3
\end{array}
$$

Clearly then the advantageous strategy is for any two players to form a coalition against the third: by this the set will gain, and the third lose, one unit".[71]

[69] Halperin Interview with Albert Tucker, May 25, 1984, Princeton University, *Princeton Mathematics Community in the 1930's*, Transcript Number 18 (PMC18). Of von Neumann, Halperin later wrote: "[S]ometimes he would stand apart, deep in thought, his brown eyes staring into space, his lips moving silently and rapidly, and at such times no one ventured to disturb him" (1990), p. 16.
[70] Von Neumann, "Theory of Games I (General Foundations)", OMDU, File John von Neumann, 1940–1948.
[71] *Ibid*, p. 12.

Von Neumann then describes how the apportionments among the three players are determined by the set function just described.[72] Each member of the "winning coalition" will receive 1/2. Were either of them to insist on more, the other could profitably deflect to form a coalition with the "defeated" player. Also, no player can improve his chances of entering a winning coalition by offering to accept less than 1/2, for the other two players would compete with each other to join him, thereby eroding away the premium offered. "So we see: each of the two members of the 'winning' coalition gets 1/2 . . . and the formation of any particular one among the three possible 'winning' coalitions cannot be brought about by paying 'compensations' and the like. Which 'winning' coalition is actually formed, will be due to causes entirely outside the limits of our present discussion".[73] These external causes were those sociological or other features, not reflected in the rules of the game, that restricted or promoted the formation of particular coalitions. Here lay the limits of the theory. It carried the analysis up to the point at which such social influences entered the picture and it showed how they mattered, but could say little about where they came from.

Von Neumann begins his search for a general definition of stability with the three-person, zero-sum game described earlier, in which there are three possible outcomes, each comprising a coalition of two winners against a single loser:

(4.b) None of them can be considered a solution by itself – it is the system of all three and their relationship to each other, which really constitute a solution.

(4.c) The three apportionments possess together, in particular, a certain "stability" to which we have referred so far only very sketchily. It consists in this, that any strategic course, followed by a majority of the players, will ultimately lead to one of them. Or, that no equilibrium can be found outside of these three apportionments.

(4.d) Again it is conspicuous that this "stability" is only a characteristic of all three apportionments together. Neither one possesses it alone – each one, taken by itself, could be circumvented if a different coalition pattern should spread to the necessary majority of the players.

We will now proceed to search for an exact formulation of the heuristic principles which lead us to our solution . . .

A more precise statement of the intuitive "stability" of the above system of three apportionments may be made in this form: If we had any other possible apportionment, then some group of players would be able and willing to exchange it for one of the three already offered, but within the system of given apportionments

---

[72] This function would later become the "characteristic function" in the book by von Neumann and Morgenstern.

[73] Von Neumann, "Theory of Games I", p. 13.

we cannot find a group of players who find it both desirable and possible to exchange one scheme for another . . . [74]

To extend the solution concept to the case of the general n-person game, he develops further notation and terminology. A coalition is effective for a particular valuation (later called an imputation) if, by forming a coalition, members may find it possible to get as much as the valuation offers them. Thus, it becomes possible to speak of a valuation, α *dominating* another, β, if there exists a non-empty set, S, effective for α, for which $\alpha_i > \beta_i$ for all members of S. Von Neumann discusses the dominance relation, $>$, noting its similarity to the order relation but also its particular curious properties: lack of completeness, intransitivity, the possibility that both $\alpha > \beta$ and $\beta > \alpha$. For the n-person game, the solution can then be defined as a collection of valuations, ν, such that:

(i) for, $\alpha, \beta \in \nu$, it is never the case that $\alpha > \beta$, (that is, no imputation in the solution is dominated by any other member imputation) and

(ii) for every $\alpha' \notin \nu$ there exists an $\alpha \in \nu$ for which $\alpha > \alpha'$ (that is, every imputation outside the solution is dominated by at least one imputation inside).

He proceeds to discuss the properties, in a manner quite different from that done earlier with the three-person game. He notes that the definition of a solution has not ruled out the existence of a α' where $\alpha' > \alpha$ – that is, the existence of an imputation lying outside the solution that dominates at least one of the member imputations, and therefore would be preferred by some effective coalition. His defence of the definition of solution in the face of such a possibility is most interesting.

If the solution ν, i.e., the system of valuations, is "accepted" by the players $1, \ldots n$, then it must impress upon their minds the idea that only the valuations $\beta \in \nu$ are "sound" ways of apportionment. An $\alpha \notin \nu$ with $\alpha' > \beta$, will, although preferable to β, fail to attract them, because it is "unsound". [For the three-person game, he refers here to the earlier explanation of why a player will be averse to accepting more than $1/2$ in a coalition.] The view on the "unsoundness" of α' may also be supported by the existence of an $\alpha \in \nu$ with $\alpha > \alpha'$. [that is, the mere presence in the solution of a third imputation that dominates the "dominating" nonmember, α', may be sufficient to deter players from seeking α']. All of these arguments are, of course, circular in a sense, and again dependent on the selection of ν as "standard of behaviour", i.e., as a criterion of soundness. But this sort of circularity is not unfamiliar in everyday considerations dealing with "soundness".

[74] *Ibid*, p. 14.

If the players have accepted ν as a "standard of behaviour", then it is necessary, in order to maintain their faith in ν, to be able to discredit with the help of ν any valuation not in ν. Indeed for every outside α' ($\in \nu$) there must exist an $\alpha \in \nu$ with $\alpha > \alpha'$.

... The above considerations make it even more clear that only ν in its entirety is a solution and possesses any kind of stability – but none of its elements individually. The circular character stressed [above] makes it also plausible that several solutions ν may exist for the same game – i.e., several stable "standards of behaviour" in the same factual situation. Each one of these would, of course, be stable and consistent in itself, but conflict with all others.[75]

In the same manuscript, von Neumann then devotes several pages to a graphical illustration of the solutions to the three-person, zero-sum, normalized game, which he uses to illustrate the distinction between proper and improper solutions, the first being a solution set that is finite, the latter being one that is infinite. The theme of social discrimination remains central.

The example 7.B also indicates one of the major reasons which lead to improper solutions. There one player – it happens to be 2 – is being discriminated against, for no intrinsic reason, i.e., for no reason suggested by the rules of the game itself, which are perfectly symmetrical. Yet a "stable standard of behavior", i.e., a solution ν can be built up on such a principle. This player has a – rather arbitrary – value assigned to him: $\alpha_2 = b_0$ for all valuations $(\alpha_1, \alpha_2, \alpha_3) \in \nu$. He is excluded from the competitory (sic) part of the game, which takes place between the other players exclusively -1 and 3.

This discrimination, however, need not be clearly disadvantageous to the player who is affected. It is disadvantageous if $b_0 = -1$. But we can also choose $b_0 > -1$, as long as $b_0 < 1/2$. At any rate, however, it amounts to an arbitrary segregration of one of the players from the general competitive negotiations for coalitions, an arbitrary assignment of a fixed – uncompetitive – value for this player in all valuations of the solution, and all this causes an indefiniteness of apportionment between the other players.

He closes by noting that subsequent discussions will show that there may be other causes of improper solutions, all of which "can be interpreted as expressing some arbitrary restriction on the competitive negotiations for coalitions which does nevertheless permit the definition of a 'stable standard of behavior'".[76]

The "stable set" would become the central solution concept of *The Theory of Games and Economic Behaviour*, with two-thirds of the book devoted to its exploration in games of three players and more. That exploration is

[75] *Ibid*, pp. 17–18.
[76] *Ibid*, pp. 26–27.

enormously ramified and complex, given the combinatorial complexity of certain games, but the importance of social norms in determining equilibrium outcomes remains fundamental throughout.

While von Neumann worked on this mathematics of arbitrary social restriction, it was precisely the latter that remained critical in Hungary. If the Téléki government believed that the laws of 1938 and 1939 were satisfactory in restraining Jewish participation, the Germans did not, accusing the Hungarians of not going far enough. Anxious to preserve Hungarian–German relations, in November 1940, the prime minister, Téléki, endorsed the Tripartite Pact signed by Germany, Italy, and Japan. He then visited Hitler in Vienna. The latter, at that point, was considering segregating Europe's Jews by sending them to the French colonies, all of which he discussed with Téléki, who apparently agreed that Europe should be free of the Jewish presence.[77] Having aligned itself with the Axis, Hungary was now no longer neutral. Part of its purpose here lay in its revisionist designs to regain territories lost to Yugoslavia after Trianon. By March 1941, however, Hitler had decided to invade Yugoslavia as well as Greece. Téléki conceded on the use of Hungary for passage of German troops through to Yugoslavia. This, in turn, brought a threat of reprisal from Britain. Under the intense pressure, at the beginning of April, Téléki committed suicide. The Germans attacked Yugoslavia and the Hungarians followed through, annexing their old territories, including the Délvidék, in the Yugoslavian northwest. Téléki was replaced by his foreign minister, László Bárdossy, whose tenure would show the 1938–1939 bid for stability to have been futile, and prove disastrous for the Jews of Hungary.

In Germany, the madness continued. In 1936, Bieberbach had defended *Deutsche Mathematik*, his new journal, against the threat of reduced funding by denouncing other journals: "One (*Mathematische Annalen*) is edited by a Jew". This was a reference to Otto Blumenthal. "In another (*Mathematische Zeitschrift*) there appear papers dedicated to female Jewish communists". This meant Emmy Noether. "In a third (Crelle's Journal) papers by emigrants are printed". This was Richard von Mises. "A fourth (*Quellen und Studien*) is led by a Jew and an emigrant half-breed".[78] These were Otto Toeplitz and Otto Neugebauer, respectively.

In 1939, in a public lecture in Heidelberg, Bieberbach elaborated upon his theory of psychological types in mathematics, presenting finer distinctions made by Jaensch. Thus the desirable J-type had three subdivisions, J_1,

---

[77] Braham (1981), p. 177.
[78] Mehrtens (1987), p. 223.

J_2, and J_3. The J_1-type, of which Felix Klein was typical, "does not turn the world into a problem, rather the problem comes to him out of the world . . . He is attracted by the colorful richness of reality; he is interested in coherence, in the grand scale of events; while thinking he has to see or feel the relation to reality" (p. 229). As for the J_2, of which Gauss was representative, he "does not so much long for a wealth of knowledge but rather for its meaning and range. He approaches reality with fixed values and ideals. He tries to form cognitions into a world-view. His aim of work is a perfect harmonious construction. He loves truth for its beauty" *(ibid)*. The third type was illustrated by Berlin's own Weierstrass, "It is a type for whom knowledge must have command over the world . . . The scientist and his subject matter are standing face to face like fighters struggling for power. Cognition is a struggle with a reality. In pure mathematics these are the critics, the systematists, who carve out clear rules for the control of the subject, who clarify the basic concepts and deprive them of their mysteries, who form the accumulated results into a system" (p. 230). For each of the above, the main criterion was the "inner coherence of the edifice of his thought". Each was concerned with "the natural place of things in science" (*ibid*). However, this could not be said of the unitary S-type, or *Strahltypus*:

who beams his autistic thought into reality.[79] At best he tries to recover his ideas within reality, but not as a confirmation of his thinking, rather as an *epitheton ornans* of reality. Among the great mathematicians – I emphasize the word "great" – no case of this intellectualist type can be found. Among Germans, however, juveniles and also mathematicians frequently remind one of this type. Namely, among the strangers who took up certain studies of Hilbert, there are some who belong to this intellectualist type" (p. 230).

One can only imagine how Bieberbach might have viewed von Neumann and his new, Hilbertian mathematics of social reality.

Conclusion

A significant stimulus in bringing von Neumann back to game theory was his engagement with the political tumult of the late 1930s. In the course of his correspondence with Ortvay and others about European politics, and as his family and friends were increasingly caught up in the swirl of events, he gravitated towards the consideration of rationality and social configuration. Unlike at Göttingen a decade previously, where he had been concerned with the behaviour of the chess- or poker-player, his concern

[79] *Strahl:* a "ray" or "beam".

was now with the rationality of the social actor or unit, yet with the same relatively simple conception of psychology as before. His concern became that of understanding social coalitions, and a key element of his theory – the dependence of stable equilibria upon social norms – bears a striking resemblance to what was then taking place in Hungary and other countries, where seismic social shifts were being brought about by changing attitudes towards some groups, and the Jews in particular, codified in legislation. Building upon this central idea, von Neumann launched into the creation of a new social mathematics, game theory, providing analytical insight into the exercise of power and social discrimination. The fact that von Neumann did this explicitly emphasising his use of the formalist mathematics of the Hilbert school – in the late 1930's still the object of Bieberbach's Nazi vitriol – lent potent symbolism to his effort. As we shall see, von Neumann's approach here involved what he termed a form of creative opportunism: a distinctly modern, and yet thoroughly human dialectics of discovery, creation, and coercion. In this manner, he ensured the mutual adjustment of the mathematics and the world it purported to describe.

TEN

Ars Combinatoria

Creating the *Theory of Games*

Introduction

From the moment he began to consider leaving Vienna, Morgenstern had his eye on the Institute for Advanced Study. In 1935, in the belief that economic research would have social benefits, Institute director Flexner had added a School of Economics and Politics, bringing in economists Edward Mead Earle, David Mitrany, and Winfield Riefler. In a community that privileged research in pure mathematics and mathematical physics, this overture to social science, which many considered not even good applied mathematics, provoked outrage. Indeed, it ultimately led to Flexner's ouster and replacement in 1939 by Frank Aydelotte.

Morgenstern was probably unaware of these internal troubles when, in June 1937, in anticipation of his visit to the United States the following year as Carnegie Professor, he sent Flexner a copy of *The Limits of Economics*, and said that he looked forward to meeting him and Riefler.[1] With Wagemann's arrival in Vienna occurring in the interim, Morgenstern never returned to Austria and he accepted a position in the nearby Princeton Economics Department.[2]

Once there, he found himself in a situation not dissimilar to the one he had left behind: amongst economists, of whom he was critical, looking elsewhere for stimulus. By November 1938, unsatisfied with his lecturer's salary at Princeton, he was viewing the Institute increasingly keenly – "the

---

[1] OM to Abraham Flexner, June 2, 1937, Archives, I.A.S., General 'Mi -Mz'.

[2] Shortly after arriving at Princeton, Morgenstern had written to Bertil Ohlin in Stockholm. During his tour as Visiting Carnegie Professor, he said, he had had offers from several universities, but he was quite happy to have chosen Princeton. Not only was it close to New York, but they had already shown an interest in him prior to the political tumult in Austria, something which gave it "a very nice psychological side" (OM to Bertil Ohlin, July 21, 1938, OMDU, Box 8, Corresp. Chronological 1925–1938).

double income etc. and no worries" – and it was in this context that he ingratiated himself with Weyl, von Neumann, and others.[3] He gave a talk at the Nassau Club early the next year, which drew Flexner, von Neumann, and Bohr, and where the latter spoke about the difficulty of the social sciences, in which "physics was taken too much as a model for our philosophy of science, physics being much too simple for that. The social sciences are so much more complicated (this incidentally is also what Planck thinks)".[4] Morgenstern now wrote about being on good terms with several of the Institute people, including "John von Neumann".[5]

Their acquaintance growing steadily in the interim, by July 1941 Morgenstern wrote: "The Hicks essay has been published, 33 pages. The Maxims are still lying around and have to be worked on. In the meantime, have started a treatise with Johnny about games, minimax, bilateral monopoly, and duopoly. What fun. We will probably get it finished before September, and the treatise will surely be of far reaching significance, especially because it touches upon the foundation of the subjective theory of value (i.e. Robinson [Crusoe] = max[imum] problem; Individual in social science = Min Max problem!) I want to publish that in the J.P.E. After close study of his manuscript, I am rewriting the Introduction"[6]

This entry contains many of the important elements of their collaboration: the Hicks essay, the Maxims paper, von Neumann's "manuscript". What were they, and how did they relate to what became the *Theory of Games?*

The Collaboration

When Morgenstern completed his review of Hicks's 1939 *Value and Capital* in mid-1941, he had already been several months under von Neumann's

---

[3] See Diary, Nov. 18, 1938, OMDU. "Perhaps they want to invite me there. That would be very pleasant.... With that everything would be decided ... And I could work on whatever I wanted. Riefler told me before his departure that Flexner really liked me and that I should visit Flexner again in the next few days" (*ibid*). The previous month, he had, in a manner of speaking, tried to "move in", himself, with Flexner agreeing that he could have an office for his own use. However, when Morgenstern indicated that it would be used by his assistant to store "gradually accumulating statistical material", the project was nipped in the bud. (See Letters, Morgenstern to Flexner, I.A.S. Archives, General 'Mi-Mz').

[4] Diary Feb. 15, 1939, OMDU.

[5] "If I only had a position at the Institute. Perhaps in time something can be done. I am on good terms with Walt Stewart, amongst others. Now also with Weyl, John von Neumann, Lowe etc. But that doesn't mean I will be able to get something", Feb. 15, 1939, OMDU.

[6] Diary July 12, 1941, OMDU.

spell, and his harsh critique of the book reflected the mathematician's influence: among "the most unreadable works...on economic theory" (1941a, p. 364). Hicks is criticised for continuing to assume that the counting of equations and unknowns guarantees the determinateness of a linear system, and for ignoring Wald's (1935) and von Neumann's (1937) existence proofs, both of which have overcome this difficulty. The "indiscriminate use of the word 'equilibrium'", writes Morgenstern, "is also objectionable, if not often entirely misleading.... If the respective equilibrium is not qualified further as being either stable, labile or indifferent, the whole statement hangs in the air, adding to the vagueness of the usual procedure.... Some of these equilibrium conditions need not at all conform to the ordinary simple maximums or minimums. They are more likely of the so-called 'minimax' type, the analysis of which requires instruments of great subtlety".[7]

Keynes is once again impugned for his "casual use of notions about expectations" (*ibid*, pp. 381–82), and it is probably to Hayek that Morgenstern is referring when he condemns the unclear use of the concept of "consistency of plans" as a condition of equilibrium. Hicks, too, had failed to give "consistency" any concrete meaning: "It is obvious that...it has not been decided whether there exists only one single grouping of plans which is compatible with equilibrium or whether there are many possible ones, each of which would be 'consistent'. In order to decide a problem of this kind it is, naturally, necessary to be more specific about the character of the plans or, in other words, to define them more specifically. The problems involved are of quite exceptional difficulty and resemble closely those of the theory of games" [*ibid*, p. 380].

At this stage Morgenstern was poised between Menger and von Neumann, actively trying to reconcile his own concerns with the former's work on social theory and the latter's new ideas on games.[8] His thinking was still dominated by a familiar range of concerns: interdependence, knowledge,

---

[7] 1941a, p. 374–75, footnote. Morgenstern had offered several criticisms of Hicks in a discursive, nonmathematical class he taught on "Advanced Problems in Economic Theory" in January–June 1939, notes from which were kindly provided to me by Dr. Burton Hallowell of East Orleans, Massachussetts, who was a student in that class. There, Morgenstern mentions the neglect of Wald by Hicks, whose *Value and Capital* was the most recent item on the reading list. In October 1940, Morgenstern discussed Hicks with von Neumann, and, several months later, wrote of "a long discussion with Johnny to whom I read parts of my Hicks manuscript, all the parts that have to do with equations. He was in complete agreement and made 1–2 suggestions. Very stimulating" (Diary, Feb. 12, 1941, OMDU).

[8] "It is interesting to compare him to Menger. Menger takes himself much more seriously. Neumann practically always excuses himself when he has to say that he has worked on something" (Diary, Sept. 1, 1940, OMDU).

and time.[9] Reinforced by his encounter with von Neumann, Morgenstern's grip on the Viennese questions had tightened, yet the links had not been worked out. What von Neumann was saying seemed to be important, yet what it meant exactly for foresight and dynamics was unclear. And it was all as difficult as it was fascinating:

Yesterday at Johnny's, who gave me a long lecture about quantum logic. Quick as always, often assuming more than I could contribute, but still to such a point that I now know what is happening. And that disturbs me extraordinarily, because it means that far-reaching epistemological conclusions must be drawn. The beautiful, comfortable division of the sciences into logical and empirical sciences falls. . . . Everything comes from quantum mechanics, and it again leads back to the foundations debate. . . . Since I had read [d'Abro's *Decline of Mechanism*] I was in a better position to follow Johnny yesterday. . . . I have the suspicion that it would also be necessary to introduce new thought forms in economics. For example, a logic of wishing (it leads back to Menger). The bad thing for me is that I see all this, and feel that it is necessary, but suspect darkly that it escapes me. . . . All this probably because I never had the necessary wide mathematical training. Sad.[10]

With the Hicks critique completed, Morgenstern turned to the "Maxims", with which he had been preoccupied on and off since leaving Vienna.[11] Completed in May 1941 under von Neumann's stimulus, "Quantitative Implications of Maxims of Behavior" (1941b) is a particularly interesting document insofar as it captures Morgenstern at a crossroads. Unlike virtually everything he had written up to then, the paper moves forward from critique to construction; it is entirely nontechnical; it draws directly on Menger; and it makes direct appeals to von Neumann for clarification. Indeed, it is fair to say that it is more truly representative of Morgenstern's own thinking than is the *Theory of Games*.

[9] The review closes with the suggestion that it would be better "to start building dynamic theory on the basis of the observations of economic fluctuations rather than to try to cramp them into a helpless and vague dynamic theory whose empirical foundations lie in limbo" (p. 393).

[10] Diary, Jan. 22, 1941, OMDU. A month earlier: "Saturday went to a lecture by Johnny. Again a ray of light. But at the same time it made me sad, because I can't handle such questions. At the same time, I know even more fundamental questions" (Diary, Dec. 10, 1940, OMDU).

[11] Morgenstern's concern with maxims had preceded his meeting with von Neumann. In December 1938, he wrote: "I keep coming back to the maxims. I must think about this keenly. Also about prediction" (OMDU, Dec. 17, 1938). For the next two and a half years, he referred to the issue frequently. Finally, in May 1941, he wrote: "Had a discussion with Johnny. . . . I should absolutely write the Maxims. He wants to read the manuscript very closely. I will start with it tomorrow, and look forward to it. . . . I have a pile of notes and want to write a short treatise, perhaps 5000 words" (*ibid*, May 17, 1941).

The issue, Morgenstern says at the outset, is the theory of *society*, economics being a branch of same. In an abundant appropriation of an unnamed Menger, the principles governing behaviour are to be called "maxims", and we are "concerned only with the formal aspects of these principles" (p. 1) and not with the larger philosophical problems connected with them.[12] Maxims may be classed into two types: restricted and unrestricted. The former can be followed regardless of whether or not others do the same (for example, "Thou shalt not steal"), whereas observing restricted maxims *will* depend on whether or not others do the same. For example, the maxim to "withdraw bank deposits when a danger threatens" is practicable only if some, and not all, follow it. Its feasibility will depend on quantitative limits such as knowing how others perceive and will react to the same "danger".

Connected with the concept of maxims is the Austrian notion of 'subjective rationality': the pursuit of aims given the individual's knowledge of facts and the intelligence with which they interpret those facts. A certain minimum level of subjective rationality of individuals is assumed, but the degree is clearly subject to change. This notion matters in the context of restricted maxims: "The individual will have to make an appraisal in his mind as to what the consequences of his acts will be and whether the consequences are still compatible with the aims which gave rise to the maxim he follows. In the case of unrestricted maxims the answer is simple because there the subjective rationality will not be impaired by the behavior of others according to the same behavior" (p. 7).

In this framework, many of the topics previously raised by Morgenstern come up naturally. The knowledge relevant to the adoption of maxims may be called 'foresight': in the case of bank withdrawals, this would refer to the knowledge of where one stood in the queue, so to speak, for the bank. This will depend on the behaviour of others in respect of the maxim and, in turn, on the particular individual's understanding of that behaviour. Clearly, "very great requirements are made as far as the intelligence of the acting individuals is concerned if they are supposed to understand precisely where the restrictions of maxims set in" (p. 8).

He then devotes a large proportion of the paper to considering the policy implications of it all, and here the politics of Mises and Hayek are once again targeted. Collective regulation may be intended to substitute for

---

[12] As in Menger (1934), the traditional philosophical questions in ethics concerning, for example, the nature of the good and the essence of evil, are deliberately ignored. Also, like Menger, Morgenstern invokes Kant, who took "one of the most famous steps in this direction . . . when he formulated his categorical imperative" (p. 1).

individual subjective rationality. For example, in the case of the withdrawal of bank deposits, the government may declare a moratorium and thereby bring about the protection of deposits that all individuals would have liked to secure voluntarily had they had "sufficient information and assurance that their own action would be followed by appropriate behavior of the others concerned" (p. 16). This regulation concerning upholding restricted maxims suggests a type of *positive interventionism*, which is "not exposed to any of the criticisms which are voiced against every intervention by the adherents of a purely laisser-faire attitude" for it "leaves the maxim by which people are moved entirely untouched": it is "merely a substitute for the corrective which superior information and intelligence would offer" (p. 17).

Finally, he turns to the issue of the compatibility and coexistence of maxims of different kinds, again following Menger. For example, are there policies that would allow maximum fulfillment of a given set of maxims? He says the coexistence of unrestricted maxims requires "very interesting methods of a mathematical nature: It is sufficient here to refer to the writings of Karl Menger [*Moral*, etc.] where a fairly exhaustive first treatment has been offered for the first time on the basis of exact methods. Space does not permit going into the matter here which also lies outside the lines of this paper which is concerned with somewhat wider aspects of the question of compatibility" (p. 20).

The key difference, of course, between Menger's and Morgenstern's analysis is that the former deals with ethical maxims that are unrestricted in Morgenstern's sense. Thus, Menger is concerned with the implications for compatibility *once the choices have been made*. Morgenstern, on the other hand, faces the added complication of considering choices that are *a priori interdependent*. Whereas Menger was interested in how smokers and nonsmokers might, so to speak, get along, Morgenstern wanted to know how strategic factors might influence one's choice to smoke or not. The restrictiveness of certain maxims meant that the individual's problem of choice was now much more complicated than traditionally suggested in economics. It "is not only a matter of making preferences between aims in a purely qualitative manner in order to establish a line of action of maximum subjective rationality": "the further problem of compatibility has to be solved by each single individual where such maxims occur. If the individuals are not aware of the existence of this problem then we clearly get a different kind of behavior than if they are aware of it and accordingly the structure of theory would have to be modified. It does not need many words to make clear that this opens up new fields of investigations" (p. 21).

Unfortunately, the character of the mathematical problems involved is such as to make it exceedingly difficult even when the use of the most advanced mathematics is envisaged; it may even be argued that the necessary mathematics do not, as yet, exist. Consequently, it will be necessary to leave it to the professional mathematicians to work out solutions and it is not exaggerated to think that they will have a very hard nut to crack. Reference should be made to the studies on the theory of games by J. von Neumann which was [sic] recently extended in a series of lectures. However, the cases treated there are as yet of a restricted nature and do not take the problems into consideration which have been described above. It is, even, not certain that they can be translated into the schemes devised in his latest researches (p. 22).

He ensured that von Neumann got the message: "[T]he Maxims are at Johnny's for eight days, but he has only read a part up to now. He questions whether one can make such a sharp division of the two classes. I believe so. Have checked up on everything: Kant, Menger, Schlick and M. Weber etc. and found nothing. If I'm right, then it is important. Even if the division cannot remain, the quantitative implications which are important to me stay. On Friday, Johnny wants to discuss it all with me".[13]

The third element in the collaboration is von Neumann's "manuscript", the first part of which was "Theory of Games I, General Foundations", completed by October 1940, after von Neumann's summer stay in Seattle.[14] Morgenstern appears not to have turned his full attention to this until July 1941, nine months after it was written, having been occupied with the Hicks review and his "Maxims" paper in the interim.[15] When he did turn to it, he wrote an introduction for economists to its first part, the "essay

---

[13] Diary, June 2, 1941, OMDU. Hence the modifications to Morgenstern's Maxims paper. Pages 6a, b, c, d, 21a, and others unnumbered, appear to have been inserted following comments by von Neumann. For example, on the division of the two classes into restricted and unrestricted, a less plodding, more tightly argued two-page insertion questions, following von Neumann, "whether the distinction is at all a clear cut one" (p. 6a). Some unrestricted maxims may be the reciprocal of restricted ones. For example, acceptance of the unrestricted maxim "I will not steal" may be the result of the fact that the maxim "I will steal" would be a restricted one since its universal observance would defeat it. The Maxims paper remained unpublished. A note in Morgenstern's handwriting, jotted in the margins years later, reads "both JvN and Gödel urged me to work this out & publish it: (never done!)" (*ibid*, p. 1).

[14] The manuscript was in two parts: "Theory of Games I", dated October 1940, mentioned in the previous chapter, and "Theory of Games II", dated January 1941.

[15] In April 1940, Morgenstern had written: "We spent nearly four hours in discussion. Maxims of behavior, (I understand perfectly what it's about and how difficult this is) about games, and about foundational questions in mathematics" (Diary, Apr. 5, 1940, OMDU). A few days later, he reported that von Neumann had read his Time Moment paper, which he found to be interesting and full of "difficult problems" (Diary, Apr. 9, 1940, OMDU). In August, when von Neumann returned from Seattle, Morgenstern wrote: "Neumann came for three hours and we had a discussion. First, my [maxims] essay, with

with Johnny", which he mentions often and considered to be of great importance, and which von Neumann read at intervals, suggesting revisions.[16] By September, they agreed, at Morgenstern's suggestion, that it should become a book for Princeton University Press: "Will perhaps be 100 pages long and full of dynamite. The consequences are becoming clearer and clearer to me. I do not exaggerate when I say to myself that it is something that goes into the deepest depths of the theory, overturns many things and introduces many new and positive things. It will be a difficult but totally clear book. Johnny is a fantastic scientist!"[17]

Von Neumann, in the meantime, ploughed ahead with the second part of his "manuscript", "Theory of Games II", completed in January 1941. Here, the theory is extended to cover non–zero-sum situations and, adopting a more formal set-theoretic notation, he proves some short theorems on stability and discusses decomposition of games. Throughout, the presentation

which he was generally in agreement (I only gave him a general outline), and then 4-person games, where he has solved nearly all the questions. There is only one catch. These combinatorical things are complicated and not very clear. He hopes that a stimulus from me will make an arithmetic treatment possible, which would give one fantastic tools" (*ibid*, Aug. 20, 1940). Two months later, in October 1940, von Neumann completed the first part of his manuscript on "Theory of Games". Nine months later, in July 1941, having completed the Hicks review, Morgenstern speaks of rewriting the introduction to the manuscript, having given it "close study" (*ibid*, July 12, 1941). In the interim, in November 1940, Morgenstern reported his being "finally able to go one of his lectures about games. Most of it I already knew (not all proofs). Yesterday I showed him the contract curve. That ties into his games... He wants to think about all this" (Nov. 4, 1940).

[16] "The essay with Johnny is preoccupying me a lot. I hope he is going to stick to it. There are going to be great representational difficulties, because the mathematics that is being used is very disparate (set theory, vector theory, higher algebra (bilinear forms etc.)). I am absolutely convinced that it is a matter of great importance. It must be clearly shown what a minimax problem is. Johnny thinks one will elicit 99% stupid objections, but one just has to put up with it. The essay will probably only be a beginning. It's also going to start something in the whole discussion about duopoly etc.; and also as far as consistency of plans is concerned, but it will take a lot of time until the present, usual prattle about 'dynamic theory' stops" (Diary, July 12, 1941, OMDU). Again, "The essay is advancing. I have finished the presentation of game theory and will write the introduction anew and enlarge it significantly" (*ibid*, July 30, 1941). And, "Today gave Johnny my manuscript, about 50 pages... He will write a commentary which I will integrate etc.... We'll certainly have a hundred pages. What we will do then remains to be seen. Perhaps in two parts in the JPE" (*ibid*, Aug. 4, 1941). "[Johnny] wants elaborations. He is in agreement with all general comments but thinks a few things would be dealt with in greater detail because, while we would understand, others perhaps not. He also liked the presentation of game theory, and wants a few additions. I have already written a number of things, made a new comment (on the supposed inapplicability of physics and its methods as a model)" (*ibid*, Aug. 14, 1941).
[17] Diary, Sept. 22, 1941, OMDU.

is dense and rigorous, without any discussion of applications. Once again, this was delivered to Morgenstern after nine months' gestation, in October 1941.[18] The second part of von Neumann's manuscript was integrated into the book after this point, and it quickly became clear that it would be much larger than expected. Morgenstern still hoped that the theory would have something to say about dynamics, but as the collaboration wore on, it became evident that it would not.[19]

At one point in 1942, von Neumann rapidly constructed his axiomatisation of measurable utility and, by December, they were discussing possible titles, the penultimate of which was "Theory of Games – and its applications to Economics & Sociology".[20] The collaboration soon found itself in competition with von Neumann's wartime consulting, as Klari von Neumann later remembered:

Johnny had started on his travels. Almost continuous: from Princeton to Boston, from Boston to Washington, from Washington to New York – a short stop in Princeton, then to Aberdeen, Maryland, the Army Proving Grounds – back to Washington, maybe one night at home, then starting on the rounds again, not necessarily in the same order, but up and down the Eastern seaboard with occasional forays further inland – not to the West yet- that came later.

In between the little time he spent at home, he was trying feverishly to finish writing the mammoth book that he and Oscar Morgenstern had started in late 1940 on the "Theory of Games". This again was an operation of a really impressive mental gymnastics and flexibility. Johnny would get home in the evening after having zig-zagged through a number of meetings up and down the Coast. As soon as he got in, he called Oscar and then they would spend the better half of the night

---

[18] Morgenstern wrote: "Thursday talked twice with Johnny at his place and he showed me the case (3,1) according to the latest development.... Yesterday in the afternoon again, he gave me MS II (his) and he discussed it through with me yesterday. It will give me a lot of work but it is not insurmountable. It already covers the [non-zero sum] case". And he continues: "Aydelotte wrote to him and also said how important he finds the whole matter. In the coming week, A. will tell the board meeting of the Institute about it! (Perhaps he'll get the idea after all that I should be at the Institute!?)" (Diary, Oct. 12, 1941, OMDU).

[19] "One of my points for [Theory of Games II] we have not thought about yet: what the difference is between static and dynamic games" (Diary, Oct. 31, 1941, OMDU). This issue is discussed in the book (von Neumann and Morgenstern [1947], p. 44).

[20] "Today at Johnny's: axiomatization of measurable utility together with the numbers. It developed slowly, more and more quickly, and at the end, after two hours (!) it was nearly completely finished. It gave me great satisfaction, and moved me so much that afterwards I could not think about anything else.... We ... will not print the whole proof. There are still some serious questions there, especially the question of risk and pleasure of gambling. But that doesn't belong in the book; perhaps later" (Diary, Apr. 14, 1942, OMDU). The proposed title was 'Theory of Rational Behavior and of Exchange'. By the end of October, it had become 'Principles of Rational Economics'. Then in December 1942, it changed, for a second last time, to the title referring to 'Economics & Sociology'.

writing the book. Next morning, fresh as a daisy, Johnny was ready for his next trip – this went on for nearly two years, with continuous interruptions of one kind or another. Sometimes they could not get together for a couple of weeks, but the moment Johnny got back, he was ready to pick up right where they stopped, as if nothing had happened since the last session.[21]

In January 1943, there began a particularly long delay when von Neumann headed off to England for a six-month stint as mathematical advisor to the Navy. The 600-page manuscript went to the publisher in April.

Between Empirics and Aesthetics: The Psychology of Mathematical Creativity

Insight into the way in which von Neumann created a mathematics of games can be gained from "The Mathematician", a lecture delivered shortly after the war and published in 1947. It was concerned with the nature of mathematical creativity and dealt with the standard issues in the debates on the foundations of mathematics, including many of those already treated, albeit in a very different manner, by Bieberbach.

Von Neumann sought to elaborate upon the relationship between intuitively given subjects and the "multiple phenomenon which is mathematics" (1947, p. 2). On the question of whether mathematics was an empirical subject, he said its best inspirations indeed came from the natural world, the two monumental examples here being Euclidean geometry, which was clearly of empirical inspiration, and the calculus, which developed as a result of the explicitly physical interests of Kepler, Newton, and Leibniz. The notion that mathematics was some sort of pure realm, untouched by empirical experience, he said, was untenable.

Empirical considerations had even helped provide a resolution of sorts to the twentieth century debates on the foundations of mathematics, regarding the fact that the criteria of mathematical rigour had varied radically from one quarter to the next. After Brouwer, some accepted the intuitionist critique, but continued on as if nothing had happened. Others, such as Hilbert, turned to an intuitionistic justification of the consistency of pre-intuitionistic mathematics, which would have been fine, said von Neumann, had Gödel not shown Hilbert's aim to be hopeless. But, even after Gödel, most continued to use classical mathematics anyway. After all, "it was producing results which were both elegant and useful, and even though one could never again be absolutely certain of its reliability, it stood on at

---

[21] Von Neumann-Eckhart, unpublished draft autobiography, pp. 26–27, KEMNW.

least as sound a foundation as, for example, the existence of the electron" (*op cit*, p. 6). Von Neumann himself was one such mathematician, and although Gödel's results saw him abandon work in theoretical logic, he persisted with the use of classical mathematics – even mathematics of whose foundations one could not be sure – given its demonstrated utility in accounting for empirical phenomena.[22]

The *Theory of Games* was an example of such work by von Neumann. Unlike the 1928 paper, the theory is now explicitly presented as a piece of modern, Hilbertian axiomatic mathematics. Using the concepts of sets and partitions (which are presented in the form of nothing less than an introductory lesson) and invoking Hilbert's *Grundlagen der Geometrie*, Chapter II provides an axiomatic description of the game as a formal mathematical entity, complete with discussion of the completeness, freedom from contradiction, and independence of the axioms.[23] The treatment is explicitly "modern" in the sense that, although the axioms are stimulated by the common-sense features of games, the intuitive interpretation is allowed to recede into the background and the theory pursued in a spirit of relative abstraction. While the mathematics is being followed through, the empirical is held at arm's length and everyday terms are introduced cautiously, in inverted commas. Hence, for example, throughout the book: "class", "discrimination", and "exploitation". Only during periodical returns to the heuristics is the vocabulary of the everyday reinvoked freely. According to von Neumann, the axiomatics "follows the classical lines of obtaining an exact formulation for intuitively – empirically – given ideas" (von Neumann and Morgenstern [1947], p. 76), and shows that it "is possible to describe and discuss mathematically human actions in which the main emphasis lies on the psychological side" (*ibid*, pp. 76–77). This was in the same spirit as Hilbert's earlier axiomatic forays into such other "human" realms as insurance mathematics and psychophysics, and it was also a response to the doubts expressed by Karl Menger in his

---

[22] In a talk given in the mid-1950s, he asked why, after Gödel, shouldn't less-than-completely rigorous mathematics continue to be used anyway? After all it had served theoretical physics perfectly well. "This [view] may sound odd, as well as a bad debasement of standards, but it was believed in by a large group of people for whom I have some sympathy, for I'm one of them" (1954a, p. 481).

[23] Of these, the first and third, but not the second, hold. The reason why completeness does not hold is because, as the authors state on p. 76 and in footnote 3, the axioms are intended to describe, not a unique object, as in Euclidean geometry, but a class of objects, as in rational mechanics or group theory – that is, there are many different games that fulfill the axioms.

paper on the St. Petersburg Paradox, as to the possibility of constructing a mathematics of choice behaviour.

In *The Mathematician*, von Neumann insisted that mathematics had a "double face"; it presented a certain duplicity. In addition to depending on empirical inspiration, it had large sections, amongst them topology, set theory and real function theory, that were "very much self-contained and aesthetical and free (or nearly free) of empirical connections" (1947, p. 7).

As to the criteria for success in empirical sciences and in mathematics, there were strong similarities. Modern physics classified and correlated phenomena; it didn't "explain" them. Although the impetus for physics lay with the needs of experimental physics, the criteria of theoretical success were, to a great extent, aesthetic: was the theory simple and yet able to cover many phenomena? In mathematics, such aesthetic considerations, he said, were absolutely predominant. A mathematical theory was expected to describe and classify disparate cases and to be elegant in its structural makeup. What did one look for in mathematical practice?

Ease in stating the problem, great difficulty in getting hold of it and in all attempts at approaching it, then again some very surprising twist by which the approach, or some part of the approach, becomes easy, etc. Also, if the deductions are lengthy or complicated, there should be some general principle involved, which 'explains' the complications and detours, reduces the apparent arbitrariness to a few simple guiding motivations, etc. These criteria are clearly those of any creative art, and the existence of some underlying empirical, worldly motif in the background – often in a very remote background – overgrown by aestheticizing developments and followed into a multitude of labyrinthine variants – all this is much more akin to the atmosphere of art pure and simple than to that of the empirical sciences (p. 9).

In its elaborate interweaving of analytical, empirical, and aesthetic considerations, the *Theory of Games* exemplifies perfectly this view of mathematics and provides great insight into von Neumann's particular type of creativity. The chapters of the book, in their very unwieldiness, show a definite unfolding of ideas, an intelligence at work. At times, one feels that von Neumann is justifying the theory as much for himself as for the reader.

Following the introduction and presentation of the axiomatic description of the concept of the game, the rest of the book explores the particularities of games of different size. Chapter III presents the zero-sum, two-person game, and the minimax theorem is proved, using, not von Neumann's earlier proof, but a modification of the elementary 1938 proof by Ville, based on the theory of convex sets. An application to a simplified two-person poker shows the game's solution to correspond with the empirically familiar ploy of "bluffing".[24] From here on, von Neumann systematically goes through the zero-sum game for three, four, and more players, exploring their combinatorial possibilities for coalition-formation and compensations (side payments). Each game is described in terms of its characteristic function, which shows the maximal payoff available to each possible coalition of the game, assuming that the coalition plays minimax against its complement and that utility is transferable among players. In Chapter IX, the concept of strategic equivalence is introduced to show how the move from the zero-sum restriction to a constant sum retains the basic features of the game, thus allowing it to be solved by the same means. In the eleventh chapter, von Neumann drops the zero (or constant) sum restriction, moving to the "general game".

We have already seen how von Neumann related the notion of social "discrimination" to the analysis of the zero-sum, three-person game. When it comes to the zero-sum, four-person game (Chapter VII), he enters an even more subtle discussion linking mathematical features and social organisation. The characteristic function for the generic four-person game is given by:

$$
\begin{aligned}
v(S) &= 0 &&\text{when S has} &&0 \text{ elements} \\
&= -1 &&&&1 \text{ elt.} \\
&= 1 &&&&3 \text{ elts.} \\
&= 0 &&&&4 \text{ elts.}
\end{aligned}
$$

And
$$
\begin{aligned}
v((1,4)) &= 2x_1 & v((2,3)) &= 2x_1 \\
v((2,4)) &= 2x_2 & v((1,3)) &= 2x_2 \\
v((3,4)) &= 2x_3 & v((1,2)) &= 2x_3 \\
\end{aligned}
$$
$$
\text{with} -1 \le x_1, x_2, x_3 \le 1
$$

Showing how each possible game corresponds to a point in the cube the coordinates of which are described by the variables x_1, x_2, x_3, von Neumann

---

[24]   Ville's proof is in J. Ville (1938). The same chapter contains a significant moment, when Morgenstern's 1928 example of Moriarty's pursuit of Sherlock Holmes is resurrected and shown, this time, to be reducible to a two-person, zero-sum game, similar to matching pennies. Depending on the values assigned to the matrix, the game has a solution in mixed strategies, so that in the example used by von Neumann and Morgenstern, the minimax solution has Moriarty continuing on to Dover in his pursuit of Holmes, with a probability of sixty percent, the same probability with which Holmes should descend from the train at the intermediate stop. Moriarty, used in 1928 to illustrate the impossibility of economic prediction and, in 1935, the impossibility of general equilibrium, has come a long way indeed . . . In a footnote, Morgenstern quietly rejects the pessimism of his earlier paper.

(p. 314) considers the particular game in which $(x_1, x_2, x_3) = (0, 0, 0)$ – that is, any majority coalition of three players wins, but the formation of coalitions of two members simply leads to a tie, with no payment being made. The characteristic function is thus:

$$
\begin{aligned}
v(S) &= 0 && \text{when S has} && 0 \text{ elements} \\
&= -1 && && 1 \text{ elt.} \\
&= 0 && && 2 \text{ elts.} \\
&= 1 && && 3 \text{ elts.} \\
&= 0 && && 4 \text{ elts.}
\end{aligned}
$$

Providing heuristic motivation, von Neumann says that, in the formation of the coalition of three, it is likely that an alliance of two will form first, followed by their searching for a third member (p. 314). In the division of the payoff of 1 between the three members, there are two possibilities: $(1/3, 1/3, 1/3)$ or $(1/3 + \varepsilon, 1/3 + \varepsilon, 1/3 - 2\varepsilon)$, with $\varepsilon > 0$. In the first case, the third to enter to coalition is treated equally. In the second, he is not. Convincing arguments could be construed for either possibility, and von Neumann considers both cases. Taking the second (asymmetrical) alternative first, one extreme solution is given by:

$$\alpha' = (1/2, 1/2, 0, -1)$$

(and imputations given by the permutation of these players)

$$\alpha'' = (0, 0, 0, 0)$$

Taking the first (symmetrical) alternative, the resulting imputations are:

$$\alpha''' = (1/3, 1/3, 1/3, -1) \quad \text{(and permutations etc.).}$$

Do the α''' constitute a solution in the technical sense of the term? No, von Neumann demonstrates: α''' do not respect the condition that all imputations outside the solution be dominated by some member of it. What needs to be added to the α''', in order to form a solution, is

$$\alpha^{IV} = (1/3, 1/3, -1/3, -1/3)$$

Thus the solution becomes:

$$
\begin{aligned}
\alpha''' &= (1/3, 1/3, 1/3, -1) \quad \text{(and permutations etc.)} \\
\alpha^{IV} &= (1/3, 1/3, -1/3, -1/3) \qquad (")
\end{aligned}
$$

Von Neumann says that if heuristic explanations are required, the added set "seems to be some kind of compromise between a part (two members) of a possible victorious coalition and the other two players" (p. 318). We do not attempt to find a full heuristic interpretation of these solutions, he concedes, as "it may well be that this part of the exact theory is already beyond such possibilities"(*ibid*).

· Note the subtle and opportunistic manner in which von Neumann argues, alternating back and forth between the empirical heuristics and the mathematical formalism: Heuristic consideration of the game leads to the suggestion of imputations. These, however, are shown to be (mathematically) insufficient to constitute a solution, which leads to his addition of imputations technically necessary to complete it. Keen to retain the solution, which, as a formal entity, has power for him, von Neumann feels compelled to provide heuristic motivation for the additional imputations. Thus it is now the *mathematics* that generates the plausibility considerations. Although he acknowledges that he is not too convinced of his explanation, that it may have been "left behind", so to speak, by the mathematics, he falls back on the defence that this free flight of the mathematics beyond plausibility arguments is "a well-known occurrence in mathematical-physical theories, even if they originate in heuristic considerations" (p. 318, n. 1). He places his faith in the formalism or in the heuristics, according to what suits him. This is what is meant by respecting the autonomy of the mathematics in "modern" fashion, and it seems to correspond to what Ingrao and Israel (1991) describe as letting the mathematics function as analogy.[25]

In this particular game, therefore, the two solutions are:

$$(1) \; \alpha' = (1/2, 1/2, 0, -1) \text{ (and permutations etc.)}$$
$$\alpha'' = (0, 0, 0, 0)$$

and:

$$(2) \; \alpha''' = (1/3, 1/3, 1/3, -1) \text{ (and permutations etc.)}$$
$$\alpha^{IV} = (1/3, 1/3, -1/3, -1/3) \quad (")$$

The two solutions correspond to different ways of treating the third member of the winning coalition, and "therefore seem to correspond to two perfectly normal principles of social organisation" (p. 318). Probing further, he says that the first solution can be shown to be surrounded by "peculiar heuristic phenomena", not discernible in the original discussion, yet suggestive of the "possibilities and interpretations" of the theory. He isolates a solution

[25] *Op cit*, pp. 181ff.

that is closely related to α' and α'', and that treats players 1, 2, and 3 symmetrically, but holds player 4 in a special position (p. 319):

$$\beta^{I} = (1/2, 1/2, 0, -1) \text{ (and imputations given by}$$
$$\text{permutations of players } 1, 2, 3)$$
$$\beta^{II} = (1/2, 1/2, -1, 0) \qquad (")$$
$$\beta^{III} = (1/2, 0, -1, 1/2) \qquad (") \qquad \text{and}$$
$$\beta^{IV} = (0, 0, 0, 0)$$

An asymmetric solution that contains some of the $(1/2, 1/2, 0, -1)$ (and permutations) is:

$$\beta^{I} = (1/2, 1/2, 0, -1) \text{ (and permutations of players } 1, 2, 3)$$
$$\beta^{II} = (1/2, 1/2, -1, 0) \qquad (")$$
$$\beta^{IV} = (0, 0, 0, 0)$$
$$\text{and} \quad \beta^{V} = (1/2, 0, -1/2, 0)$$

That is, β^{III} in this case has been replaced by β^{V}, which is to say that player 4 can no longer be one of the first two members of the winning coalition. According to von Neumann, the first three players – 1, 2, and 3 – form a "privileged group": no one from outside them can enter the "first" coalition of two, but, between them, the wrangling over which two will get the winnings continues to take place. Note, he says, that the third member may indeed be completely defeated, as in β^{II}, but "only by a majority of his 'class' who form the 'first' coalition and who may admit the 'underprivileged' player 4 to the third membership of the 'final' coalition, to which he is eligible".[26]

The reader will note that this describes a perfectly possible form of social organization. This form is discriminatory to be sure, although not in the simple way of "discriminatory" solutions of the three-person game. *It describes a more complex and a more delicate type of social inter-relation, due to the solution rather than to the game itself.* One may think it somewhat arbitrary, but since we are considering a "society" of very small size, all possible standards of behavior must be adjusted rather precisely and delicately to the narrowness of its possibilities (p. 320, emphasis added).

Once again, von Neumann's to-ing and fro-ing between the mathematics and the world was on display, and if there was one feature of the work that would appear foreign to his economist readers it was the implicit suggestion

[26] In a footnote, von Neumann notes the parallel between β^{V} and α^{IV} in the asymmetric solution (α^{III}, α^{IV}), but says that he can find no similar heuristic motivation.

that features of social relations might be uncovered, not by experience, but rather by the logical exploration of the mathematics being created to describe the domain. The chapter closes with further discussion of how the mathematics can generate the phenomenon of the less-than-complete exploitation of a defeated player, something that von Neumann says is also quite characteristic of social organisations.

Truth and the Ficitious Player

Further displays of this "opportunism" appear in Chapter XI, where von Neumann extends the stable set solution concept from zero-sum to non–zero-sum, or general, *n*-person games (pp. 504ff), the kind appropriate to socioeconomic questions, in which there are possibilities for net social gains.[27] How, he asks, might one use the zero-sum theory developed so far for these conceptually different situations? It would be "very discouraging" if one had to start out fresh all over again (p. 505). Von Neumann thus conjures up a ghost at the table, the "ficitious player", whose identity, depending on the arguments required to sustain the construction of the mathematics, wavers between that of an active character, with his own beliefs and desires, and a passive, inanimate, formal device.

Von Neumann's procedure is to interpret the *n*-person general game as an $n + 1$-person zero-sum game, to which one can then apply the existing zero-sum theory. The $(n + 1)$th person is a fictitious player who loses the amount won by the other *n* players and vice versa but who has no direct influence on the game. His payoff is:

$$\aleph_{n+1}(\tau_1, \ldots, \tau_n) \equiv -\sum_{k=1}^{n} \aleph_k(\tau_1, \ldots, \tau_n)$$

We thus have the zero-sum extension of the game Γ, denoted by $\overline{\Gamma}$. The argument is that by applying the existing (zero-sum) theory to the extended game, $\overline{\Gamma}$, we can "solve" the original (non–zero-sum) game. Central to this strategy, of course, is the need to ensure that the fictitious player does not materially influence the outcome. After all, he was introduced by the mathematician; he is not present in any game. Von Neumann's defence of the fictitious player, his "surprising twist", is a complex piece of casuistry.

---

[27] The authors earlier consider constant-sum games, in which the sum of gains and losses always equals some constant, and show by means of the notion of strategic equivalence how these games are structurally isomorphic to the zero-sum type, and therefore require no conceptually different approach (see *op cit*, pp. 345–48).

The analysis, he says, cannot be "purely mathematical" because we have temporarily abandoned the solid mathematical theory used in the previous part of the book; it must mainly be "in the nature of plausibility arguments" (p. 506). Although the fictitious player must have "no influence whatever" on the game, we must nevertheless ask ourselves "whether the fictitious player is absolutely excluded from all transactions connected with the game" (p. 507).

If we are to use the solutions of $\overline{\Gamma}$ for Γ, then we can only use that subset of solutions to the former in which the dummy plays no active role – that is, has no concrete effect on the play of the remaining n, real, players. Although the dummy has no moves in the game and therefore cannot be seen as a desirable coalition partner by the others, he himself, says von Neumann, "may have an interest in finding allies" (p. 508). In particular, the dummy may have an interest in paying compensation to other players so as to prevent the formation of coalitions unfavourable to him.

It is important not to misunderstand this: As long as Γ is played, i.e. as long as the fictitious player is really a formalistic fiction, no such thing will happen; but if the game really played is $\overline{\Gamma}$, i.e. if the fictitious player behaves as a real player would in his position, then his offer of compensations to the others must be expected (p. 508).

As soon as we are dealing with the imaginary $\overline{\Gamma}$, then the dummy, with his ability to pay compensations, becomes a force to be reckoned with. He "gets into the game in spite of his inability to influence its course directly by moves of his own" (*ibid*). Von Neumann clarifies with an example, a coordination game. There are two players, 1 and 2, each of which must choose the numbers 1 or 2. If both choose 1, each wins $1/2$; otherwise, each gets -1. The characteristic function is thus:

$$v((1)) = v((2)) = -1$$
$$v((1, 2)) = 1$$

Introducing to the characteristic function the fictitious player, 3, who wins the residual of the others' earnings, we add:

$$v((3)) = -v((1, 2)) = -1$$
$$v((1, 3)) = -v((2)) = 1$$
$$v((2, 3)) = -v((1)) = 1$$
$$v((1, 2, 3)) = -v((\emptyset)) = 0$$

The game is thus a competition amongst three players for coalitions, and the dummy can be expected to try to enter coalitions just like players 1 or 2. We might also argue that he will be willing to pay either 1 or 2 in order to induce them not to form a coalition together, in which case he wins 2 by default. (He will be willing to pay up to 3/2 for this, the amount necessary to compensate either 1 or 2 for not entering a coalition). If we interpret the zero-sum extension literally and rigorously, therefore, we find ourselves forced, says von Neumann, to attribute to the fictitious player abilities "conflicting with the spirit in which he was introduced" (p. 510). Thus the extended game cannot be unqualifiedly applied *in toto* to the original game. What is to be done?

Von Neumann's way out is to *restrict* the scope of the characteristic function in the extended game. The aforementioned complications concerning the efficacy of the dummy player lay in our application, to the extended two-person game, of the complete set of (nondiscriminatory) solutions to the normal three-person, zero-sum, game. If the dummy is not to wreak philosophical havoc, he must be given a restricted role. This von Neumann does by considering only the subset of solutions that "discriminate" against the fictitious player – that is that restrict his coalition-competing capabilities. Specifically, one might consider only those solutions that assign to the "excluded" fictitious player a fixed amount, c, which may be as low as − 1, the amount he could obtain on his own, or may be higher, depending on the social standard of behavior then obtaining, as manifested in considerations of prejudice and privilege. Might it be reasonable that the fictitious player *not* be held at the absolute minimum possible, that society might not exploit all possible gains and deny itself something? The possibility is not so absurd, says von Neumann: there may be forms of social organisation that are inefficient – in the sense that all possibilities for collective gain are not exploited – yet stable.[28]

Conclusion

It is difficult to accept the assertion that von Neumann's return to game theory was related to his rejection of the "Hilbert program", inspired by the

---

[28] Although he says, without giving references, that discussion of this point in the sociological literature is far from concluded, he invokes an argument based on dominance to conclude that c, the amount received by the fictitious player, must be restricted to its minimum value: there will always be some set of players for whom outcomes that do not exploit the whole social product will be dominated by those that do (see pp. 520–27).

Gödel impossibility theorems of 1931.[29] On the contrary, von Neumann absorbed the implications of the Gödel proofs very quickly, at Königsberg in 1931, and immediately abandoned all work in metamathematics. His return to game theory, almost a decade later, was stimulated, not by Gödel's work, but by the very concrete personal and historical circumstances of the late 1930s. As for the nature of this new mathematics, our examination of the content of *Theory of Games* suggests the importance, not so much of the failure of Formalism in *metamathematics* – to which the book makes no reference – as the success of the modern, formalist *mathematics* of Hilbert and his school.[30] Thoroughly Hilbertian, the book is a manifesto for the use of modern set theory and discrete mathematics in the social realm, and this stance is reinforced by both von Neumann's private disdain for the antiquity of Hicksian and Samuelsonian mathematical economics and Morgenstern's numerous diary references to his new "modern" reading, be it Hausdorff, Fraenkel, or van der Waerden.

Also, given the historical circumstances, it is difficult not to see a certain symbolism in this burst of mathematical creativity. As the *Theory of Games* was being written, Göttingen was no more, Hilbert was on his deathbed – he would die in February 1943 – and Nazism was destroying the Europe von Neumann had known. With volcanic energy and an extrordinary degree of obsession, von Neumann turned to the creation of a new social theory, the fine details of which he presented in precisely the modern, Hilbertian style dismissed by Bieberbach and *Deutsche Mathematik*. It is difficult not

---

[29] Mirowski repeatedly asserts that von Neumann's return to game theory was related to his rejection of the "Hilbert program". The "leitmotiv of the period of the mid-1930's was one of absorbing the implications of Gödel's results, and incoporating them into a recalibrated sense of the efficacy and aims and purposes of mathematics, and reconciling them with his newfound alliance with the military. What could rigor and formalization do for us after Gödel?" (2002, p. 119). Mirowski summarily dismisses "The Mathematician" as "an apologia for the years spent wandering in the desert" (p. 119).

[30] Following Leo Corry, Weintraub (2002, pp. 72–100) properly distinguishes between Hilbert's programme in the foundations of mathematics and his promotion of the axiomatic method in mathematics and science, arguing that Gödel's work, although it ended the former, left the latter intact. This is in keeping with the view expressed by von Neumann in "The Mathematician". Philippe Mongin (2003) attempts to detect the possible influence of Gödel on von Neumann's economics, and suggests that the minimalist style of axiomatics employed by von Neumann in the *Theory of Games* is reflective of a changed view of axiomatics after Gödel. However, Mongin admits that this applies only to the appendix containing the axiomatic treatment of utility theory (TGEB, pp. 617–32) and not to the book's central axiomatic treatment of the concept of the game (TGEB, pp. 73–79), the style of which is "pre-Gödelian" by Mongin's own criteria (Communication, Mongin to author, May 29, 2007).

to see in von Neumann's work here an element of symbolic re-affirmation of loyalty to a world now lost, a riposte to the forces confronting him.

In his absorbing account of the dialectic between creation and discovery in mathematics, Ulam protégé and M.I.T. combinatorics specialist Gian-Carlo Rota describes the field of mathematics as the ultimate escape from reality: "All other escapes... are ephemeral by comparison. The mathematician's feeling of triumph, as he forces the world to obey the laws his imagination has freely created, feeds on its own success. The world is permanently changed by the workings of his mind and the certainty that his creations will endure renews his confidence as no other pursuit".[31] Perhaps this is particularly salient in the case of a mathematics intended to capture psychological or social features of the mathematician's own world. It is difficult not to see in von Neumann's efforts an element of perhaps subconscious resistance to the conditions of the time; an almost defiant willingness to see order beyond the disorder, equilibrium beyond the confusion, an inevitable return to normality once the present transition, with its "abnormal spiritual tensions", was over.

Even at the risk of getting ahead of our story, it is worth noting that, to the very end, von Neumann spoke of the stable set in terms consonant with this account. In 1953, young Princeton mathematician Harold Kuhn wrote to him, asking him about the possibility of testing the stable set solution using the experimental methods then beginning at the RAND Corporation. Von Neumann replied in the negative: "I think that nothing smaller than a complete social system will give a reasonable 'empirical' picture [of the stable set solution]. Here, over relatively long periods of time, one can meaningfully assert that the 'system' has not changed, while the positions of various participants within it may have changed many times. This would seem to me to be the analogue of a single solution and an 'exploration' of the imputations that belong to it. After relatively long times, there occur discontinuous changes, 'revolutions' which produce a different 'system'".[32]

---

[31] Rota (1997), p. 70. In an moment of unwitting poetry, Rota goes on to illustrate the "monstrosity" of the mathematician's view of the world by comparing him to none other than Nabokov's Luzhin, "who eventually sees all life as subordinate to the game of chess".

[32] Von Neumann to Harold Kuhn, Apr. 14, 1953, Container 24, File: Kuhn, H. W., Von Neumann Papers, Library of Congress. In a striking coincidence, the metaphor likening society to a game recurs in the last book written by Emanuel Lasker. In 1938, he had left Stalin's Russia for New York, where he tried, with increasing difficulty, to make ends meet. He nonetheless found the time to write *The Community of the Future*, a 300-page book about the establishment of a noncompetitive community as a way to absorb the unemployed Jews of Europe (Lasker 1941). It was an attempt to "try the method of the

By the same token, we can understand von Neumann's dismissal of John Nash's 1950 proof of the existence of an equilibrium point in a game without coalitions.[33] Given everything we have observed about him, it seems that to von Neumann, the formation of alliances and coalitions was *sine qua non* in any theory of social organisation. It is easy to understand why the idea of noncooperation would have appeared artificial to him, the elegance of Nash's proof notwithstanding. At a Princeton conference in 1955, he defended, against the criticism of Nash himself, the multiplicity of solutions permitted by the stable set: "[T]his result", he said, "was not surprising in view of the correspondingly enormous variety of observed stable social structures; many differing conventions can endure, existing today for no better reason than that they were here yesterday".[34]

---

chess master on a political problem ... : that of unemployment" (1941, p. 12). An idea pervading the book is that of the parallel between social life and games. Society was like chess played on an "infinite board" (p. 66). Questions of stability and balance, and ethics and power, were central. The study of chess shed light on the analysis of social power, allowing one to observe, for example, that "the alliance of weak powers is enduring but not that of the strong", or "in a state of balance every piece has political authority proportionate to its intrinsic power" (p. 138). The first of these maxims was an application of the concept of balance, which, said Lasker, lay at the root of every compromise. The second determined the authority due to each force when the game was in a state of balance. Were it not respected, there could be no peace. In chess, as soon as the state of balance is disturbed, "a tension arises which seeks an outlet". Likewise in society: "As long as no new needs or aspirations arise, the old established parties maintain their hold. They have laid down the rules of the game which have stood the test of experience ... But at a period of distress which all feel, or of injustice ... or of great creations, which set novel problems, the written and the unwritten law become the object of criticism backed by ethical force, and the ensuing struggle is apt to lead to changes in the prevailing mode of life" (pp. 140–41). That last sentence could have been uttered by von Neumann.

Perhaps Albert Einstein had not read that book on the noncompetitive community when, a year later in 1942, he wrote the Foreword to Lasker's biography. Although he expressed admiration for someone he regarded as a true Renaissance man, Einstein felt that the game of chess exercised too strong a hold over Lasker's imagination: "I am not a chess expert and therefore not in a position to marvel at the force of mind revealed in his greatest intellectual accomplishment – in the field of chess. I must even confess that the struggle for power and the competitive spirit expressed in the form of an ingenious game have always been repugnant to me ... To my mind, there was a tragic note in his personality ... The enormous psychological tension, without which nobody can be a chess master, was so deeply interwoven with chess that he could never entirely rid himself of the spirit of the game, even when he was occupied with philosophic and human problems". Einstein in the Preface to Hannak (1959).

[33] This is the contribution for which Nash would be awarded the Nobel Prize in Economics in 1994. On von Neumann's dismissive attitude towards the Nash Equilibrium, see Leonard (1994) and Nasar (1998).

[34] Wolfe, Philip (1955), p. 25.

Coda

Ortvay's letters from Budapest trickled to a halt in 1941. In January, he was sending three separate copies to be sure that they reached von Neumann in Princeton. He appealed for help in raising funds for the beleagured Mathematical and Physical Society. He continued to write about the application of mathematics to the realm of "spiritual" states. He was now reading Kurt Lewin's *Principles of Topological Psychology*,[35] and felt it likely that the areas most ripe for such mathematical treatment were those that included sharp distinctions, such as music, or juridical systems.[36] His last letter, in February 1941, was a brief summary of the previous one, written as though he thought von Neumann had not received it the first time.

In 1944, by which time von Neumann and Morgenstern's book had appeared, matters had worsened in Budapest. Because of their services to the State, von Neumann's teachers, Fejér and Riesz, were granted special status and each allowed to spend part of the war in the one of the protected houses in Budapest's "little ghetto", around Pozsonyi and Szent István streets.[37] These houses were under the diplomatic protection of various countries, and it was here that the Swedish diplomat Raoul Wallenberg managed to save many Hungarian Jews. Von Neumann's teachers Fejér and Riesz appear to have been housed in the hospital of the Swedish Embassy at 14–16 Tátra St.: Riesz early in 1944, when the Jews of Szeged and the provinces were being deported; Fejér later in the year. There, although crowded, in terrible conditions, with up to fifteen in a room, they were at least safe from deportation, and they survived the war.

So, too, albeit with greater precarity, did their student, Paul Turán. On the eve of his death many years later, Turán remembered his time in the camps, where he conserved his spirit by working on mathematics. In September 1940, he had been making a living as private tutor in Budapest when he was called to labour camp service. A friend in Shanghai had recently written him about a problem in graph theory: what is the maximum number of edges in a graph with n vertices not containing a complete subgraph

---

[35] Lewin (1936) took set theory, topology, and Karl Menger's dimension theory and used them to recast psychological situations.

[36] Ortvay to von Neumann, Jan. 29, 1941. It should be possible to achieve in these areas, Ortvay felt, that which had been achieved in the science of heredity, where natural selection became something that could be discussed rigorously once the essentials were properly treated. Here he continued to describe what he saw as the appropriate way to model the functioning of the neural system.

[37] See Frojimovics et al (1999), pp. 402–03.

with k vertices? In the camp, Turán was recognized by the commandant, a Hungarian engineer with mathematical training. The commandant took pity on Turán's weak physique and gave him an easy job, directing visitors to piles of wooden logs of different sizes. In this "serene setting", Turán recalled, he was able to work on the extremal problem in his head and solve it: "I cannot properly describe my feelings during the next few days. The pleasure of dealing with a quite unusual type of problem, the beauty of it, the gradual nearing of the solution, and finally the complete solution made these days really ecstatic. The feeling of some intellectual freedom and being, to a certain extent, spiritually free of oppression only added to this ecstasy".[38] In the context of our story, Turán's confrontation with the camp commandant had a certain poignance, for the latter was one Joszéf Winkler, erstwhile contributor to *KöMaL* and joint winner of the Eotvös Competition eighteen years previously – in 1926, the year with none other than König's question about the knight's move on the infinite chessboard.

In July 1944, by which time the threat of deportation was real, Turán was working in a brick factory near Budapest. There, all the kilns were connected by rail to all the storage yards, but, at the crossings, the moving trucks tended to jump the tracks. He began to work on the graph-theoretic problem of minimizing the number of crossings in a yard with m kilns and n storage yards. This time, however, his thinking was stifled by fears for his family. By late 1944, there was no work to do and Turán and other Jews expected to be deported from one day to the next. He began to think about another problem, concerning the maximal size of subgraph in a graph of given size. He conjectured a solution, for which he had no support other than "the symmetry and some dim feeling of beauty; perhaps the ugly reality was what made me believe in the strong connection of beauty and truth. But this unsuccessful fight gave me strength, hence, when it was necessary, I could act properly".[39]

Others were not so sustained. Dénes König's elder brother, the literary scholar, György, took his life after the German occupation of Budapest on March 19. Then, when the *Nyilas* took over on October 16, König himself, the very one who had introduced the young von Neumann to the mathematics of chess in the 1920s, also committed suicide. Under the Arrow Cross gangs of the *Nyilas*, Hungary entered its darkest period, with Jews being

---

[38] Turán (1977), p. 8. Incidentally, questions of the psychological ambiguity and ambivalence of life in confinement are central to the work of Hungarian writer, Imré Kertész, recipient of the Nobel Prize in Literature in 2002. See his *Fateless* (1992) or *Kaddish for a Child not Born* (1997).

[39] *Ibid*, p. 9.

tortured and shot, their bodies dumped in the Danube. At one point, in late December, Fejér and the occupants of the Swedish hospital were marched by night to the river's edge by the Arrow Crossers, but saved by the last-minute intervention of an army officer.[40] The forced labour, deportation to the camps, and the local attacks saw the destruction of Hungarian Jewry, with 600,000 perishing within a few short months. On January 2, 1945, when the Germans were fleeing and the Russians about to enter Budapest, von Neumann's friend, Ortvay, took his own life, apparently fearing revenge by the "liberators".[41] Neither he, König, nor others close to its genesis would get to read the *Theory of Games.*

[40] See Turán (1960), p. 1205.

[41] In the Ulam papers of the American Philosophical Society, in a document describing "Family Memorabilia of Nicholas A. Vonneuman relevant for John von Neumann biography donated to the A.P.S.", September 15, 1994, Nicholas Vonneuman writes that Ortvay had been able to retain his position at the Physics Department of the University of Budapest by virtue of his having "qualified" ancestors. Whether this implies that Ortvay was Jewish is unclear. Dr. László Filep claims that Ortvay was not. Filep also notes that, being unassociated with the *Nyilas*, Ortvay's fear of the Russians was unnecessary, but it illustrates well the fear and tension abroad in Hungary at this time. In a letter to his brother, Marcel, in Sweden, written in July 1945, Frigyes Riesz wrote of the König and Ortvay suicides, and of Fejér's sufferings during the war. He also said that Szeged mathematician István Lipka had been fired the previous day from his university position, having been discovered to have joined the Nazi party as early as 1939. (Riesz, F. to M. Riesz, July 18, 1945, Marcel Riesz Papers, Lund, Sweden). I am grateful to Dr. Filep for providing me with a copy of this letter.

ELEVEN

Morgenstern's Catharsis

The "Introduction"

Morgenstern was energized during this time by writing the part of the *Theory of Games* that was to be most widely read – his accessible introduction, the "Formulation of the Economic Problem". A manifesto for a new mathematical economics, this chapter represents the confluence of the two different intellectual traditions of the two authors, bearing traces of Austrian criticism and yet pervaded with the influence of von Neumann.

The essay defends the use of a certain kind of mathematics in economics and discusses rational behaviour, utility, and the new concepts of solution and standard of behaviour. The primitive state of economics, it is repeatedly emphasised, has been the result of an inadequate understanding of the basic categories and facts, onto which an unsophisticated and imitative mathematics has been grafted. The simple application to social interaction of the physical metaphors of rational mechanics or conservative systems is rejected.[1]

So, too, is the idea that the natural and social worlds constitute two different domains, to be treated differently in science. This rejection of the distinction between *Geisteswissenschaft* and *Naturwissenschaft*, a rebellious stance already taken by Karl Menger in the mid-1930s, continued Morgenstern's break with the Austrian economic tradition. Rather, following von Neumann, the empirical world, as a heterogeneous collection of phenomena, be they natural or social, is seen as a repository of potential bodies of mathematics, and the prominence of the phenomena is interpreted as a sign of their promise in this regard. That new mathematics will be necessary in the social domain is suggested by the "importance of the social phenomena,

---

[1] This point is discussed in Mirowski (1992).

249

the wealth and multiplicity of their manifestations, and the complexity of their structure" (p. 6).

Any ambiguity regarding the use of mathematics previously shown by Morgenstern has now disappeared. A section is devoted to dismissing the arguments typically raised against the mathematical analysis of human behaviour – psychological factors, measurement difficulties, and so on. Similar arguments were raised against the use of mathematics in sixteenth-century physics and eighteenth-century chemistry and biology, all of which now depend completely on mathematical techniques. Furthermore, the mathematical development *facilitated* the conceptual development: in the theory of heat, for example, energy and temperature emerged from the mathematics; they did not precede it.

The reason why individual rationality has never been properly treated lies in the lack of mathematical methods capable of quantitatively accounting for all the elements in the usual qualitative description. The "significant relationships are much more complicated than the popular and the 'philosophical' use of the word 'rational' indicates" (p. 9). Central to the complexity is the transition from Robinson Crusoe's world to the social economy, where the individual's optimisation depends on "live variables" – that is, the decisions of others. This is a conceptually new problem, "nowhere dealt with in classical mathematics ... no conditional maximum problem, no problem of the calculus of variations, of functional analysis, etc." (p. 11). In the social economy, as one moves from two to three to more participants, the complexity increases insofar as each individual has to take account of an increasing number of other players *and* the number of choice variables for each player increases. The "combinatorial complications of the problem ... increase tremendously with every increase in the number of players" (*ibid*).

Morgenstern dwells on the relationship between the new theory and the current theory of competition. Unlike in mechanics, in which the theory for two, three, four, and more bodies is well known and provides a foundation for the statistical theory of the case of large numbers, the same cannot be said of the social exchange economy. Heretofore, there has been no theory for small numbers of participants. Unlike in mechanics, in economics it cannot be said whether or not the movement from small to large numbers of agents represents a simplification. The "current assertions concerning free competition appear to be very valuable surmises and inspiring anticipations of results", but, without a theory of small numbers, they are not scientific results. "The problem must be formulated, solved and understood for small numbers of participants before anything can be proved about the changes of its character in any limiting case of large numbers such as free competition" (p. 14). This is particularly important because of the widespread belief that increasing the number of participants inevitably leads to free competition. The current theories of interlocking individual plans, which was a veiled reference to Hayek, of the Lausanne School and others, all make far-reaching restrictions: sometimes free competition is assumed; other times, coalition-formation is simply ruled out. These amount to a *petitio principii*, avoiding the real difficulty. If, however, there form coalitions of participants acting together, then the situation changes dramatically. The new theory emphasises the importance of coalition formation, and a footnote here refers to trade unions, consumers' cooperatives, cartels, and "conceivably some organizations more in the political sphere" (p. 15, n. 1).

Several pages are devoted to defending the use of numerically measurable, cardinal utilities, especially given the perception that this represents a retrograde step relative to the theory of indifference curves and the presumed impossibility of interpersonal comparison of utility. The current situation, Morgenstern argues, is very similar to the early stage of the theory of heat, when one object was seen to be warmer or colder than another but no quantitative comparison was possible. It turned out that comparison was posssible in terms of both the quantity of heat and temperature. The former is directly numerical and additive, and related to a numerical mechanical energy. The latter is more subtly numerical, not being additive, but standing nonetheless on a "rigidly numerical scale" that emerged from the study of the behaviour of ideal gases and the role of temperature in relation to the entropy theorem. The historical experience of the theory of heat, therefore, should provide some encouragement regarding a theory of utility.

The step from indifference theory to cardinal utility is actually a relatively small one. If it is plausible to believe that an individual can choose between two items, A and B, then it is also plausible to believe that he can choose between an item, A, and a combination (B and C, each with a probability of fifty percent). If he prefers A to both B and C, then he will choose A over the bundle. If, however, he prefers A to B and C to A, then if he chooses A over the bundle, we can assume that he prefers A to B by *more* than he prefers C to A. Specifically, if the individual is indifferent between A and the combination (B with probability a, C with probability $1-a$), then a may be used as a numerical estimate of the ratio of preference for A over B to that of C over B. The indifference curve theory, therefore, either assumes too much or too little: the former if preferences are not at all comparable, the latter if indeed they are, in which case a numerical utility can be obtained, making the indifference curve superfluous.

To the objection that the individual usually operates in a vague and hazy environment, Morgenstern replies that the same could be said of the individual's conduct "regarding light, heat, muscular effort, etc." (p. 20). "But in order to build a science of physics, these phenomena had to be measured. And subsequently the individal has come to use the results of such measurements – directly or indirectly – even in his everyday life" (*ibid*). Perhaps the material economic life of the individual may one day be affected in a similar manner, thanks to such measurement, and there follows a detailed discussion of the principles of measurement and of operations on such measures as mass, distance, and position. In the case of utility, indifference curve theory assumes that one can speak only of the relation "greater" – that is, that utilities are numerical up to a monotone transformation. There follows von Neumann's demonstration that the range of transformations may be narrowed to the set of all linear transformations. In the new theory, Morgenstern concludes, the concept of marginal utility may no longer have any meaning at all.[2]

The end of the introduction forsakes broad philosophical considerations for an encapsulative description of the theory to which the rest of the book is devoted. The central theoretical idea, apart, obviously, from the belief that social situations should be described in terms of coalitional games, is the identification of the range of possible outcomes in a game. These outcomes are solutions, in the sense that they constitute equilibria from which there is no tendency to depart. A solution is plausibly a "set of rules for each participant which tell him how to behave in every situation which may conceivably arise" (*ibid*). The solution must allow for the fact that some events facing the participants may be determined only statistically and that

---

[2] If u and v represent two utilities to an individual, $u > v$ expresses the preference relation, and $\alpha u + (1 - \alpha)v$ the weighted combination or bundle, then it is possible to attach a number $v(u)$ to the utility u, such that $u > v$ implies $v(u) > v(v)$ and $v(\alpha u + (1 - \alpha)v) = \alpha v(u) + (1 - \alpha)v(v)$. Attaching such a number requires that this relation and operation satisfy certain properties or axioms, which the authors describe and discuss in detail over the next five pages. In particular, they address the question of the possibility that an individual may derive utility from gambling – that is, may have a positive preference for risk. The axiom that comes closest to eliminating this was the one that stated that an individual is indifferent between obtaining a combination of two constituents in two successive steps or in one operation. Allowance for specific utility of gambling could not be made without contradiction in the current system of axioms. Also, such a modification would require a much more refined system of psychology than is currently available to economics. Reference is made here to Menger's 1934 paper on the Petersburg Paradox, which indicated that the intricacies of using methods of mathematical expectation are far from completely analysed. The present method, the authors say, is a direct, but plausible, attempt at simplification.

some individuals may act irrationally. Thus the concept of the game fulfills the same function for economic and social problems as do "geometrico-mathematical models" for the physical sciences: precise, exhaustive, not unduly complicated, and similar to reality, just like Newton's simplifying interpretation of the solar system as a small number of 'masspoints'. The solution is a "combinatorial enumeration of enormous complexity", indicating the minimum the participant under consideration can get if he behaves rationally. He may get more if the others behave "irrationally" – that is, make mistakes. Just as in the development of general relativity and quantum mechanics, these are pretheoretical "desiderata", features the theory may be expected to show if it is to be deemed satisfactory.

For society as a whole, the ideal "solution" would be one that describes the amount obtainable by each player, playing rationally. Were such a single imputation available, then the "structure of the society under consideration would be extremely simple: There would exist an absolute state of equilibrium in which the quantitative share of every participant would be precisely determined" (p. 34). However, such a simple solution does not generally exist, so the notion needs to be broadened – something which "is closely connected with certain inherent features of social organization that are well known from a 'common sense' point of view but thus far have not been viewed in proper perspective" (*ibid*).

The set of games for which this simple determinate solution holds are those of the two-person, zero-sum type, which, although not typical of "major economic processes", contain universally important traits of all games and form the basis for the general theory. The simplest divergences from this basic case are where the zero-sum requirement no longer holds, the problem of bilateral monopoly, or, when another player enters the picture – the three-person, zero-sum game. In the second case, because there are three possibilities for coalition formation, of two players against the remaining one, one finds three possible imputations, depending on which coalition obtains. The solution becomes a *set* of possible imputations, rather than a single one, and this is something that characterises games in general.

Any particular alliance describes only one particular consideration which enters the minds of the participants when they plan their behavior. Even if a particular alliance is ultimately formed, the division of the proceeds between the allies will be decisively infuenced by the other alliances which each one might alternatively have entered. Thus only the three alliances and the imputations together form a rational whole which determines all of its details and possesses a stability of its own. It is, indeed, this whole which is the really significant entity, more so than its constituent imputations. Even if one of these is actually applied, i.e., if one particular alliance is actually formed, the others are present in a 'virtual' existence: Although they have not materialized, they have contributed essentially to shaping and determining the actual reality (p. 36).

In an n-person game, a "solution should be a system of imputations possessing in its entirety some kind of balance and stability the nature of which we will try to determine. We emphasize that this stability – whatever it may turn out to be – will be a property of the system as a whole and not of the single imputations of which it is composed" (*ibid*).

There follows a systematic presentation of the elements sketched in von Neumann's earlier manuscript of 1940. One imputation or outcome, x, is said to dominate another, y, "when there exists a group of participants each one of which prefers his individual situation in x to that in y, and who are convinced that they are able, as a group – that is, as an alliance – to enforce their preferences" (p. 38). Were domination a transitive ordering, then a single best outcome would be the imputation that dominated all others and was dominated by none. However, it is not transitive: x may dominate y; y may dominate z, yet z may dominate x. This is because the "effective set", the preferring coalition, may not be the same in each case, something that is "a most typical phenomenon in all social organizations. The domination relationships between various imputations . . . i.e. between various states of society – correspond to the various ways in which these can unstabilize – i.e. upset – each other" (p. 39).

A "solution", predicated, as we have already seen, upon the concept of dominance, is related to "standards of behaviour" – that is, the particular set of rules, customs, or institutions governing social organisation at a particular time. To understand the analogy, the reader is advised to "temporarily forget the analogy with games and think entirely in terms of social organization" (p. 41, n. 1):

Let the physical basis of a social economy be given, – or to take a broader view of the matter, of a society. According to all tradition and experience human beings have a characteristic way of adjusting themselves to such a background. This consists of not setting up one rigid system of apportionment, i.e. of imputation, but rather a variety of alternatives, which will probably express some general principles but nevertheless differ among themselves in many particular respects. This system of imputations describes the 'established order of society' or 'accepted standard of behavior' (p. 41).

The circular nature of the definition, based on dominance characteristics of the member imputations, lends each solution a kind of inner stability. However, a solution will be adopted only to the extent that the underlying standard of behavior commands general acceptance: the theory predicts neither which solution will be observed nor which imputation within any solution will obtain. This approach to social theory contrasts with those that normally involve some concept of "social purpose". The latter is usually based on principles concerning distribution or other overall aims, which are inevitably arbitrary and are usually supported by reference to arguments concerning inner stability or the desirability of the resulting distribution: "Little can be said concerning about the latter type of motivation. Our problem is not to determine what ought to happen in pursuance of any set of – necessarily arbitrary – *a priori* principles, but to investigate where the equilibrium of forces lies. As far as the first motivation is concerned, it has been our aim to give just those arguments precise and satisfactory form, concerning both global aims and individual apportionments... A theory which is consistent at this point cannot fail to give a precise account of the entire interplay of economic interest, influence and power " (p. 43).

Catharsis

The collaboration with von Neumann was a huge event in the life of Morgenstern, carrying him onto unfamiliar ground. Although he saw how the minimax theorem resolved Holmes and Moriarty's problem, it is fair to say that the mathematical intricacies of the stable set remained beyond him, as they did for many a reader. If he had earlier sought a logic of foresight and interaction, he now had more than he could handle. Issues that several years previously would have troubled him as an Austrian economist – that the agent of game theory was all-knowing, a "demi-god" of sorts, that the theory evacuated the passage of time – these now paled against the 600 pages of analysis with which he was publicly associated. It was a strange position for a co-author: "One of these days, I have to write down a few things about the story of the book (and my minimal share; but I seem to have acted as a kind of catalytic factor)".[3]

For all his public posturing, Morgenstern privately reveals a fine sensitivity to the human dimensions of intellectual collaboration. The more he got to know von Neumann, the more he realised he would never know him: "He is a mysterious man. The moment he touches something scientific, he is totally enthusiastic, clear, alive, then he sinks, dreams, talks superficially,

---

[3] Diary, Jan.1, 1943, OMDU.

in a strange mixture".[4] Unlike her husband, Morgenstern shared Klari's love of music, something they indulged in while von Neumann sat off to one side. "Yesterday for dinner at the Neumann's, . . . We then went to my place, and Clari and I listened to Brahms, while Johnny flipped through Sorokin with surprise".[5] In time, he found himself becoming something of an intermediary between the couple. Von Neumann spoke to him about the difficulties of marital life, the tensions of parenthood. And Morgenstern confided in Klari as in no one beforehand.

Time and time again, he returned to the change he was undergoing: "I have a clear feeling of freedom from prejudices and ties to theories and general views, as if I were shedding my skin. Hopefully a few things will remain. On the emotional level I am more open to small joys, and I see how often I was a complete donkey. Took everything too tragically. That must come from my education in the First World War".[6] He felt as though his original habits of thought and expression were withering away. But he welcomed it: "behind the break I see the light of a new science".[7]

The closer he drew to von Neumann and mathematics, the less comfortable he felt with himself. "I neither can nor wish to let go of set theory . . . I was an idiot not to have studied math. even as a sideline at the University of Vienna, instead of this silly philosophy, which took so much of my time and of which so little is left. After Fraenkel, I will read Hausdorff".[8] This, in turn, reminded him of where his priorities should lie. It was pleasant to have female company, but it did not touch him as deeply as his work, which now interested him more than anything else: "That's all over" (*ibid*).

He had lunch with mathematician Shizuo Kakutani, who said that he regarded von Neumann as being on a par with Hilbert or Poincaré. Kakutani recalled having recently presented von Neumann with a problem about which he had been thinking for years, only to have the latter phone back an

---

[4] Diary, Sept. 1, 1940, OMDU. Then a year later, "The more I get to know Johnny's mind, the more I must admire him. One is presented with the incomprehensible (*ibid*, Jan. 10, 1941). "[Johnny] was working so hard that he was totally absent from everything else. I saw that while they were here. We played 2–3 records, and he made pained faces, deeply in thought. What a strange process. How it overcomes these people, and they cannot defend themselves against it at all" (*ibid*, Jan. 22, 1941). And later, "Johnny called me; he likes my manuscript. . . . I am very happy about this . . . After all, is wasn't easy for me to simplify his mathematical theory, and to represent it correctly. He is working continuously without a break; it is nearly eerie" (*ibid*, Aug. 7, 1941).

[5] Diary, July 16, 1941, OMDU.

[6] And then, as if to check himself, "I hope that this liberation will not lead to uninhibitedness". Diary, Jan. 27, 1941, OMDU.

[7] Diary, Feb. 22, 1941, OMDU.

[8] Diary, Aug. 7, 1941, OMDU.

hour and a half later with a solution – apparently discovered over lunch. The same afternoon, Kakutani marvelled as von Neumann presented a proof in Operator Theory over the course of several hours, without notes and without having previously written anything on paper.

Von Neumann was increasingly immersed in war work, slipping away frequently to New York, Washington, and elsewhere to provide mathematical advice on military problems. Morgenstern was an enemy alien, but this fact, and, if his journal is any indication, the war itself, paled into insignificance against the personal catharsis he was undergoing. At one point, in January 1942, von Neumann dashed off to the Aberdeen Proving Ground, leaving Morgenstern with a "great headache" – that is, ten pages on the analysis of poker.[9] He agreed with von Neumann that it was unlikely that economists would build on the book before a long time, ill-equipped as they were to absorb its contents.

The exposure to mathematical work cast the social sciences in a different light: "Yesterday I saw Weyl while he was thinking about a new mathematical work. This total evident concentration. Recently I asked Johnny why in his opinion the creative power of mathematicians decreases very often so early. He thought it must be increasing difficulties in concentration. What a difference in the work of the social sciences. How much looser, easier the mental processes. Sometimes I think that it would be valuable to write an autobiography containing the scientific undercurrent of such an analysis, especially if I could connect it to the real sciences. I really have no illusions about the social sciences".[10] Thus, when it came to his disciplinary peers, there was little room for even private indulgence. Fortified by von Neumann's claim that, in the future, one would regard twentieth-century mathematical economics as being contemporaneous with Newton, Morgenstern became "very angry with those terrible textbook scribblers. There is nothing solid. Everything is third class, full of mistakes, lack of knowledge of the literature etc.".[11] Withering remarks were directed towards Hayek, Keynes, and others.[12] A visit from Eve Burns, then involved in her work on

---

[9] Diary, Jan. 27, 1942, OMDU.

[10] Diary, Oct. 26, 1940, OMDU.

[11] Diary, Sept. 8, 1941, OMDU.

[12] For example, "Last Thursday a report in the graduate course about Hayek's 'Pure Theory of Capital'. It is higher nonsense . . . They keep talking about 'Investment Function', but there is no question of stating the concept with any precision. He does not seem to know what a function really is. This type of 'economic theory' must vanish (Diary, Mar. 15, 1942, OMDU). "I am more and more of the opinion that Keynes is a scientific charlatan, and his followers not even that. People like Lerner, for example, are simply out of their minds and don't have a clue what science is all about" (*ibid*, May 2, 1943).

the social security system, convinced Morgenstern that he was now "living outside the world of most economists. It seems that I have come in contact much more with the truly mathematical spirit, and once one has [had] one's eyes opened, it stays that way".[13]

In May 1943, he lectured to his own department on the theory of games, drawing a mixed reaction. His biggest critic was F. A. Lutz, whose own work on Bentham Morgenstern so admired, but whose vested interest in the current theory apparently left him "decidedly hostile and ironical" towards the theory of games.[14] When it turned out that the upset Lutz later approached Weyl, wondering what was wrong with the use of calculus in economics, Morgenstern chortled, and reported all to an amused von Neumann. From a meeting with British reformer William Beveridge, he came away seething. The claims by economists that they could define and achieve full employment were a sham, similar to the claims of alchemists in the Middle Ages, for which some of *them* were beheaded. "Yesterday, when I saw Johnny again, it again struck me that in the social sciences there are hardly any people of such intellectual eminence as in mathematics and the natural sciences. (Perhaps Böhm-Bawerk, Menger, surely not Marshall!). That's why young talented potential economists cannot really look up to anybody, or get a feeling for the spirit of the science. There is a completely different atmosphere in economics from in these other circles. The economists don't seem to know this; otherwise, they would show much greater humility, which is completely missing in their character".[15]

Morgenstern's reflections on Keynes are unremittingly negative and continuous with his earlier criticism of his lack of precision in the treatment of expectations. He felt one of the reasons students were attracted to Keynes, as to John Hicks, was because the criticism by the Institutionalists or Clark was not on the same intellectual level. A letter from Haberler provoked the reflection that Keynes was "one of the biggest charlatans who has ever appeared on the economic scene. And everybody is on their belly before him... Someone should really face up to Keynes one of these days, he is brilliant, very intelligent, and his opponents are no match for him".[16] Several months later, he was reading more of Keynes: "a strange mixture of good knowledge and pseudoscientific thinking, but intelligent in its mode of expression and therefore (for those three reasons) so dangerous".[17] In

[13] Diary, Nov. 8, 1942, OMDU.
[14] Diary, May 2, 1943, OMDU.
[15] Diary, June 29, 1943, OMDU.
[16] Diary, Oct. 10, 1943, OMDU.
[17] Diary, Mar. 12, 1944, OMDU.

September 1943, he received a letter from Hayek, who "hated science as he always has", and claimed to have heard curious rumours about the book. "He is going to find it even more 'curious' when he sees it... He is in a dead end. The Pure Theory of Capital is not worth reading".[18] A year later, he read *The Road to Serfdom*, for which he managed grudging praise: "It's not worth a lot. Only wonderful in its love of freedom. He should look up what is said in our book about symmetry and fairness! There is nothing profound in it. Naturally I don't like planning either, but the intellectual situation is much more complicated".[19]

The Princeton economics department he now felt to be even more provincial and unsatisfying: "There are only 4 or 5 graduate courses, no seminars yet, no discussions. The students don't like it either. [X] is totally unsuitable: the mathematicians have weekly colloquia, as do the physicists, psychologists and chemists. We have to have somebody with a truly scientific spirit... something must happen".[20] At the Institute, he attended lectures and seminars in a range of subjects, the most recent at this point being on probability by Princeton statistician Sam Wilks, who said he had little esteem for what he encountered in "statistics and mathematical economics a la Econometrica".[21]

In March 1943, just before the book went to press, Bertrand Russell visited Princeton, to Morgenstern's delight but, also, disappointment. Russell made remarks concerning politics and culture, which, Morgenstern felt, were inaccurate and poorly defended. For example, he spoke of the "unalterable and deep seated hatred of England among the common people" of the United States, which Morgenstern regarded as completely exaggerated. It was also clear that Russell had somewhat lost contact with mathematics, and Morgenstern felt that he was surpassed, as a personality, by Weyl, Johnny, and Siegel, and, as a logician, by Gödel. In October, Morgenstern refused an invitation from the American Philosophical Society to speak about the influence philosophy had had on him: "I am afraid it would be

---

[18] Diary, Sept. 1, 1943, OMDU. The previous year, he had written: "Last Thursday a report in the graduate course about Hayek's 'Pure Theory of Capital'. It is higher nonsense... They keep talking about 'Investment Function', but there is no question of stating the concept with any precision. He does not seem to know what a function really is. This type of 'economic theory' *must* vanish" (Diary, Mar. 15, 1942, OMDU). "Talked about Kalecki in the graduate course yesterday, with the book fresh in my mind. Economists simply don't know what science means. I am quite disgusted with all of this garbage" (*ibid*, April 14, 1942).

[19] Diary, Oct. 25, 1943, OMDU.

[20] Diary, Oct. 7, 1941, OMDU.

[21] Diary, Jan. 2, 1944, OMDU.

too much work, and would be no fun for them if I said openly what I think now . . . ".[22] Nineteen forty-three closed with dinner with Russell, the von Neumanns, and Siegel at the Nassau Tavern, where the conversation was "interesting, and difficult", and Russell stimulating, as always, but not up to date. "He said that quantum mechanics is inexact, to which Johnny responded immediately, and one saw that R. was not informed. He is also not conversant with the newer mathematics . . . [Nonetheless] I have much to thank him for, and I told him that. I only wish I had read more of him earlier. And that brings me back to the old theme; time wasted on 'philosophy'".[23] A later conversation with Herman Weyl turned to Kant, by whom the mathematician, in his youth, had been impressed, until he discovered Hilbert's geometry. "That not only brought him to mathematics, but cleared the whole of philosophy as a pseudoscience out of his way".

Reception

Von Neumann's theory of games brought together empirical, mathematical, and aesthetic considerations in a way quite unlike anything seen in social theory beforehand. This mystified some and rankled others. In an early review of the book for the *American Journal of Sociology*, and then in a letter to Morgenstern, Herbert Simon expressed doubts about the stability concept. The set of solutions might be narrowed down, he suggested, by accepting that "a player will not help to maintain an imputation if he receives the smallest share from that imputation".[24] Whether or not this restriction was acceptable, he said, was ultimately an empirical question: "I am not certain that I understand Professor von Neumann's position on this point, but I got the distinct impression from discussion with him that his preference for the definition you used was based largely on aesthetic and formal grounds. Being a social scientist rather than a mathematician, I am not quite willing for (sic) the formal theory to lead the facts around by the nose to quite this extent – although I recognize that similar methods have been very fruitful both in relativity theory and in quantum mechanics" (*ibid*).

I have further difficulty in pinning down the operational meaning of the term "solution". It is clear that only one imputation can follow from a single play. If a single imputation were observed, it could only be concluded that the solution which held for the players of the game was one of all those solutions of which

[22] Diary, Oct. 25, 1943, OMDU.
[23] Diary, Jan. 2, 1944, OMDU.
[24] Simon to OM, August 20, 1945, OMDU, Container 32, File 90.

the particular observed imputation was a member. There might be an infinity of such solutions. If the society were 'stable' over a period of time then successive observations of imputations might successively narrow the set of possible solutions. Since an assumption of stability would be necessary to determine the solution, the stability of the solution could not be empirically tested (*ibid*).

In 1951, in New York, at an interdisciplinary meeting to discuss circular causal and feedback mechanisms in biological and social systems, the Eighth "Cybernetics" Conference, statistician Leonard "Jimmie" Savage emphasised this feature of game theory. He had been von Neumann's assistant at the Institute in 1941, while the book was being written.

[S]o far as I have been able to understand, Von Neumann's theory simply doesn't make testable predictions about many-person games. Though a lot of mathematical machinery is constructed in this connection, Von Neumann, neither in writing nor in conversation, seems to me to make at all clear what empirical consequences this machinery may suggest. The situation is totally different from that for two-person games, about which Von Neumann's writing suggests quite definite consequences. Thus, though such an experimenter as Bavelas can and does test the consequences of the two-person zero-sum theory, I think he would not even know what to look for in the case of many-person games.[25]

Other opinions were mixed.[26] One E. J. Gumbel at the New School for Social Research felt that there was a long way to go before the theory would furnish "tangible results for the rational solution of economic problems" (1945, p. 210). He doubted whether the method, which was based on a capitalist form of production, could cover "all rational economics", and, like many social science readers, was discouraged by "page after page full of formulae". Writing in *Science and Society*, Louis Weisner of Hunter College curiously found the authors' vision to be "bounded on all sides by the doctrine of marginal utility" (1945, p. 368). He said "they explicitly exclude the economy of a communist society from their program because their theory demands that combat and competition prevail in the distribution

---

[25] In von Foerster et al (eds.) (1950), p. 38. Luce and Raiffa (1957), in the landmark book on game theory of the postwar period, express a similar mystification concerning the nebulousness of the stable set.

[26] A partial list of publications in which the book was reviewed includes, in chronological order, *Psychological Abstracts*, *View*, *American Journal of Sociology*, *Annals of American Academy of Political and Social Science*, *The Times Literary Supplement*, *Bulletin of the American Mathematical Society*, *Mathematical Gazette*, *Journal of Farm Economics*, *Mathematical Reviews*, *American Economic Review*, *Journal of American Statistical Association*, *Science and Society*, *Commentary*, *Journal of Philosophy*. "Reviews of Theory of Games and Economic Behavior by John von Neumann and Oskar Morgenstern", from I.A.S. Archives, Faculty Files, von N. John, folder: "Morgenstern Book".

of the social product". Because of this, they were silent on the problems of economic production, a reticence that was "not peculiar to the authors but to the marginal utility school of economics to which they belong". Their theory had nothing to say about why mercantilism gave way to industrial capitalism, and why the latter gave way to imperialism. To such questions, their theory, "being static and unhistorical, cannot provide any answers at all".

Writing in *View, the Modern Magazine,* alongside André Breton, Man Ray, and Meyer Schapiro, the Princeton algebraist, Claude Chevalley, welcomed the appearance of set and group theories in social science. He said applications of mathematics to date had been unsuccessful because, too often, one has tried to follow "the pattern indicated by the mechanical or physical theories, where the spotlight is taken by differential equations expressing the immediate future of a system in terms of its present condition" (1945, p. 43). Chevalley gives a succinct, verbal presentation of the mathematical theory, noting that the question of the existence of a solution for all *n*-person games remained "unfortunately open". He praised the book, hoping that it would help economics "emerge from its actual condition of vagueness and confusion to the rank of a body of precise statements bearing on precisely defined situations".

In the *Journal of Political Economy,* Jacob Marschak of the University of Chicago wrote an extensive review, which also appeared as a Cowles Commission Paper. "All is not well with static economics", he opened. The first part of the review is devoted to a masterful exposition of the general three-person game of two buyers and one seller. Which solution will emerge, and which imputation within it, will depend on standards of behavior, "legal or moral codes" (1946, p. 104). In the two-person zero-sum game, the assumed symmetry of intelligence of the players is contrasted with Keynes' analysis in Chapter 10 of the *General Theory,* which describes the importance to the stock market of that "third degree where we devote our intelligence to anticipating what average opinion expects the average opinion to be" (p. 106). The main achievement of von Neumann's theory, Marschak felt, was not so much in its concrete results as in its having "intoduced into economics the tools of modern logic and in using them with an astounding power of generalization" (p. 114). He underlined the procedure by which empirical concepts generated by experience were formulated into theoretical concepts that were then detached from experience until the final conclusions were reached. At this point, they were "materialized", transformed again into the language of the concrete field, prepared for empirical test. He agreed with von Neumann that mathematical economics had hitherto illegitimately borrowed its tools from older physics, "without more scrutiny of logical foundations than was then usual in physics itself" (p. 115). Not surprisingly, however, Marschak felt that the authors underestimated the value of existing work in mathematical economics, especially when compared with the "vague propositions of premathematical economics" (*ibid*). He closed with high praise for the book's "simplicity, clarity, and patience", a "spectacle of vigorous thinking". He concluded, "Ten more such books, and the progress of economics is assured" (*ibid*).

Another Russian, well-trained in mathematics and statistics, was then on leave from Iowa State on a Guggenheim Fellowship at the Cowles Commission, where he wrote a review of the book. After a systematic account, Leonid Hurwicz, too, expressed dismay, even greater than Marschak's, at the indiscriminate attacks by von Neumann and Morgenstern on the present techniques used in economics. The discipline, he said, could not afford the luxury of developing in the most "logical" manner when results were needed to attend to employment fluctuations and the like. One did not always need a mathematical proof to know that a cartel was likely, nor was it clear that the ultimate results of the theory of games would be so different from those obtainable by current methods as to justify the harshness of the opening chapter. These "vague criticisms", he felt, were hardly "worthy of the constructive achievements of the rest of the book" – which must have stung Morgenstern. He also noted the curious absence of reference to earlier writers: one might be led to conclude that only Böhm-Bawerk and Pareto were known to the authors. Nonetheless, the work was a spectacular achievement, "remarkably lucid and fascinating", a "rare event" (1945, p. 925).

Writing in the *Economic Journal*, Cambridge's Richard Stone was sensitive to the breadth of the work, noting that it would interest not only mathematicians, economists, and students of games *per se*, but also sociologists, political theorists, and students of diplomatic and military strategy. Harvard's Carl Kaysen wrote a detailed and impressive review in the *Review of Economic Studies*. In his words, it is an "act of some temerity for a mere economist to pass judgment on a great work in which a writer who is a brilliant logician, mathematician and physicist, with the aid of an economist, offers all economists a new foundation for their thinking, a foundation which is at once vast and finely wrought, made with infinite labour" (1946–47, p. 12). However, he found the utility axioms to be a weak point. Whereas von Neumann was content to gloss over this issue with the observations that it was impossible to construct a consistent system of axioms that allowed for a utility from gambling, and that at least an axiomatisation of a psychological domain had been achieved, Kaysen emphasised the fact that people bought lottery tickets, purchased insurance policies, and acted as entrepreneurs. Kaysen noted, if von Neumann and Morgenstern found it "impossible to formulate the concept of the utility of gambling in a consistent way, so much the worse for their postulate systems" (p. 13). The major difficulty, however, was not with utility, but with the indeterminacy of the solution concept, which often permitted an infinite range of values to be achieved by a single player. To be of "practical value in really complex cases", the theory would have to say something about what forces would be operational in a given situation, which coalitions would be formed, what compensation payments made. For the moment, Kaysen concluded, the theory of games promised no revolution in economics.

Conclusion

In the short run, Kaysen proved to be quite right. Although the book found isolated, thorough readers such as Gerard Debreu, Herbert Simon, and John Harsanyi, and although von Neumann's treatment of expected utility caught the interest of economists–statisticians such as Friedman and Savage, the work as a whole was not widely embraced by the community of economists. As predicted by von Neumann and heralded in some of the reviews, the book's mathematical apparatus, difficulty of access, and sheer strangeness of subject combined to make it appear rebarbative in the eyes of many economist readers. What, therefore, happened to it? If academic economists did not care about game theory in the immediate postwar period, who did?

To answer that question, we have to turn away, yet again, from economics, and consider what was happening during World War II, when von Neumann and Morgenstern were, albeit increasingly sporadically, writing together. In a nutshell, the war saw the development of operations research and, with that, the application of a small, relatively minor part of game theory to the analysis of military engagements. This was where game theory made its entry into the world, so to speak. That wartime experience, in turn, shaped the postwar period, with the continued promotion of operations research and, in time, the development of Cold War social science. In this isolated world, centred on the RAND Corporation, game theory became a constitutive element, naturally passing out of von Neumann's hands in the process. Throughout the war and afterwards, whether in battlefield applications or in the social self-scrutiny of the Cold War laboratory, the demands made of game theory were new. In its application, be it to bomber–fighter duels or experimental games, there was a distinct collapse of scope, leaving the theory far removed from the broad theme of social configuration and reconfiguration that had originally fired von Neumann. Amidst the accidents of history, his aim to create an abstract mathematics of ambitious social explanation was quietly forgotten.

Plate 19. Von Neumann party in 1930s' Princeton. From left Angela Robertson, Mariette von Neumann, Eugene Wigner and his wife, von Neumann, Edward Teller, and H. P. "Bob" Robertson. *Credit:* Courtesy of Marina von Neumann Whitman.

Plate 20. Rudolf Ortvay. *Credit:* Courtesy of Professor Gyula Radnai, Budapest.

Plate 21. Oskar Morgenstern and John von Neumann, Sea Girt, New Jersey, 1949.
Credit: Courtesy of Dorothy Morgenstern-Thomas.

Plate 22. RAND Corporation Headquarters. *Credit:* Courtesy of Getty Images.

Plate 23. Oskar Morgenstern, his son Carl, and John von Neumann, California, 1953. *Credit:* Courtesy of Dorothy Morgenstern-Thomas.

Plate 24. Willy Fellner, Klari von Neumann, Oskar Morgenstern, John von Neumann, Pacific Palisades, California, 1953. *Credit:* Courtesy of Dorothy Morgenstern-Thomas.

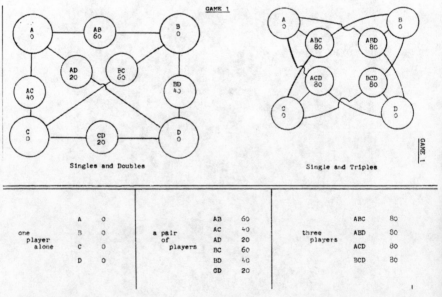

one player alone			a pair of players			three players		
	A	0		AB	60		ABC	80
	B	0		AC	40		ABD	80
	C	0		AD	20		ACD	80
	D	0		BC	60		BCD	80
				BD	40			
				CD	20			

Plate 25. Game 1 from Kalisch, G., J. Milnor, J. Nash, and E. D. Nering (1952) "Some Experimental n-Person Games", RAND RM-948, August 25, reprinted with permission.

Plate 26. Playing war games at RAND, 1955. Clockwise, from left: John D. Williams, Philip Moseley, Lee DuBridge, Philip M. Morse, and Robert F. Bacher. *Credit:* From Philip M. Morse, *In at the Beginnings: A Physicist's Life,* © 1976, Massachusetts Institute of Technology, by permission of The MIT Press.

Plate 27. Close-up of New England during Strategic War Planning (SWAP) play. *Credit:* Reprinted with permission from Helmer, Olaf (1960), "Strategic Gaming", RAND P-1902, February 10.

TWELVE

Von Neumann's War

Although Morgenstern at first was completely absorbed by the writing of the *Theory of Games* with von Neumann, it soon became one activity amongst several. We have portrayed his return to game theory as partly an early reaction to upheaval and war; an attempt to construct a mathematics of social configuration. As reactions go, it was an abstract, intellectual one – the theoretical analysis of coalitions and compensations, with the "distant hope of some application to social phenomena", as von Neumann described to Ulam[1]. By 1941, however, the very events that had earlier fired von Neumann's imagination had culminated in American involvement in war, and he was drawn from the distant observation of political drama to actual participation in the conflict. He became directly, physically involved, interacting with military officers, incessantly moving about, boarding ships and submarines.

With that involvement came a significant, unforeseeable moment in the history of game theory: this new mathematics made its wartime entrance into the world, not as the abstract theory of social order central to the book, but as a problem-solving technique. In the course of World War II, a small, narrow part of the mathematics, far removed from the stable set in either content or philosophy, was adopted in the emerging field of operations research. What might be called the theory's social integration occurred in this context.

Science and War

Late in the evening of July 14, 1943, from his room in the Cosmos Club in Washington, D.C., von Neumann wrote to Ulam, saying that he was "not

---

[1] Von N to Ulam, April 2, 1942, Stan Ulam Papers, American Philosophical Society (hereafter, SUAP).

quite dead mathematically yet", but was doing mostly gas dynamics, and "even experiments, and, horribile dictu, find it amusing".[2] The mock apology for his enjoyment is revealing, for it speaks to a distinction, important to von Neumann and Ulam, between being a "pure" and an "applied" mathematician. Applied mathematicians were viewed by purists with a certain disdain: they had turned to applications, it was felt, because they were not up to the rigours of abstract mathematics. To work on the axioms of set theory was an elevated pursuit; to model the growth of locust populations or calculate the trajectory of artillery shells was footwork. Thus, during the war, when von Neumann's interests were changing, Ulam could poke good-natured fun at him: "When it comes to the applications of mathematics to dentistry, maybe you'll stop".[3]

Having laid down the foundations of the book with Morgenstern, von Neumann became increasingly involved in military consulting to the Ballistic Research Laboratory (BRL) at Aberdeen Proving Ground, Maryland; the National Defense Research Committee (NDRC); and the Navy's Section for Mine Warfare. This saw him spend the last part of 1942 in Washington, D.C., and the first half of the next year in England.[4] As we shall see, he also became deeply involved in the Manhattan Project at Los Alamos. A decade later, at a ceremony in honor of General Robert Kent, who, in the late 1930s, had been a senior official at Aberdeen, von Neumann said: "It was through him that I was introduced to military science, and it was through military science that I was introduced to applied sciences. Before this I was, apart from some lesser infidelities, essentially a pure mathematician, or at least a very pure theoretician. Whatever else may have happened in the meantime, I have certainly succeeded in losing my purity".[5]

The loss of purity trumpeted by von Neumann had begun in the late 1930s, when he was introduced by Oswald Veblen to the BRL.[6] This project, in which Veblen had been involved during World War I, was renewed in 1937 with the heightening tensions in Europe, so that, from 1937 to 1940, von Neumann worked occasionally for the BRL, mostly in mathematical statistics, on the procedures used in testing ammunition. As of 1940, along with physicists I. I. Rabi and Theodor von Karman, von Neumann

---

[2] Von N to Ulam, July 14, 1943, SUAP.

[3] 1976, p. 245.

[4] See von Neumann to Frank Aydelotte, Oct. 11, 1945, VNIAS, Faculty Files, John von Neumann, Folder: 1941–.

[5] Kent Symposium, December 1955, quoted in Aspray (1990), p. 26.

[6] See Aspray (1990), p. 26ff. On Veblen's involvement in fundraising for mathematical research after World War I, see Feffer (1998).

continued to act as adviser to the BRL, now mostly on aerodynamics. That year marked what his assistant Paul Halmos (1973) describes as a "discontinuous break" in von Neumann's publications: up to then, he was a "topflight mathematician who understood physics", says Halmos; after that, he was "an applied mathematician who remembered his pure work" (p. 391). From that point onwards, as one involvement led to another, von Neumann became steeped in successive wartime applications: statistics, shock waves, flow problems, hydrodynamics, ballistics, computation, and problems of atomic detonation.

In a confidential letter in October 1945 to Frank Aydelotte, Flexner's successor at the Institute for Advanced Study, von Neumann described that wartime work.[7] Following his BRL induction, in 1941 and 1942 he was a member of Division 8 of the National Defense Research Committee (NDRC), working mainly on high explosives, in particular the shaping of charges: "In these devices the precise geometrical shape of an explosive charge is used to modify, concentrate, or direct, the physical effects of detonation". In 1942 and 1943, he worked for the Section for Mine Warfare of the U.S. Navy's Bureau of Ordnance, Research and Development Division, on "operational research". In this connection, he had spent the last part of 1942 in Washington, and from January to July 1943 in England. It was there that he developed what he described as an "obscene" interest in calculation and computing. As of late 1943, he had continued to consult with the Army Ordnance Department and the Navy Bureau of Ordnance on aerodynamics and the application of high explosives.

Beginning in early 1944, at the Institute at Princeton, he had directed a project, initiated by the Applied Mathematics Panel and then taken over by the U.S. Navy Bureau of Ordnance, the objective of which was to develop new computing methods suited to "very-high-speed computing devices that [would] become available in the near future". Starting in late 1944, he had been a member of Division 2 of the NDRC. He had previously advised them on "the interaction and reflection of blast waves" and was now working on "the effects of explosive blast and of projectile impact on various structures", work he thought would likely continue to be supported, at Princeton, by the government after the war: "The conclusions reached in the course of this work led to direct military applications in determining the conditions under which large and extremely large bombs have to be detonated; the subject is too highly classified to be discussed here".

---

[7] Von Neumann to Aydelotte, Oct. 11, 1945, VNIAS, Faculty Files; von Neumann, J., Folder 1941–.

Since September 1943, he had been spending almost a third of his time as consultant with the Manhattan District, United States Engineers, on a project at Los Alamos, New Mexico, devoted "to the properly military and explosive aspects of nuclear fission". His work there, he continued, "was of a triple character: theoretical, in certain phases engineering, and operational research. It is still highly classified . . . so that I cannot discuss it here". Finally, since early 1945, again at the BRL in Aberdeen, he had worked increasingly on the development of "various high-speed computing devices, and quite particularly in planning a new electronic machine" (*ibid*).

Von Neumann's wartime role, one associate later wrote, "was unique; for he was a consultant or other participant in so many government . . . activities that his influence was very broadly felt".[8]

His numerous and tangled involvements reflect the changing relationship between the American military and scientific establishments during the war. In the space of a few years, physicists and mathematicians especially, but also economists and humanists, gained an unprecedented role in advising various branches of the military and the government, and their advice was seen as significant in defeating the Germans and Japanese. In what follows, we consider the wartime engagement of mathematicians, in general, and see, in particular, how game theory was used. All of this, in turn, will allow us to understand many features of the postwar, Cold War, world, in which game theory played a distinctive role.[9]

The wartime mathematical activity of interest to us fell under the control of the NDRC, which was established in June 1940 by an order of President Roosevelt to "correlate governmental and civil research in the fields of military importance outside of aeronautics" and was chaired by Massachussetts Institute of Technology engineer Vannevar Bush.[10] The NDRC administered contracts for defence research with universities and other institutions. For example, the key contribution of the Radiation Laboratory at MIT was the development of shorter wave radar that yielded better resolution and clarity on the radar screen. At its peak, it employed more than

---

[8] Rees (1980), p. 609.

[9] For a detailed account of the use of economics, statistics, and operations research during World War II and the Cold War, and the ensuing impact on the theories and techniques of modern economics, see the work of Judy Klein, including (2000) and her forthcoming book on the subject.

[10] Baxter (1946), p. 15. Aeronautics was in the domain of the National Advisory Committee for Aeronautics (NACA) established in 1938. The NDRC had five internal divisions catering to perceived needs in armor and ordnance; bombs; fuels and chemicals; communication and transportation; detection, controls, and instruments; patents and inventions.

4,000 people from all over the country. Other large laboratories, developing underwater sound, were operated in California, at Columbia, Harvard, and the Woods Hole Oceanographic Institute.

Under the NDRC umbrella, mathematicians were involved in improving the effectiveness of weapons, with some early work on the design of new types, but by and large through better use of existing ones.[11] It was here that von Neumann's wartime influence began to take hold, as he became active in advising on applications of mathematics to operations research and related problems. The wartime centres of activity of particular interest to us were the Anti-Submarine Warfare Operations Research Group (ASWORG), located in Boston and attached to the Navy, and the Applied Mathematics Panel, the research of which was used primarily by the Army Air Force.

ASWORG

I have just been informed that Joh. v. Neumann is in this country and will visit me next week accompanying a man from the Admiralty for whom he will do some war research.

Max Born (Edinburgh) to Albert Einstein, May 10, 1943[12]

In March 1942, MIT physicist Philip Morse was given the role of directing a group of civilian scientists who would advise the Navy's new Anti-Submarine Warfare Unit.[13] German submarines were now active off the East Coast and there was a need to protect American coastal traffic and transatlantic convoys.

Morse was already familiar with military work. An acoustics specialist, he had joined the Radiation Laboratory at MIT in early 1941 and then set up an Army Air Force noise-control project at Harvard.[14] It was then

[11] The mathematicians had shown an early keenness to become involved, with the American Mathematical Society and the Mathematics Association of America, in 1940, jointly appointing a War Preparedness Committee, with subcommittees on research, preparation for war, and education for service. Amongst the consultants to the research group were von Neumann (Ballistics), Norbert Wiener (Computation), and Samuel S. Wilks (Probability & Statistics). See Morse and Hart (1941), p. 296.

[12] Quoted in Born (1978), p. 142.

[13] See Morse (1977), p. 172ff, and (1948), his Josiah Willard Gibbs lecture to the American Mathematical Society. See also Morse and Kimball (1951), the declassified version of a report by the same authors written in 1946 (see Tidman [1984], pp. 102–03).

[14] The project had two components: a physiological–psychological study of the effects of noise on the crew to decide whether cabin noise reduction in the bomber was necessary, and measuring and devising methods to reduce noise. The engineering part was headed by Leo Beranek and the psychological part by S. S. "Sid" Stevens, both Harvard faculty members.

that Morse was approached by the Navy, which was facing a new problem with mines in the English Channel, being laid by the Germans in the shallow parts of the shipping lanes. Previously they had used a magnetic detonation system for the mines, which the British had learned to counter by neutralising the magnetic field of their ships through the use of a system of coils. Now, however, they had switched to noise detonation, with the mine being triggered by the sound of a passing ship. The result was a Navy contract with MIT, which saw Morse and his group working in a tug boat off the Massachussetts coast testing and building an underwater microphone. The idea was to use the recordings to make a device that would imitate a ship's noise and could be dragged over the mines so as to detonate them. With test trips on the sea and involving a great deal of *bricolage*, it was all a far cry from theoretical acoustics.[15]

Morse's next problem was the Navy anti-submarine project. By early 1942, with German U-boats attacking along the East coast, the matter had become urgent. Using his contacts to secure personnel in what had become a very competitive climate, within a few months, Morse had assembled a mixed group of physicists, mathematicians, and even actuaries and geneticists. Morse and his group started to think about how to find and destroy German submarines, the "search problem". In this case, given the detection technology, this required considering the *pattern* of search. If submarines were submerged, they could be detected using expensive, specially equipped surface vessels. However, they actually spent more than half their time on the surface, where they were able to travel more quickly, charge their batteries, and send coded messages on shortwave radio. Therefore, less expensive airplane search could be used and Morse developed a technique for optimal search depending on distance, angle of vision, and general visibility conditions (see Figure 12.1).[16]

---

[15] Morse (1977), pp. 162–63.

[16] The airplane proceeds by making uniform 360° visual or radar "glimpses" of the horizon. The probability of the airplane making contact with the target in a period dt, can be written as $\gamma(r)$ dt, where γ is usually independent of the angle between the line of sight and the direction of motion of observer or target, and depends on both r, the horizontal distance between observer and target, and general visibility conditions (haziness, quality of the radar set, etc.). In particular, within narrow limits, says Morse, the probability of detection will depend on ω, the solid angle subtended by the object at the eye. If the target is a wake of constant area on the surface of the ocean, then, if $r \gg h$, we can write $\gamma = k\,\omega = (B/s^2)\sin a = (Bh/s^3) \cong (Bh/r^3)$, where the quantity B depends on the area and the contrast of the wake, the atmospheric conditions, and so on. This is the inverse cube law for contact rate, which provides a guide for searching for small objects in conditions of good visibility, and can be handled analytically. Morse shows how the analysis can be extended to take account of the motion of the observer with respect to the target. See Morse (1948, p. 607).

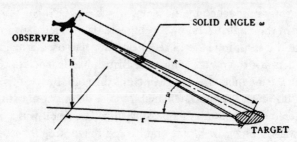

Figure 12.1. Visual detection of wake on ocean surface by observer in aircraft. *Credit:* From Philip Morse (1948), "Mathematical Problems in Operations Research", *Bulletin of the American Mathematical Society*, Vol. 54, pp. 602–21. With permission from the American Mathematical Society.

Locating a passive target is one thing. What happens if measure is met with countermeasure – if A's choice of search tactic depends on B's choice of evasion or defense tactic, and vice versa? "Problems of this sort, are discussed by von Neumann and Morgenstern in their *Theory of games*", says Morse.[17]

Morse had known the von Neumanns since the late 1920s, when he was completing his doctorate at Princeton. He particularly liked Mariette, von Neumann's first wife, and both she and Desmond Kuper, the physicist for whom she had left von Neumann in 1937, would be amongst the first of Morse's recruits in 1946 when he became director of the Brookhaven nuclear laboratory on Long Island. He also liked von Neumann, whom he regarded as smarter than Oppenheimer and more versatile than Einstein.[18] Morse was thus being advised on the use of minimax theory in the search problem, in Washington in late 1942, before the *Theory of Games* even appeared: "Johnny von Neumann was round, genial, helpful, and interested in everything. No one I even knew could as quickly grasp the essence of one's halting explanations, point out the crux of the difficulty, and suggest ways to solve it. He would really listen, while simultaneously thinking through the implications of what he was hearing. It was at times discouraging to have him come out with the answer before I got through explaining the problem".[19]

The problem in this case was that of a submarine attempting to pass through a channel, undetected by an airplane that is trying to spot it.

---

[17] Morse, *op cit*, p. 613.
[18] Morse (1977), p. 87.
[19] Morse (1977), p. 98.

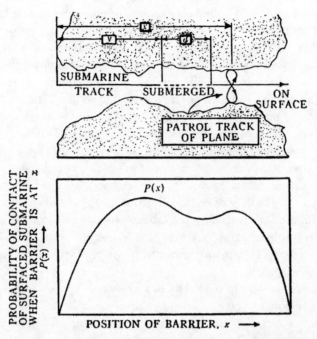

Figure 12.2. Barrier patrol versus diving submarine. $P(x)$ = probability of contact if barrier is at x and submarine is surfaced. *Credit:* From Philip Morse (1948), "Mathematical Problems in Operations Research", *Bulletin of the American Mathematical Society*, Vol. 54, pp. 602–21. With permission from the American Mathematical Society.

Whether in sections or all at once, the submarine can only travel submerged a total distance, a: the rest of the time it is visible. The plane must choose a point along the channel, the width of which varies, at which it will traverse in surveillance (Figure 12.2).

$P(x)$ is the probability of detection if the plane patrols at a point at distance x from the channel entry, given that the submarine is visible at that point. The narrower the channel at any point, the higher the probability of detection, because the average time spent over the water is inversely proportional to channel width. If the plane chose the point that maximises $P(x)$, the submarine would choose never to surface at that point. The plane, therefore, should choose a density function, $\phi(x)$, which will allow it to choose its patrol point in some *random*, but optimal, manner. Similarly, if it is to avoid detection, the submarine must choose a submergence probability $\psi(x)$, which will be conditional upon $P(x)$ and physical submergence constraints.

Morse shows how the submarine commander can ensure that the probability of contact never exceeds a certain amount, H, regardless of whether or not $\psi(x)$ is known to the plane pilot. The submergence probability density $\psi(x)$ thus chosen will require the submarine commander to concentrate his resources where they are needed: depending on the shape of $P(x)$, he may even be advised not to dive at all in certain parts, to disregard less promising regions completely.[20] "The same result . . . turns up in many problems discussed by von Neumann and Morgenstern" (1948, p. 617).[21]

For his part, the pilot flies a random patrol along x, in such a way that even if its density, $\phi(x)$, is known by the submarine commander, it will be of no advantage to him. Symmetrical to the submarine, he can guarantee a minimum probability of contact, K. And it can be shown that K = H, that the optimum probability of contact generated by both sides is the same. "The problem considered here, says Morse, is an extremely simplifed example of a large class of problems which turn up continuously in tactical studies" (*ibid*, p. 619).

As a continuation of the search problem, Morse's group worked on the question of how to best place air-carried depth charges, and apparently managed to increase submarine casualties by changing the depth at which charges exploded. Solutions of this kind were directly useful during the war, says Morse, with casualties increasing significantly after certain search and attack procedures were implemented. Morse travelled to blacked-out London in late 1942, where he consulted with P.M.S. Blackett, founder of operational research and adviser to the British Admiralty on radar detection. He and William Shockley also visited the various other English and

---

[20] In short, let:

$$\text{Density of patrol} = \phi(x) \int \phi(x)\mathrm{d}(x) = 1$$
$$\text{Probability of submergence} = \psi(x) \quad \int \psi(x)\mathrm{d}(x) = 1$$
$$\text{Probability of detection } G = \int P(x)\phi(x)[1 - \psi(x)]\mathrm{d}(x)$$

Given $\psi(x)$, the probability density of submergence chosen by the submarine commander, the patrol-squadron commander wishes to maximise $P(x)[1 - \psi(x)]$, and the submarine commander wishes to keep it as low as possible, subject to the constraint that he must spend some time at the surface. His best tactic is to ensure that $P(x)[1 - \psi(x)]$ is a constant over as great a range of x as possible. He sets $P(x)[1 - \psi(x)] = H$, i.e.:

$$\psi(x) = 1H P(x) \quad \text{when } P(x) > H \text{ (over range B of } x)$$
$$0 \qquad \qquad \text{when } P(x) < H \text{ (over range B of } x)$$

where H is determined by the submergence limitations of the submarine.

[21] As it happens, it turns up relatively little in the *Theory of Games*, but such a casual remark by Morse is quite consistent with his having been advised by von Neumann prior to the book's publication.

American operations research groups at work in the country: the Coastal Command group dealing with submarines in the Bay of Biscay, and the group attached to the U.S. strategic bombing command. Here, he became aware of the emerging conflict over the choice of German targets. Following the Blitz, the British were in favour of extracting an eye for an eye, with general high-altitude bombing likely to result in many civilian casualties. Some of the Americans, including Bob Robertson, Princeton mathematician and friend of von Neumann, believed that this was not only excessive but would have the contrary effect of strengthening rather than diminishing German resolve. However, Robertson found it impossible to convey his opinions to his group leader or above, and was generally frustrated by this gap in communication between the various levels.[22] It was also in late 1942 that von Neumann became involved in Morse's group on the anti-mine warfare, the work that took him to England for the first half of 1943.

The Applied Mathematics Panel

In the fall of 1942, Vannevar Bush reorganised the NDRC to include a new unit that would coordinate the activities of mathematicians – the Applied Mathematics Panel (AMP). Warren Weaver was appointed director and his advisory committee included, amongst others, Richard Courant, Griffith Evans, Oswald Veblen, Marston Morse, and Sam Wilks. Von Neumann also provided counsel.[23]

Originally an applied mathematician from Wisconsin, Warren Weaver had been an officer with the Rockefeller Foundation since 1934, where he headed the Natural Sciences Division.[24] Based initially in Paris, he travelled

[22] Robertson had become involved in operations research relatively early. Following initial contacts on Blackett's operations research between the British and Conant, and then Columbia's Shirley Quimby, Robertson and J. E. Burchard, for Division 2 of the NDRC, visited Britain in the early fall of 1941, consulting with several of those involved in military consulting, including Solly Zuckerman. According to Fortun and Schweber (1994), Robertson then became deeply interested in Blackett's work, and became "a prime force in the development of operational analysis" in the United States (*op cit*, p. 603). It might also be mentioned that Robertson was also von Neumann's partner in crime in the composition of dirty limericks, for which they were both renowned (see Morse [1977], p. 98).

[23] Courant had left Göttingen in 1933, moving to NYU to head the newly established Courant Institute. Griffith Evans, who incidentally was one of the few mathematicians in interwar America actively working in economics, was at Rice. See Evans (1930) and Weintraub (1998). Marston Morse, like Veblen and von Neumann, was at the Institute for Advanced Study. See Rees (1980), p. 609.

[24] See Weaver (1970).

extensively to American and European universities, and was significantly involved in the extension of Rockefeller support to the sciences, and to biology in particular. Important beneficiaries included the laboratory at Cold Spring Harbor, Long Island, that of Linus Pauling at California Institute of Technology, and a range of other projects, many of them in molecular biology. In his slim, selective autobiography, Weaver recalls being appointed by Vannevar Bush, around 1940, as director of the Fire Control division of the Office of Scientific Research and Development. He writes, "Fire control refers to all the devices and procedures used to assure that any 'projectile' (a shell fired from an anti-aircraft gun, a bomb dropped from an airplane or a torpedo launched from a ship) will in fact hit the desired target" (pp. 77–78). The key project here, from mid-1940 to early 1942, was the development of the ground-to-air M-9 anti-aircraft director, designed to improve the accuracy with which German bombers and buzz bombs could be shot down as they attacked London.[25] In addition to the M-9 director, Weaver's Fire Control Division was involved in other similar projects involving the improvement of sighting and tracking procedures, such as directing guns from an airplane against other aircraft, or controlling fire in low-altitude attacks on submarines.

The establishment of the AMP in 1942 was not a straightforward matter because, at the NDRC, there had been a history of resistance towards mathematicians. This had its roots, in part, in a pre-war political clash at Harvard, where Conant had disagreed with mathematician Marshall Stone over the tightening-up of academic appointment procedures.[26] In 1940, when the American mathematicians set up a War Preparedness Committee, which included von Neumann, Wiener, and Wilks, they met with little interest from Bush and Conant. Amidst increasing unhappiness, spokesmen Stone and Marston Morse approached the NDRC to present their case, and in early 1942 presented a memorandum, "Mathematics in War", which bemoaned the fact that the mathematical intelligence of the country was being ignored by the czars of wartime science, especially given the fact that the Germans carried their calculations to "the fourth significant figure".[27] In response, Frank Jewett formed a Joint Committee on Mathematics and

---

[25] See Galison (1994).

[26] See MacLane (1989).

[27] Dunham Jackson, G. A. Bliss, G. C. Evans, and M. Stone, "Mathematics in War", National Academy of Science Archives, Administration, Organization, NAS-Committee on Mathematics, Joint with NRC 1942: General, quoted in Owens (1989), p. 302. I found Owens' discussion of this period to be very helpful.

included Stone, Morse, and Evans on the executive committee.[28] Nobody was happy with the Committee, Owens (1989) suggests. Stone was upset that he was not chairman, and Weaver felt excluded from the centre of action – little action though there may have been. Attempts to reconcile the engineering view of Weaver with the mathematical perspective of Stone came to little, so that even after the AMP was established in late 1942, under Weaver, the conflict continued, with letters flying back and forth between him and the marginalised Stone.[29] Weaver initially sought to hide behind the authority of Bush and Conant, suggesting that he was merely following orders in establishing the Panel, but in the face of Stone's antagonism he soon made explicit his "applied" leanings and his antipathy towards the demigods of "pure" mathematics. The war demanded a certain kind of mathematician, he told Stone, one who was familiar with the NDRC, comfortable with military officers, and knowledgeable concerning weapons procedures. The circumstances of war, he said, were different and were very particular: "It is unfortunately true, he continued, that these conditions exclude a good many mathematicians, the dreamy moonchildren, the prima donnas, the a-social geniuses. Many of them are ornaments of a peaceful civilization; some of them are very good or even great mathematicians, but they are certainly a severe pain in the neck in this kind of situation".[30] Weaver felt that mathematics did not form a unity, that the pure and the applied were quite different. There was even something, he felt, in the training of the applied mathematician, unlike that of the purist, that was conducive to service to one's country. The letters went back and forth for some time, but by then Stone had effectively been excluded and Weaver firmly held the reins. He would hold them for a long time.

Weaver ran the AMP from its head office at New York's Rockefeller Centre with the help of aides Ed Paxson and Mina Rees, two figures who would be important in the postwar period.[31] Contracts were set up with

---

[28] Other members included George Birkhoff, Veblen, Weaver, Bob Robertson, Walter Bartky, Harry Bateman, and Dunham Jackson.

[29] See Owens *op cit*, pp. 295–96.

[30] Weaver to Stone, Dec. 6, 1943, OSRD, Record Group 227, General Records, Organization-Applied Mathematics, National Archives, quoted in Owens, *op cit*, p. 296.

[31] Rees (1902–1997) obtained a doctorate in Mathematics at the University of Chicago in 1931 and taught at Hunter College until 1943. After the dissolution of the AMP in 1945, she became head of the Mathematics Branch of the Office for Naval Research, and was thus responsible for a great deal of research support of mathematical activity, including game theory, in universities all over the United States. See "Mina Rees" in Albers and Anderson (eds.) (1985). Edwin Paxson was a mathematician with a doctorate from Caltech.

eleven universities.[32] At Brown, for example, a group worked on classical dynamics and the mechanics of deformable media. At New York University, Courant's group worked on gas dynamics and, in particular, studied the shock fronts associated with explosions in the air and underwater.[33]

In Probability and Statistics, work was done by Statistical Research Groups (SRG) at Berkeley, Columbia, and Princeton. The Columbia group, the largest, was run by Allen Wallis with Harold Hotelling as principal investigator.[34] Wallis's group occupied a portion of a building at W. 118th St. New York, which also had as tenants the Strategic Bombing section of the Princeton SRG and Columbia's Applied Mathematics Group, discussed later. At the SRG-Columbia, Wallis and Hotelling gathered around them a group of capable statisticians, including Vienna émigré Abraham Wald, J. Wolfowitz, Milton Friedman, Jimmie Savage, Abe Girschick, Fred Mosteller, and George Stigler.[35]

The study of aerial combat was central to the work of the Columbia SRG. For example, an early study of alternative ways of placing machine guns on a fighter involved analysing the geometry and tactics of aerial combat. This, in turn, led to work on anti-aircraft weapons, aircraft turret sights, and dispersion of aircraft machine guns. A second broad area was the design of the optimal lead angles of aircraft torpedo salvos. This involved the interpretation of photographs of Japanese destroyers in order to glean information on speed and turning radius. In 1887, Lord Kelvin had established that, when a ship is moving at constant speed, the waves behind it remain in a sector of semi-angle $19° 28'$, regardless of the craft's size and speed, and that the actual speed is indicated by the intervals between the cusps along the bow waves. These facts, used in conjunction with the battle photographs,

---

[32] They included Brown, Berkeley, Columbia, Harvard, NYU, Northwestern, and Princeton (see Rees, *op cit*; AMP, Summary Technical Report, NCRC, Washington, D.C., 1946, 3 volumes). The main areas involved were Classical Dynamics, Mechanics of Deformable Media, Fluid Mechanics, Probability and Statistics, and Air Warfare.

[33] This resulted in *Supersonic Flow and Shock Waves* (1948) by Courant and Kurt Friedrichs.

[34] Wallis had left Stanford for the Office of Price Administration (OPA) in Washington (Wallis 1980). At Hotelling's suggestion, Weaver approached Wallis in mid-1942, suggesting that he lead the Columbia SRG, which he did from then until its dissolution in 1945.

[35] Amongst these, Wolfowitz, Friedman, and Girschick had all been Hotelling's students during the 1930s. Savage, after a doctorate from Michigan in 1941 and a year as von Neumann's assistant, had gone to Cornell and then Brown. See Wallis, *op cit*. Girschick, in fact, only spent 1944–45 at the SRG: the remainder of 1939–46 being spent as principal statistician for the Bureau of Agricultural Economics. Affected by his work with Wald, however, after a brief stay at the Bureau of the Census, he went to RAND, soon after its creation in 1946 (see Kruskal and Tanur (eds.) (1978), pp. 398–99).

allowed the mathematicians to make conjectures about the capabilities and flexibility of Japanese destroyers, and thus provide a better rationale for the arrangement of torpedo barrages.

The third field involved the development of inspection and testing procedures.[36] In early 1943, Navy Captain Schuyler wanted some way of evaluating the probability of a hit by anti-aircraft fire on a directly approaching dive bomber. Wallis and Ed Paulson worked out the statistical requirements for the (destructive) testing of the anti-aircraft artillery shells in the experimental development of such an evaluation. Schuyler was dismayed at the large samples involved and observed that, at the Naval Proving Ground at Dahlgren, Virginia, an experienced military officer would likely be able to make a judgement well before the prescribed number of shells had been test-fired. Perhaps there was some more formal way, he suggested, of using the sequential information provided by the testing so that the exercise could be conducted more economically? What was the point of destroying thousands of shells if it was "obvious" after a few hundred that further testing would not lead to rejection, or acceptance, as the case may be, of the hypothesis? Wallis approached Friedman and the two then approached Abraham Wald.

After his stay at the Cowles Commission, Wald had moved in the fall of 1938 to teach at Columbia, with the help of Hotelling.[37] Within a year, he had become professor of mathematical statistics and Fellow of the Econometric Society. During the war, however, he had still not been granted American citizenship and was technically an enemy alien, even when doing wartime research that was classified: the legend soon emerged that his work was snatched away from him as he completed it.[38] Although initially pessimistic before Wallis's and Friedman's problem, Wald soon turned to it and, not long afterwards, developed the sequential probability ratio test, a formal procedure corresponding to Schuyler's initial intuition. The ratification and formalisation of this idea became sequential analysis, and the use of these procedures apparently yielded significant economies in inspection for the Navy. The techniques were brought together in Wald's 1947 *Sequential Analysis*, and sequential testing tables became common on postwar industry assembly lines. Beginning in 1939, Wald had also integrated von Neumann's minimax theorem into his work on statistical decision theory, and while he was at the SRG during the war, he completed two related

---

[36] See Rees, *op cit*, pp. 614–15; Wallis, *op cit*.
[37] See Freeman (1968).
[38] See Wallis, *op cit*, p. 329.

papers, one reinterpreting statistical decision as a game against an inanimate nature and deriving decision functions that minimise the maximum risk; the other generalizing von Neumann's 1928 theorem to games with continua of strategies (Wald 1945a, b).[39] In general, the period saw many innovations in statistics, to many of which Wald was central. Looking back later, his student Harold Freeman wrote: "Statistical interest and relevance were high. Testing hypotheses, estimation theory were alive. The Fisher–Neyman controversies were unsettled. Sequential theory, decision theory, game theory, Bayesian notions, all exciting innovations, were underway. Those were effective years for statistical theory and application . . . Ghastly that such progress should have involved, even depended on, the deaths of millions of people thousands of miles away".[40]

The third group of statisticians was at Princeton, directed by Sam Wilks, dominant figure in mathematical statistics at that university's mathematics department.[41] Involved in the work of the NDRC as of 1941, Wilks stands out, along with Weaver and von Neumann, as a crucial figure in military research. It was at his recommendation that Wallis looked to Hotelling at the inception of the SRG-Columbia. At Princeton, Wilks directed statistical analysis of the Atlantic convoy problem and anti-submarine warfare, and he was involved in the statistical work done by the Columbia group. Like von Neumann, he was a peripatetic influence.[42] Of wartime mathematics, he remarked:

The methodology of research varied from formal mathematical analysis, at one extreme, to synthetic processes and statistical experiments or models at the other. Formal analysis is the more precise and hence satisfying process, but the difficulties of formulating the problem in analytical terms and then (worse) of finding numerical solutions increase rapidly with the complexity of the bombing situation. For example, it is very easy to deduce almost all the probability consequences regarding the problem of aiming a single bomb at a rectangular target, but very few deductions can be made directly from the equations which describe the dropping of a train of as few as three bombs on a rectangular target. Since the problem

---

[39] Wallis recalled: "[At lunch one day], Wald discussed some of his ideas on decision theory and Savage . . . remarked that he knew a rather obscure paper that would interest Wald, namely, von Neumann's 1928 paper on games. Wald laughed and said that some of his ideas were based on that paper" (*op cit*, p. 334). In 1947, Wald reviewed von Neumann and Morgenstern's book in the *Review of Economics Statistics*.

[40] Quoted in Wallis, *op cit*, p. 329.

[41] A Texan, with a doctorate from Iowa, Wilks came to New York in the early 1930s to work with Hotelling, and then spent a year in London and Cambridge, interacting with Karl Pearson, Egon Pearson, R. A. Fisher, and Jerzy Neyman. He joined Princeton's mathematics department in 1933, and assumed the editorship of the *Annals of Mathematical Statistics* in 1938. See Mosteller (1968).

[42] He was awarded a Presidential Certificate of Merit in 1947.

of dropping a train of three bombs is itself extremely simple, compared to many common bombing operations, it is apparent that formal mathematical processes cannot alone be depended upon to carry the burden but they are powerful when used in conjunction with synthetic methods and statistical models.[43]

Other Princeton figures close to this work included Wilks' colleague in mathematics, John Tukey, and their graduate students Frederick Mosteller, Merrill Flood and John Williams.[44] The latter two are important to our account.

Merrill Flood had completed a doctorate in mathematics in 1935 at Princeton, where he then taught for several years while also acting as consultant mathematician externally.[45] During the war, he was at Princeton when he became involved in the mathematical research division of the Fire Control Research Office, located on the campus and working under contract to Weaver at the AMP. At Weaver's initiative, John Tukey assigned Flood in 1944 to work on Project AC-92 and prepare a study on the aerial bombing of Japan with the B-29 bomber. Flood's report was concerned with the general tactical problem of aerial bombing. If the analysis was to be adequate, precise, and efficient, it was stated at the outset, it must be mathematical: "the strategic judgments have to be made in quantitative terms" (1944, p. iv). The result was "Aerial Bombing Tactics: General Considerations", which analysed bombing methods in terms of "gain" – i.e., the difference between an operation's "worth" and its "costs". This was another wartime study that made use of game theory.[46]

---

[43] Wilks in AMP, 1946, quoted in Rees, *op cit*, p. 613.

[44] Fred Mosteller was a graduate student of Wilks' who had arrived at Princeton, from Carnegie Institute of Technology, in 1939. Mosteller (b. 1916) returned to Princeton after the war to complete his doctorate in 1945. He was appointed to the Department of Social Relations at Harvard in 1946 and moved to the Department of Statistics in 1951, where he became chairman. In 1977, he shifted to the Department of Biostatistics at the Harvard School of Public Health, and several years later became chair of the latter's Department of Health Policy and Management. His many books include *Stochastic Models for Learning*, with Robert R. Bush (New York: Wiley, 1955), *Probability with Statistical Applications* (Reading MA: Addison-Wesley, 1961). See Fienberg et al (eds.) (1990).

[45] See the entry for Flood, Merrill in *Who's Who in America*, Vol. 34.

[46] The report was written in late 1944 and appeared as Memorandum No. 8 of the Fire Control Research Office at Princeton. It was distributed as a working paper through the AMP and declassified in 1948. It was later published as a Project RAND memorandum (*RM-913*, "Aerial Bombing Tactics: General Considerations, (A World War II Study)"), dated Sept. 2, 1952. Flood notes that, during the war, secrecy requirements meant that he knew absolutely nothing about the parallel work in operations research by P. M. S. Blackett in England. Incidentally, it may well have been here, in Flood's work, that the term "military worth" was coined. It was later adopted by Williams at RAND when naming his research group of mathematicians, game theorists, and economists.

The report opened with a general discussion of the tactical factors that determined the outcome of an operation. Some were important ("dominant"), such as number of planes, flight formation, altitude. Others were not: for example, there was no point in harmonizing a bomber's Central Fire Control system to within 1 millimetre accuracy if flexures of the plane in flight introduced aligning errors of 5 millimetres. As for Japan, the possible types of activity for the B-29 bomber included precision bombing of small targets such as gasoline refineries, area bombing of large targets such as industrial areas, area bombing with incendiaries, mine laying, area bombing with poison gas or with propaganda leaflets, and photographic reconnaissance.

The worth of an operation might include destruction of ground installations, mining water areas, casualties, interrupted production, erosion of enemy morale, destruction of enemy aircraft, enemy casualties in the air, and so on. These had to be measured and aggregated in terms of some common unit, such as man-hours necessary to repair the various forms of damage; or in terms of a vector, with values for, say, property destruction (measured in, say, lost production or the cost of clearing mines plus their value as obstacles to shipping); personnel losses; or intelligence information obtained. Flood felt that the easiest thing to do would be to avoid any such construction of vectors and lump all into a single worth index, $W(Z)$ – man-hours.

The cost of an operation included bombers lost, casualties, bomber maintenance and repair, personnel training, materiel consumed, base operation costs, and so on. Again, these had to be expressed in terms of a unit such as man-hours, and aggregated. The comparison of worth and cost to produce some index of gain was not straightforward; for example, American man-hours were not equivalent in any simple sense to Japanese man-hours. The aim was the construction of a gain function $G(Z)$, capturing the relationship between the tactical parameters (controllable and uncontrollable) and the gain of the operation. The use of photographs taken from above the bomber formation just before, and during, the bombing run – the sort of material Williams and Mosteller were working with at the Columbia office – might help ensure that there were no discrepancies attributable to the neglect of dominant parameters. Flood thought that, with sufficient vigilance, it would be possible to estimate a reliable gain function.

By way of simplified example, Flood considered the choice of size of bombing sortie and flight altitude, in the case of a "bombing operation to be conducted at long range by a large force of bombers against a single important target" (1944, p. 14). Initially, a fixed defensive strategy by the Japanese was assumed. Flood simplified further by assuming that worth depends only on the damage inflicted, and cost, on bombers lost in the operation. It was assumed, further, that a common unit of measurement was available for all components.[47] Given assumptions about average accuracy and average expected losses from enemy flak and fighters, Flood showed how a gain function could be constructed in terms of sortie size and flight altitude. For example, for a given sortie size, higher altitude would reduce attrition due to flak but also reduce accuracy. For a given altitude, low sortie size would reduce attrition by presenting a less compact target but would reduce the number of strikes. Using data extrapolated from the activities of the Eighth Air Force, Flood constructed worth, cost, and gain tables, and used these to construct contour charts of the gain as a function of sortie size and altitude.

A further section then considered the mathematics of "maximizing gain under competition", where "competition", in fact, denotes strategic counteraction by the Japanese. Here, Flood introduced game theory. The simple example given was that of a single attacking bomber, which had to choose between two widely separated targets, defended by a lone fighter. Making the assumption that there existed values of the targets common to both sides (see Figure 12.3), and given other parameters concerning probability of a hit by either side, Flood showed how each side's "perfect tactics" – that is, optimal distribution of effort (attack and defense) between the two targets – may require the use of tactical uncertainty – that is, a mixed strategy – so as to hide one's intentions from the opponent (see pp. 21–26). "The essential point shown . . . is that each opponent is faced with tactical uncertainties, some of which are uncontrollable by either opponent and *some of which are controllable by only one opponent.* The competitive objective of the attacker (defender) is to keep the defender (attacker) systematically uncertain concerning the *individual actions* within the general tactical pattern already settled upon, but this need not imply uncertainty concerning the tactical pattern itself, which will usually be impossible" (p. 25).[48] Because of wartime security, Flood did not know exactly how this work would be

---

[47] For example, the worth of the operation was the total value of the target (in, say, man-hours) multiplied by the fraction of the target destroyed, where the latter equalled $1 - e^{-kn}$, k being a constant, .0006, and n the number of bombs falling within 1,000 ft. of the target's centre.

[48] Flood attributes his use of minimax methods to his attending one of von Neumann's pre-war Princeton lectures in game theory. In 1944, when it came to writing the report, Flood was apparently unaware that von Neumann and Morgenstern actually had a book in press until informed of such by Princeton mathematician Albert Tucker (see *op cit*, p. ii).

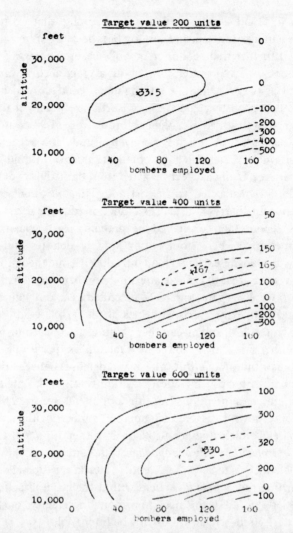

Figure 12.3. Contour Charts of the Gain Function Based on Three Different Assignments of Target Value. *Credit:* Reprinted with permission from Flood, Merrill M. (1952), "Aerial Bombing Tactics: General Considerations", RAND RM-913, Sept. 2.

used, and Tukey was not allowed to tell him. Apparently, Tukey at one point hinted to him that it had "something to with a story in the paper about a mysterious flash being reported in the New Mexican desert".[49]

---

[49] Poundstone (1992), p. 68.

Even more important to our story was John Williams. A trained astronomer, Williams had come to Princeton from Arizona to pursue graduate studies in mathematics under the supervision of Wilks. Before the outbreak of war, he published four papers on statistics.[50] He became involved in wartime statistical research, and, from 1940, was involved in what he later aptly described as a "series of baffling administrative arrangements", involving the Princeton SRG, headed by Wilks, Division 5 of the OSRD, and the 20th Air Force. For these groups and others, he produced some forty wartime memoranda on the statistical analysis of conflict. For example, for the BRL at the Aberdeen Proving Ground, he completed reports on such topics as "The Zeros of the Polynomial P_n". For the Field Artillery Board, he wrote on "Firing Pattern for Use with Fragmentation Projectiles" and on the effects of air bursts on targets in shelter trenches. He spent part of the war at the Strategic Bombing Section of the Princeton SRG, which was located in the same building on Morningside Heights as the Columbia SRG. Here, he worked on such topics as the sources of systematic error in bombing, and with Fred Mosteller he analysed "before-and-after" photographic evidence, in order to aid recommendations in bombing strategy, and assessed flak risk to controlled-missile bombers. His other efforts included analyses of "Train bombing" with Jerzy Neyman, of "Pattern Bombing with Azon" with Jimmie Savage, and the "Estimation of Balloon Barrage Probabilities and Similar Problems" with overseer Warren Weaver, of whom he became something of a *protégé*. Most of his work was done alone, however, and he later said that he tended to get the work that did not fit elsewhere. It was in this web of activities that he interacted with von Neumann and Robertson, and Williams was known to stand out in his devotion to the former. He became very involved in the war work, being, as he said, "one of the last to tidy up for the NDRC and put out the cat". Unlike Mosteller and many others who returned to academia after 1945, Williams stayed with military consulting and, as we shall see, became a crucial, if unspoken, postwar figure in the development of game theory.

Turning to air warfare, the largest Applied Mathematics Group was that at Columbia, headed by Saunders MacLane and located in the W. 118th

---

[50] Williams completed an undergraduate degree in astronomy in 1937, having already published a paper on meteor observation (Williams [1933]). See "John Davis Williams, 1909–1964, In Memoriam", from The Memorial Service for Williams at Santa Monica Civic Auditorium, California, December 6, 1964, a copy of which was kindly provided to me by Professor Fred Mosteller of Harvard University. In addition to CV and bibliography, this includes accounts of Williams by several RAND colleagues.

St. building already mentioned.[51] They were particularly concerned with air-to-air gunnery, which involved aeroballistics (the motion of a projectile from an airborne gun), the design of different types of weapon sights, and pursuit curve theory. Reports and material were received from Britain, circulated amongst the Group mathematicians, and stored in a safe at night. The problems were rather laborious and involved relatively simple mathematics, such as trigonometry and differential calculus. For example, Ed Paxson had had the Group at Brown University study fighter pursuit courses: the track taken by a fighter pursuing another plane. According to MacLane (1989), the Brown solution was unsatisfactory for three-dimensional space, and the Columbia Group took over the problem, with the help of, amongst others, Tukey at Princeton. They calculated equations for some thirty-three pursuit courses, and had the solutions cranked out by the "computers", young female graduates of Vassar, working on desktop Marchant calculators. In other work on airborne fire control, it was only with great difficulty that Ed Hewitt could persuade a gunner of the 8th Air Force to aim his fire towards the tail of the other plane, rather than the nose, so as to allow for the forward motion of the firing plane.

Los Alamos

If the war had "corrupted" von Neumann the pure mathematician, it did not take long before he, in turn, corrupted Ulam. In September, 1943, he wrote cryptically to his Polish friend from "the Southwest", apologising for not keeping up his correspondence. Since he had come back from England, he was in three or four different places each week. He would likely spend two days a week in Princeton, alternating with two or three other places. He didn't know how long it would last, he might even have to go to England before Christmas. It was all very vague. The war was going "reasonably well", he said, "but I would still think that Germany may last well into 1944 – say 10% probability for a political collapse before Spring, and the probable end summer or fall. Don't you agree? I don't agree that Russia will ever settle with Germany, unless there is a revolution in Germany, but otherwise I am 100% pessimistic about Russian relations. What do you think?"[52]

Ulam was fully aware that, had he not moved to the United States in time and remained in Warsaw, his fate might have been like that of so many of his Polish colleagues. He soon intimated to von Neumann that he was keen

---

[51] See Rees (1980); MacLane (1989).
[52] Von Neumann to Ulam, Sept. 2, 1943, SUAP.

to become involved in war work against Germany. Von Neumann took care of it. "I am very glad", he wrote, "that Mr. Hughes 'and all he stands for' have come through. I told them about you, because you wrote me several times in the past that you definitely wanted a war job, and because this is a very real possibility, where you would do very effective and useful work":

The project in question is exceedingly important, probably beyond all adjectives I could affix to it. It is very interesting, too, and the theoretical (and other) physicists connected with it are probably the best group existing anywhere at this moment. It does require some computational work, but there is no doubt that everybody will be most glad and give you all the encouragement you can wish in doing original research on the subject, for which there is ample opportunity. I can also assure you of my cooperation in this respect.

The secrecy requirements of the project are rather extreme, and it will probably necessitate your and your family's essentially staying on the premises (except for vacations) as long as you choose to be associated with it.

To repeat: If you want war work, this is probably a quite exceptional opportunity.[53]

Von Neumann had been brought into the Manhattan Project the previous September by Robert Oppenheimer.[54] He became a key member of a group dominated by fellow Hungarians, including physicists Edward Teller, Leo Szilard, Eugene Wigner, and Theodore von Karman. Only four mathematicians were involved in Los Alamos: von Neumann himself, Ulam, C. J. Everett, and Jack W. Calkin. As was the case elsewhere, von Neumann seems to have been well-liked, and he was one of the very few scientists with both full information on the project and permission to come and go as they pleased from the research site.

Technically, his main contribution was to help with the huge computational problems raised in the calculations connected with the placing of explosives in the bomb. The detonation system for the Hiroshima bomb, "Little Boy", consisted of a trigger system that fired a bullet into a sphere of uranium-235. This technique was relatively inefficient in comparison to the "implosion" method, which would force the sphere of explosive in upon itself to achieve a critical mass – and thereby achieve a bigger bang for

[53] Von Neumann to Ulam, Nov. 9, 1943, SUAP. Von Neumann continued: "Do you really count on a quite short war? I don't see that from a purely technical standpoint Germany need be broken before next fall. Of course a collapse may come any day from now on for moral and political reasons, but I can't see how to judge that, without knowing much more about the present state and efficiency of the Nazi political machine. And there is still a year's worth of Asiatic war after that. Anyhow, qui vivra verra . . ." (*ibid*).

[54] See Rhodes (1986); Poundstone (1992); Macrae (1992).

a given quantity of fissionable material – but which posed great technical obstacles.

Seth Neddermeyer had begun experimenting with the implosion method in the late summer of 1943, firing tests with metal pipes on Independence Day of that year.[55] Wrapping explosive around a metal pipe, the detonation created a converging wave which could reduce the pipe to a solid metal bar, or twist it out of shape, depending on the symmetry of the squeeze. At Oppenheimer's weekly colloquium, the work was ridiculed by Richard Feynman and others. One engineer named Parsons said that the project was akin to trying to implode a beer can without spilling the beer, but was even more difficult.

Von Neumann had been working on the hydrodynamics of the shock waves formed by shaped charges, research that was being used in the development of the bazooka. When he visited Los Alamos in late summer 1943, he encountered Neddermeyer's work on implosion, "another warren of hydrodynamic complexity".[56] There, he renewed his acquaintance with Edward Teller, whom he had known in his youth. In discussions with him on the compression effects of implosion on metals, around October 1943, von Neumann concluded that a solid mass of plutonium, rather than a hollow sphere, could be squeezed sufficiently to form a bomb core. Such an implosion bomb would measure five feet in diameter and nine feet long. With von Neumann's benediction, Neddermeyer now continued to test small-scale models of the bomb, trying to assure that the expanding shock waves from sets of explosive wrapped around a core converged to form a uniform squeeze. Continued irregularities in the jets travelling ahead of the main mass of the wave posed problems, and George Kistiakowsky, the Harvard chemist, was brought in. Kistiakowsky had worked on explosives since 1940 for the NDRC, and Manhattan Project historian Richard Rhodes says that it was he who had converted von Neumann, contra the military, to the belief that they could be turned into very precise instruments.

When Ulam arrived at Los Alamos in the winter of 1943, Teller tried to appropriate him for work on the Super, while Hans Bethe wanted every spare body to work on the implosion question. The Super, however, was soon demoted by the Los Alamos Governing Body, as it would have required tritium, which was difficult to manufacture. Von Neumann was thus able to draft Ulam to work on the hydrodynamics of implosion, where the problem was "to calculate the interactions of the several shock waves as they evolved

[55] See Rhodes, *op cit*, pp. 478ff.
[56] *Ibid*, p. 479.

through time, which meant trying to reduce the continuous motion of a number of moving, interacting surfaces to some workable mathematical model". Said Ulam, "[t]he hydrodynamical problem was simply stated, but very difficult to calculate – not only in detail, but even in order of magnitude".[57] IBM punchcard sorters were brought in in April 1944 to facilitate the massive calculations on various-shaped bomb cores, and it was this work that apparently stimulated von Neumann to think about how the machines could be improved. The task was to arrange the multiple detonators of explosive in the bomb in such a way that there were no irregularities in the shock waves; and that an even, uniform wave be pressed in on the uranium or plutonium tamper. The arrangements of explosives were termed lenses, after those in the eye which similarly focus light.

Going against the resigned pessimism of the group, von Neumann undertook extensive calculations to show how the explosive charges could be placed so that the variation in the velocity of the arriving shock waves was no more than five percent. The system he designed involved the circular arrangement of high explosive in interlocking, tapered blocks. Several detonators were placed along the outer edge and in the middle, surrounded by a natural uranium tamper, the plutonium core. The whole contraption measured five feet across: Fat Man. In the winter of early 1945, there were numerous tests of the moulded explosives, with small bombs.

By the spring of 1945, it was clear that the uranium bomb was going to work, and von Neumann's work on the explosive lens augured well for the plutonium version. Truman was consulting with Stimson, his Secretary of War, and General Leslie Groves, Los Alamos' military head, on the use of the bomb. Von Neumann's contribution now went beyond the theory and design of implosion. Unlike many of his physicist colleagues, he evinced a liking and admiration for military officers and he was brought into the committee appointed to deliberate on the choice of Japanese targets, which reported to Groves.[58] They met in a conference room in the Pentagon: Brigadier General Thomas F. Farrell, representing Groves; two Air Force officers; von Neumann; the British physicist William G. Penney; and three other scientists. They brought in an Air Force meteorologist, who discussed the weather in Japan, saying that the visibility would probably be best in August. The target deliberations were complicated. The Air Force's 21st Bomber Command, under General Curtis LeMay, was already firebombing Japanese cities at a fierce rate, and planned to be dropping 100,000 tons per

[57] *Ibid*, p. 544.
[58] See Rhodes, *op cit*, pp. 626ff; Macrae (1992), pp. 241–45; Groves (1962), pp. 267ff.

month by the end of the year. The members of the Target Committee were under the impression that this continued assault had priority. Further, there were very few Japanese cities that had not been damaged to some extent. They decided to consider some seventeen targets, including Tokyo Bay, Yokohama, Osaka, Hiroshima, and Nagasaki. Penney would consider the size of the explosion, the likely damage, the deaths; von Neumann would attend to the computations, and they would meet again in mid-May in Los Alamos.

Klari von Neumann later remembered a startling night with her husband at around this time:

Whenever he came home, we usually spent most of the night talking – well, he was talking and I was the "ear"; his pent-up tension was pouring out in a flow of words which, as a rule, he kept strictly to himself. One time, in early 1945, he came back from Los Alamos and proceeded to behave in the most unusual "Johnnyesque" manner. He arrived home sometime mid-morning, immediately went to bed and slept twelve hours. Nothing he could have done would have had me more worried than Johnny skipping two meals, not to speak of the fact that I have never known him to sleep that long in one stretch. Sometime late that night he woke up and started talking at a speed which, even for him, was extraordinarily fast. He stuttered a great deal – this he often did when he was under some strain; as a matter of fact, publicly this was the only indication that ever slipped through that he was tense about something or other. Except for this stutter, in the presence of any outsider, he never raised his voice, remained calm and detached under provoking circumstances, and mixed humor and jokes into the most solemnly serious occasions.

. . . In retrospect, that night in '45, Johnny's summary of the future was uncannily prophetic.

"What we are creating now is a monster whose influence is going to change history, provided there is any history left", he said, "yet it would be impossible not to see it through, not only for military reasons, but it would also be unethical from the point of view of the scientists not to do what they know is feasible, no matter what terrible consequences it may have. And this is only the beginning! The energy source which is now being made available will make scientists the most hated and also the most wanted citizens of any country".

Whenever Johnny got emotionally upset, he was likely to use big words and superlatives – and this he certainly did that night.

"The world could be conquered, but this nation of puritans will not grab its chance; we will be able to go into space way beyond the moon if only people could keep pace with what they create", he added, and then went into a lengthy discussion of how automation is going to become not only more important but indispensable. While speculating about the details of the future technical possibilities, he gradually got himself into such a dither that I finally suggested a couple of sleeping pills and a very strong drink to bring him back to the present and make him relax a little about his own predictions of inevitable doom".[59]

---

[59] Von Neumann-Eckhart, unpublished draft autobiography, pp. 29–31, KEMNW.

Von Neumann respected his own ethical injunction. At Oppenheimer's office on May 10 and 11, the Target Committee had been extended to include Oppenheimer himself, Captain William Parsons, Richard Tolman, and Norman Ramsey. The items on the agenda were the height of detonation; reports of weather and operations; "gadget" (that is, bomb) jettison and landing; the status of targets; psychological factors in target selection; use against military objectives; radiological effects; coordinated air operations; rehearsals; operating requirements for safety of planes; and coordinations with the 21st Bomber Command's conventional Japanese bombing program. Von Neumann was able to contribute to the question of the height of detonation, given the information then available at this pre-Trinity test stage, because he had worked on such questions for conventional bombing in England in 1943. They also had to consider the likelihood of an emergency: could a bomber return safely with the unexploded bomb? Could the bomb be jettisoned in the sea if necessary?

They now had three criteria for target selection: it must be a large urban area more than three miles in diameter, capable of being damaged by the blast, and unlikely to be attacked by Bomber Command before August. As to targets, the committee considered short lists provided by both intelligence sources and the Air Force. Von Neumann disapproved of virtually all the intelligence suggestions, such as the Yawata ironworks, the Asano dockyard, and the Dunlop rubber factory in Tokyo, apparently on the grounds that the 21st Bomber Command could dispose of them effectively. Of the Air Force list – Kyoto, Hiroshima, Yokohama, Tokyo's Imperial Palace, Kokura Arsenal, and Niigata – he apparently approved all, except the imperial palace and Niigata. The group demanded further study of Kyoto, Hiroshima, Yokohama, and Kokura Arsenal, and decided to meet at the Pentagon on May 28. At that meeting, they discussed recent Air Force tests, carried out in Cuba, concerning flying time for long distances with heavy bomb loads. Continued firebombing by the Air Force was reducing the number of targets, so that the final selection was composed of Kyoto, Hiroshima, and Niigata. The aiming points were to be chosen later depending on the weather, and they were be at the city centre, rather than outlying industrial areas, so as to have maximum impact. The Target Committee would remain on call. Subsequently, in the many intense meetings involving Stimson, Groves, Bush, Conant, and others, the Secretary of War ruled out Kyoto, on the grounds that it was a holy centre. There were several impassioned meetings at which the monumental nature of the atomic discovery was raised and the ethical and political aspects of the bomb discussed. It was decided that it would be best to proceed without prior warning to the Japanese, and to drop it on an industrial area with a large population of workers living nearby. The final list of alternatives was Hiroshima, Kokura Arsenal, and Nagasaki. "Little Boy", the equivalent of 12,500 tonnes of TNT, fell on the former on August 6, and bad weather spared Kokura, but not Nagasaki, from "Fat Man", 20,000 tonnes, three days later.

Conclusion

Watching the European descent into disequilibrium in the late 1930s, von Neumann had developed *n*-person game theory, an abstract meditation on social equilibrium and norms, predicated upon the mathematics of the stable set, and published as the *Theory of Games and Economic Behavior*. During World War II, a narrow, technical part of game theory, bearing little relationship to the theory of coalitional games, was appropriated as a mathematics of measure and countermeasure. This narrow application gave a new tractability to a range of search, bombing, and allocation problems. An element in the rise to power of mathematicians and physicists such as von Neumann and Morse, game theory became part of a general wartime reorientation of scientific work around the applied mathematical or engineering axis. The game-theoretic analyses of submarine search and fighter–bomber pursuit were part of this larger, complex, multifaceted transformation.

After the war, in the context of perceived Soviet threat and Cold War, important mutations continued to occur. As we shall see, operations research and the game-theoretic analysis of engagements of particular military engagements remained important. However, in the late 1940s and the 1950s, the heightened tension associated with the nuclear threat brought renewed attention to questions of social cohesion and stability. In this setting, game theory was integrated into experimental work on economics, psychology, and group dynamics. Against the background of the "present danger" and anxiety about the vulnerability of the body politic, game theory, one might say, regained its status as a theory of *social* behaviour.

THIRTEEN

Social Science and the "Present Danger"

Game Theory and Psychology at the RAND Corporation, 1946–1960[1]

[I]t was a long time that I was in what was called the "contamination ward". We tried to work on some problems and things like that, but it was mostly a matter of waiting until we did get a clearance... You know, you couldn't move around and you couldn't go to other parts of the building where these people were located, except for a short period of time into your administrative offices or things like that. Because you had problems with a lot of the walls... they do not go to the ceiling and things that are going on in one office can be heard in another.
      Robert Belzer (RAND game theorist), NASM interview, p. 9

In general, the procedure should be to discover breaking points under pressure, similar to the physical techniques of determining such for material and shapes. All this could be tied up closely with the problems of target analysis, the construction of a national redoubt, etc.
      Oskar Morgenstern (1954b), "The Compressibility of Organisations and Economic Systems", RAND *RM-1325*, Aug. 17, p. 19

Introduction

It was summer 1948, at the intersection of 4th St. and Broadway in Santa Monica, California. The building was formerly the office of the *Evening Outlook* newspaper, and one would have had to look closely to notice the new doors and locks, and the security guard. Inside, more than forty people had gathered for a seminar, one of many in the building that summer. All eyes were on von Neumann, his dark suit appearing even more conspicuous in informal California. The discussion was about the duels between two fighter aircraft, and, describing a game with a continuum of strategies and a discontinuous payoff function, somebody had asked von Neumann if he

---

[1] This chapter draws on two sets of interviews conducted with early RAND researchers: one by me and one by the National Air and Space Museum (NASM) of the Smithsonian Institution as part of their Oral History on Space, Science, and Technology.

could prove that it had no solution. After his habitual minute staring into the mid-distance, von Neumann turned to the board, talking and writing quickly, the eraser, as always, never very far behind.

"Well, players 1 and 2 each choose a number, x and y respectively, in the closed zero-one interval. Let 1's payoff function be:

$$M(x, y) = x + y, \quad \text{if } y \geq x$$
$$M(x, y) = x - y, \quad \text{if } y < x$$

Now, if we let $F(x)$ be the density function describing player 1's mixed strategy, and $G(y)$ be that for player 2, then we can write 1's expectation as the Stieltjes integral:

$$\phi(F, G) = \iint M(x, y) \, dF(x) \, dG(y)$$

Then, integrating with respect to x we obtain . . . "

"No! No!" interjected a young voice from the back of the room, "that can be done much more simply!".

You could have heard a pin drop. Even years later, Hans Speier, who had then just been appointed head of the Social Science division at RAND, remembered the moment:

Now my heart stood still, because I wasn't used to this sort of thing. Johnny von Neumann said "Come up here, young man. Show me". He goes up, takes the piece of chalk, and writes down another derivation, and Johnny von Neumann interrupts and says, 'not so fast, young man. I can't follow'.

Now . . . he was right, the young man was right. Johnny von Neumann, after this meeting, went to John Williams and said, 'Who is this boy?'

He said, 'I found him there and there and I was told he was a very promising young mathematician, so we hired him'.

He said, 'How long has he been here?'

'Oh, about six or nine months'.

'And what has he been doing?'

Only John Williams could do this marvellously. He said, 'Oh well, he has written three or four papers, each of which is the equivalent of a doctoral dissertation in mathematics'.

Which was true. Johnny von Neumann looked at that, and he gave him – I don't know – it was something quite fantastic, a special stipend to Princeton or something like that . . . He was the son of a Harvard professor. (Speier interview, pp. 39–40).[2]

---

[2] From Speier NASM interview, April 5, 1988. Speier had studied economics and sociology at Heidelberg, where Karl Mannheim, Karl Jaspers, and Emil Lederer were influential. In 1933, he left Germany for the New School in New York, joining what had become the university-in-exile led by Lederer. He spent World War II at the Foreign Broadcast Intelligence Service of the Federal Communications Commission, analysing Nazi radio propaganda.

The son was Lloyd Shapley, and the Harvard professor was Harlow Shapley, noted astronomer and public intellectual. Beginning in 1939, Shapley senior was a member of the American Committee for Democracy and Intellectual Freedom and World Peace, the American Russian Institute, and the Civil Rights Congress, all of which were soon regarded as Communist "fronts" in the McCarthy period. By 1948, he was involved in the National Council of the Arts, Sciences and Professions, and was a moving force behind its sponsorship the following year of the Cultural and Scientific Conference for World Peace at the Waldorf Hotel, which McCarthyites regarded as an attempt to reach out to the Stalinist regime.[3] Thus, notwithstanding his scientific eminence, the Harvard astronomer's conciliatory politics would probably have ruled him out of court as far as Project RAND was concerned. Yet, here was his 25-year-old son, active in what was to become America's quintessential Cold War institution.

Lloyd Shapley had joined RAND in early 1948. He had been studying mathematics at Harvard until interrupted by the war in 1943. This saw him spend the next two years with the army in China, as a private and weather officer. Airlifted into the country over the Himalayas, he was attached to a unit called the Flying Tigers, the activities of which involved distributing support to that part of China not occupied by the Japanese.[4] Shapley deciphered coded weather reports coming from Siberia. The code was relatively unsophisticated so that, even though it would change every so often, he was able to distinguish himself sufficiently to merit a "battlefield promotion" to Corporal. This work contributed to compiling Chinese weather reports, which, given the eastward flow of weather patterns, were important to making forecasts for Japan. The attack on Hiroshima saw Shapley return to Harvard in late 1945.

At RAND, beginning in 1948, month-long summer gatherings were held for the discussion of game theory, utility theory, linear programming, decision theory, and related subjects. The summer seminar series being discussed here gave rise to one of RAND's earliest papers, Bohnenblust et al (1948), "Mathematical Theory of Zero-sum Two-person Games with a Finite Number or a Continuum of Strategies". The game with the discontinuous payoff function and no solution is discussed on pp. 32–33.

For the RAND documents, RM indicates a memo, usually for internal circulation at RAND and limited external circulation. P indicates a paper, approved by the RAND administration – quickly read by the head, Frank Collbohm, to ensure that it adhered to the institution's "vision" – and distributed externally.

[3] On Harlow Shapley (1885–1972) from the McCarthy perspective, see Buckley and Bozell (1954), pp. 136–39, and pp. 347–49.

[4] The aid, Shapley later recalled, included aviation fuel and "economic fuel" – that is, Chinese banknotes printed by the ABC Banknote Company in Chicago, the Chinese government apparently having lost control of the large printing presses. "Another one of these rather strange kind of intrusions of economics into logistics", he said. Shapley interview with author, April 21, 1992, UCLA (hereafter Shapley interview).

Ability notwithstanding, his undergraduate career was undistinguished, and in later recollections he was somewhat dismissive of the period, singling out only a class taken with logician Willard van Orman Quine: "[T]hat influenced me a great deal because here was a person who was talking exactly, correctly saying things – the whole idea of mathematical logic – that was a big memory from my Harvard courses" (*ibid*). Shapley was glad to get out of university, being uninterested in further study, without particular ambitions for the future, and keen, one senses, to get away from the East Coast.

Through the Air Force connections of an older brother, who worked at the Budget Bureau in Washington, Shapley was put in touch with Project RAND, a new military research organisation then being developed at the Douglas Aircraft Company in Santa Monica. In 1948, he was recruited to RAND's mathematics division, which was then labelled the section for the "Evaluation of Military Worth". When it came to his gaining the security clearance required of all RAND employees, there was initially resistance – "I had relatives" – but, in time, the Federal Bureau of Investigation was able to determine that Shapley's suitability was not compromised by blood ties. He survived "de-contamination", and entered the inner circle. By early 1950, several months after confirmation of controlled atomic explosions in the Soviet interior, the son was sunk deep in game-theoretic models of nuclear conflict, while the father figured high on McCarthy's list of academics suspected of un-American activities.

From that point on, until its peak in mid-decade, McCarthyism infused American academic life with mistrust and suspicion.[5] Careers were cut short, individuals felt obliged to prove their political loyalties, and university administrations were drawn into battles to defend their faculty members. There were nervous breakdowns, and persecution by the F.B.I. disrupted family lives. The perception of external threat to American society bred a domestic paranoia, with fears of physical fracture being accompanied by fears for the cohesion of the body politic.

RAND was at the centre of this. Its *raison d'être* was bound up with matters of vulnerability and integrity. It was set up because of the likelihood of a bombing campaign against the USSR, and matured with the recognition of a Soviet nuclear threat against the United States. These considerations permeated the culture of the institution and the orientation of its research.

RAND's scientists focused on the vulnerability and integrity of physical structures. Physicists such as the Latter brothers considered the destructive

[5] See Halberstam (1993).

capacities of atomic bombs, while engineer Bruno Augenstein examined the permeability of various construction materials in the face of gamma rays, and his colleague Ted Barlow worked on the development of early warning systems to protect the U.S. territory against the arrival of Soviet bombers.

RAND social science gave full expression to a topical concern for individual behaviour and social integrity. Questions of individual rationality and group cohesion underscored the wargames as well as the experiments in game theory, psychology, and group dynamics, conducted by RAND researchers under Williams.[6] Together they formed a constellation of interconnected theories and practices – a "form of life" to use the sociological expression – broadly concerned with individual rationality and social order at a time of threat. The alertness and predictive abilities of soldiers were tested; students and RAND staff were invited to participate in group experiments to see whether they were rational in the game-theoretic sense; groups of Air Force personnel were observed as they carried out simulated air-defence under induced psychological stress; and RAND mathematicians and other researchers engaged in simulated conflicts over the war map.

Normative stipulations concerning normal behaviour abounded, not only in the experimental procedures themselves, but also at various removes from them. At the seminar with which we opened this chapter, the doors were closed and guarded, open to those with security clearance. In experiments conducted, John Nash humorously, but tellingly, spoke about his subjects being remarkably well-behaved, and John Williams, as experimental subject himself, laughingly referred to the psychological instability of his opponent, Armen Alchian. More seriously, the same Williams, as research director, encouraged interdisciplinary mixing at RAND in the promotion of scientific creativity, yet in 1951 deemed game theorist J. C. C. McKinsey a risk because of his homosexuality. Three years later,

---

[6] Note that the focus here is not RAND economics, which involves the story of Hitch's economists, and how cost–benefit analysis became important at RAND, underpinning systems analysis and, ultimately, the phenomenon of the "Whiz Kids" under McNamara at the Pentagon. Nor is the focus RAND's Social Science division under Hans Speier. This group, which soon included Russian scholar Nathan Leites, and sociologists Fritz Sallagar and Paul Kecksméti – the Hungarian encountered earlier who had reviewed Menger's 1934 book – took a literary-historical approach to social analysis (see, for example, Leites' work on the Politburo, or Kecksméti's "Sociological Aspects of the Information Process" [RAND P-430]). In time, this alienated them somewhat from the dominant culture at RAND, which variously privileged the formal models of the game theorists and the quantitative cost–benefit studies of the economists. Speier's unit soon shifted from Santa Monica to RAND's Washington, D.C. office.

after McKinsey's suicide, John Nash was dismissed from RAND for similar reasons.[7]

It was in this context that game theory was sustained at RAND in the postwar period, as one strand, itself variegated, in a web of activities that were given meaning, relevance and direction by the political concerns of that institution. The purpose of this chapter is to provide a portrait of RAND and the role played by game theory in its operations. It should be said at the outset that the subject is complex and refuses to be forced into a simple linear narrative. Let me, therefore, provide an outline, a guide to the wealth of seemingly disparate activities that enter our account of RAND in the decade from late 1940s.

Insofar as RAND was created to perpetuate operations research work of the kind applied during World War II, game theory was part of its *raison d'être*. At the early stages, game-theoretic models were thought likely to be useful in solving tactical military problems to be encountered in a war with the USSR. Encouraged by the successes of operations research during the war and by the presence of such architects as von Neumann and Weaver, RAND supported mathematical research in game theory in the belief that there would be direct applications. It soon became clear, however, that, tractable and elegant though such models might be, they were not going to provide precise guidance for decision-making in any of the large defence studies being undertaken at RAND. Game theory, it was acknowledged, was no "miraculous" technique in war strategy. Its value, rather, was that it helped structure thinking in a qualitative, conceptual way, highlighting the importance of strategic interaction, threats, credibility, and similar factors.

However, game theory at RAND was not simply a part of operations research. It also became a crucial point of reference in the rich social–scientific experimental culture that developed there, and a good deal of this chapter will be taken up with portraying this tapestry of experimental activities. The emergence of that culture coincided with the growth of anxiety about the possibility of a nuclear attack and the effects it might have on society. Thus, even where they started out with examining the behaviour of individuals, as in psychology and learning theory, these experimental activities invariably ended up studying individuals, not in isolation, but as they interacted in groups. The main experimental activities we will deal with lie in the fields of psychology (Kennedy and the work of the Systems Research Laboratory), learning (at the individual and then group levels), game theory

---

[7] See Nasar (1998), pp. 184–89.

itself (Prisoner's Dilemma and group-bargaining experiments), and, finally, war-games (Helmer-Shapley). These interconnected experimental activities defined a substantial part of RAND social science in the 1950s, and many of them drew upon game theory in one way or another.

We begin with the creation of RAND and the development of its heterogeneous social–science group. Then, following an account of some operations research work, we expand to consider the varied aforementioned experimental activities conducted at RAND. Throughout, we keep a close eye on von Neumann and Morgenstern, portraying the independent and quite different relationships the two of them enjoyed with the Santa Monica institution.[8]

Project RAND

At the time of that 1948 seminar involving von Neumann and Shapley, Project RAND had been two years in existence. As the war ended in 1945, there was concern amongst the War Department, the Office of Scientific Research and Development, and branches of the military that the many academics who had been involved in wartime military research were going to simply return to the campus and the possibility for future cooperation would simply be lost. The Air Force, in particular, recognized that scientific advisors would be crucial to interservice rivalry and to maintaining control of the atomic bomb, then in Air Force hands. Thus, in March 1946, following discussions between Army Air Force Chief of Staff "Hap" Arnold; Dr. Ed. Bowles, scientific consultant to Secretary of War Henry Stimson; and engineers Arthur Raymond and Frank Collbohm from the Douglas Aircraft Co., Arnold committed $10 million of research funds remaining from the war to Project RAND (that is, Research ANd Development). Initially located under the Douglas roof, before moving to the rented building mentioned earlier, and then to its own quarters, the function of this group of physicists, mathematicians and engineers, presided over by Collbohm, was to conduct a program of research on "intercontinental warfare, other

---

[8] On the history of the RAND Corporation, see Smith (1966), Kaplan (1983), Jardini (1997), Specht (1960), Hounshell (1997), Leonard (1991, 2004), and Mirowski (2002). Inspired by Edwards' (1997) treatment of the RAND 1957 spin-off, the Systems Development Corporation, Mirowski portrays RAND's adoption of game theory as a chapter in the emergence of "cyborg" thinking (pp. 207–22). The present account is deliberately intended to be more variegated, showing how the meaning and significance of game theory at RAND depended upon multiple facets of that institution's culture – material, social, and political.

than surface, with the object of advising the Army Air Forces on devices and techniques".[9]

Collbohm was close to Warren Weaver, wartime head of the Applied Mathematics Panel, with whom he had discussed strategy and operations research during the war, aboard Collbohm's yacht and in his Brentwood dining room. From the very earliest days of Project RAND, Collbohm sought Weaver's advice. Thus, when Collbohm was looking for a new director of mathematics research, Weaver recommended his wartime aide and protégé, John Williams. Once at Santa Monica, Williams relied on Weaver, Wilks, and von Neumann as key advisers and consultants. Williams, apparently, worshipped von Neumann, and although the latter darted about in little short steps while Williams was famously inert, the two were similar in personality. At his Pacific Palisades home, Williams held "high alcohol, high IQ" parties similar to those at the von Neumann household in Princeton.[10] They also enjoyed fast cars. Von Neumann's talent for writing off his car whilst escaping unscathed was well-known. As for Williams, he drove a Jaguar in which he had the RAND machinist install a Cadillac supercharger, and which he would apparently take out for midnight test-drives on the Santa Monica highway. From the beginning, at RAND, Williams sought close contact with von Neumann: "Paxson, Helmer and all the rest are going to be especially nice to me if I succeed in getting you on the team", he wrote in 1947, luring him with a \$200-per-month retainer. Von Neumann showed great interest.[11] In RAND's early years, in the late 1940s, even when

---

[9] Quotation from Smith (1966), p. 46. The engineers had advised Bowles in a successful study of the B-29 bomber. Arnold had been supportive of such work and wished to see it continued. Arnold and Donald Douglas were also connected by the marriage of their children.

[10] Poundstone (1992) recalls the house in which Williams lived. A very large home occupied by a millionaire had been put up for sale. A developer had the idea of slicing the house in five pieces, and demolishing numbers two and four to yield three smaller dwellings. Williams lived in the middle – and Deborah Kerr in one of the others.

[11] Letter, Dec. 16, 1947, von Neumann Papers, Library of Congress, Container 15, File RAND Corp. Contract Correspondence. Von Neumann felt that a group should meet twice a month (von Neumann to Williams, Jan. 9, 1948, *loc cit*), and Williams shortly indicated the probable other members of the group: Dollard, Lasswell, and Young on social science, and Stephen on statistics. A year later, Williams wrote of the crowd likely to appear at RAND that summer, including Richard Bellman, David Blackwell, Willard van Orman Quine, T. W. Anderson, Abe Girschick, Ken Arrow, Sam Karlin, Donald Young, and Hans Reichenbach. "We intend to make major efforts on applications of game theory and . . . the random sampling method in analysis" (Williams to von N., Dec. 27, 1948, *loc cit*). To which von Neumann replied in encouragement: "The work on game theory, which you have been pushing so energetically and successfully interests me greatly – I don't think that I need tell you this again" (von N. to Williams, Jan. 7, 1949, VNLC, Container 15, RAND Corresp.). Shortly before visiting that summer, von Neumann concluded a letter

he was no longer a visible presence, the Hungarian carried great authority: "Everybody knew that von Neumann was king", recalled economist Jack Hirshleifer.[12]

Amongst the first mathematicians brought by Williams to RAND was Edwin Paxson, technical aide to Weaver at the wartime AMP. He was soon joined by Olaf Helmer, a German immigré with a doctorate in mathematics from the University of Berlin (1934) and another in logic from the University of London (1936).[13] Also at RAND were "Abe" Girshick of the Columbia SRG; Melvin Dresher from the Office of Price Administration, and J. C. C. McKinsey, a philosopher of logic trained at Columbia in the 1930s.[14] The RAND–Princeton connection was maintained through von Neumann, Wilks, and the new graduates. Merrill Flood, who, following wartime Princeton, had been a researcher at the Department of the Army, also joined RAND around this time. Another Princeton student, Samuel Karlin, arrived at the Caltech mathematics department in 1947 and became consultant to RAND the following summer, along with his chairman, Henry Bohnenblust, a veteran of wartime operations research in England. These, too, maintained close connections with Princeton mathematics.[15] So did Richard Bellman, and, over the next decade, a steady stream of Princeton

"I shall also write to you in a few days about various papers on games which you have sent me. They are very interesting and I want to congratulate you on the excellent work which is being done at RAND on the subject" (von N. to Williams, May 11, 1949, *loc cit*).

[12] Hirshleifer interview with author, Feb. 27, 1990, UCLA Hirshleifer's colleague, Armen Alchian, recalled a presentation in which economist Stephen Enke was discussing the allocation of fissile material between the Air Force and the Army, using an Edgeworth box with fissile material on one axis and manpower on the other. In addition to von Neumann, the audience included Bacher and Lee Dubridge, both of whom seemed to have trouble understanding. "But then von Neumann began to nod in agreement with Enke's presentation, and very quickly Bacher and Dubridge were nodding too. The pecking order in action!" (Alchian interview with author, Feb. 6, 1990, UCLA).

[13] Before the war, Helmer had been assistant to Rudolf Carnap at the University of Chicago in 1937–1938, and had taught mathematics at the University of Illinois from 1938 to 1941, and at the City College of New York from 1941 to 1944. The following year was spent doing wartime applied mathematics under Williams at the New York office of the Princeton Statistical Research Group. He spent part of 1946 in France and Berlin teaching mathematics to the U.S. Army, and, later that year, went to Project RAND.

[14] Girschick had immigrated from Russia in 1932, the same year that Dresher had arrived from Poland (see Kaplan [1983]).

[15] Karlin would later persuade Shapley to return to Princeton for a doctorate in mathematics, which he did in the autumn of 1949 (hence the reference in the Speier interview earlier). He had already developed a reputation at Princeton before arriving there: apparently, when Solomon Lefschetz wrote to him at RAND, informing him of a generous assistantship on a project run by Lefschetz, Shapley is said to have brusquely replied "Lefschetz, the arrangements you suggested are very satisfactory. There need be no more correspondence about this" (Tucker interview with the author, Princeton, Dec. 11, 1991).

mathematics students, amongst them John Nash, Harold Kuhn, Herbert Scarf, and David Gale, spent summers and other sojourns at RAND.

The Rational Life

With RAND early staffed by engineers, mathematicians, and physicists, the idea of bringing in social scientists was pushed by University of Chicago economist Allen Wallis, who had worked at the wartime Columbia Statistical Research Group. One of the first people he directed towards Williams, in the summer of 1946, was economist Armen Alchian. He had been a student of Wallis at pre-war Stanford, where he obained his doctorate, and then served in the Air Force during the war. Alchian was now on the faculty at the University of California of Los Angeles (U.C.L.A.) close to RAND. Once a week, he recalls, he was collected by car and brought down to Santa Monica, with only the vague instruction from Williams that he should speak to the mathematicians and engineers and see where he could help.[16] Alchian had read the *Theory of Games* that summer, and arrived at RAND to find Paxson, Dresher, Weaver, and Helmer working on the subject, with von Neumann visiting every month or so.

Although the economic viewpoint did not assume immediate institutional importance at RAND, Alchian seems to have made a quick impact, and the idea of bringing in people in economics and the social sciences was being pushed further in December 1946 by Williams, Weaver, Helmer, Wilks, and Mosteller.[17] Collbohm, an engineer who concentrated on cultivating RAND's relationship with the Air Force, evinced no great enthusiasm for social science, and even less for philosophy, but he seems to have been prepared to defer to Weaver and Williams on most matters. Thus, in September 1947, Williams ran a five-day conference of social scientists at the New York Economic Club, bringing together Alchian, Wallis, Weaver, Wilks, and Mosteller, and several dozen handpicked economists, sociologists, anthropologists, and political scientists. Those present included Columbia anthropologist Ruth Benedict; political scientists Bernard Brodie, Bill Fox, and Harold Lasswell, all of Yale; sociologists Hans Speier of the New School and Franz Neumann of Columbia; and economists Charles Hitch of Oxford, Ed Shaw of Stanford, and Princeton's Jacob Viner.[18]

---

[16] Alchian interview with the author.
[17] See Jardini, *op cit.*
[18] The others included Bernard Berelson (Univ. Chicago Library School), Ansley Coale (Econ. Princeton), Chas. Dollard (Carnegie Corp.), Herbert Goldhamer (Sociology, Univ. Chicago), H. M. Gray (Public Policy, Univ. Illinois), Pendleton Herring (Carnegie Corp.),

Williams told those gathered that they were there so that RAND might essentially look them over and approach potential recruits. It was also an opportunity, he said, to discuss issues related to the "broad topics of the identification, measurement and control of factors important in (1) the occurrence of war, and (2) the winning of war if it should occur".[19] "I assume", added Weaver, "that every person in this room is fundamentally interested in and devoted to what can broadly be called the rational life. He believes fundamentally that there is something to this business of having some knowledge... and some analysis of problems, as compared with living in a state of ignorance, superstition and drifting-into-whatever-may-come". The specific rationality RAND had in mind, he said, concerned the evaluation of "military worth", discovering "to what extent it is possible to have useful quantitative indices for a gadget, a tactic or a strategy, so that one can compare it with available alternatives and guide decisions by analysis..." (*ibid*).[20]

This exploratory conference was, in many respects, run along the lines of an experiment in group dynamics. Each participant had been required to submit three projects deemed important to national security. At the meeting, these proposals were then discussed in such a way that no author was present when his project was being considered. For example, Jardini reports that a group on the "Economics of Preparedness and War", led by Chicago sociologist William Ogburn, discussed the economic and political transformation of society under war. By the end of the conference, the original panels had been regrouped[21], and they recommended that RAND social science look at a range of research areas, including psychological warfare and morale, crisis and disaster situations, American goals and values,

---

Clark H. Hull (Sociology, Yale), Abraham Kaplan (Philosophy, UCLA), Lawrence Kubie (Medicine, Columbia), Frank Lorimer (Statistics, Princeton), Donald Marquis (Psychology, Univ. Michigan), Fred Mosteller (Sociology, Harvard), Wm. F. Ogburn (Sociology, Univ. Chicago), Louis Ridenour (Physics, Univ. Illinois), Leo C. Rosten (Political Sci., unaffiliated), Ed. Shaw (Econ., Stanford), Hans Speier (Sociology, New School), Samuel Stouffer (Sociology, Harvard), Jacob Viner (Econ., Princeton), Allen Wallis (Econ., Univ. Chicago), W. Weaver (Math., Radio Corp. of America), Sam Wilks (Math., Princeton), Donald Young (Statistics, SSRC). See "New York Conference of Social Scientists", September 1947, RAND R-106.

[19] Quoted in Kaplan, *op cit*, p. 72.

[20] "Military worth", as we have seen, was the term coined by mathematician Merrill Flood in his cost–benefit and game-theoretic analysis of bombing at wartime Princeton. The "gadget" was the euphemism used to describe the atomic bomb.

[21] Jardini (*op cit*) reports that the final groups were Psychology and Sociology; Political Science; Economics; Intelligence and Military Affairs; and Research Methods, Organisation and Planning.

economic preparedness, content analysis and intelligence techniques, secrecy and disclosure issues, and methods of attitude measurement.

In fact, the attitudes of the conference participants themselves were
scrutinised, with their deliberations being monitored and recorded, and
then examined by Williams and his colleagues back at RAND. Hans Speier
remembered that: "everything that was said was recorded. Apparently they
believed in the possibility, or the desirability, of not losing any grain of
wisdom . . . That struck me as a little funny".[22] With its handpicked groups,
sound recorders, and "protocols", this conference on the anxieties and difficulties of a society facing war was a strange prefiguration of what would
emerge at RAND, where work in game theory, psychology, and wargames
would combine, at a time of anxiety, to sustain an interdisciplinary experimental form of life, an extended meditation upon individual rationality
and social stability.

At the conference, Williams approached several people as potential heads
of economic research at RAND. Neither Allen Wallis nor Ed Shaw was
interested. Neither, initially, was Charles Hitch.[23] He eventually conceded,
however, and took the position. Within a few years, with Alchian's encouragement, Hitch had brought in Jack Hirshleifer, Joseph Kershaw, Roland
McKean, Stephen Enke, David Novick, and Albert Wohlstetter.

---

[22] Speier interview – NASM, p. 23.

[23] Hitch had left the University of Arizona in the 1930s on a Rhodes Scholarship to Oxford,
where he had stayed on as lecturer. During the war, he worked with Averell Harriman at
the American Embassy, studying British controls of coal and steel for the War Production
Board (WPB) in Washington, D.C. After a stint with the WPB itself, he was drafted,
assigned to the Office of Strategic Services, forerunner of the Central Intelligence Agency,
and placed with an Anglo-American research unit of the Home Office, RE-8, located in
Princes-Risborough, near Oxford. Like Williams back at Columbia, whom he did not
know at the time, Hitch analysed before-and-after aerial photographs to assess the effects
of air raids on Germany. This work, under the supervision of a Royal Air Force office and
in collaboration with the group's various mathematicians, engineers, statisticians, and
biologists, reinforced Hitch's inclination to privilege quantifiable effects, such as buildings
burnt, for example, and to disregard speculations about factors less easily quantified. For
example, at one point, the group noticed that only a small proportion of bombs actually
hit their German target, and concluded that the damage inflicted could be heightened by
increasing the proportion of incendiary bombs as opposed to high explosives. This led
to counter-suggestions that high-explosive detonations were important because of their
psychological effect on the German population, to which Hitch and the economists could
reply that such effects could not be easily quantified. He returned to his lectureship at
Queen's College, Oxford, after the war, in the spring of 1946. Then, in 1947, he was visiting
professor at the University of Sao Paolo when he received a cablegram from Williams and
physicist Dana Bailey, inviting him to New York. See Hitch interview with the Smithsonian
Institution.

Letting Go

Confined to Princeton, Morgenstern had not been involved in any of the wartime economic analysis. No doubt he had seen aerial views of the bomb damage in the newspapers, but he was not privy to the kind of work being done by Hitch or Williams, and did not emerge from the war as part of their "crowd". In 1947, he was denied the possibility of participating in the RAND New York conference, because he had flown from that city in mid-June for his first return to Europe since the events of 1938.

His early jubilation on returning to France soon wore off. "Paris is very dark", he wrote, "cafés close at 10, 11! No business there, people look depressed, gray, women particularly. Only seldom sees one smart women (then they are often Am[erican] ...). The war has told on women quite particularly. The houses are dilapidated & groteskly [sic] in need of repairs. Sometimes there is genteelness in the poverty".[24] He visited Perroux's Institute, where he was well received, and spent an evening with economist Georges Lutfalla, and philosophers Schutz, Santillana and others. Visiting Charles Rist's Institute – "terrible; nothing goes on here" (*ibid*) – he gave an ex tempo lecture on games: "Few understood", he said, "but all terribly excited & interested". Allais, however, was opposed, saying that they "had not disproved that there is a soc. max for free competition (!). Ulmo then said that Johnny was the greatest living math. & that they should be careful in criticising. Lutfalla was very interested. Frechet did not come. Divisia is out of town. Nobody has even seen the book" (*ibid*). He later had long talks with Maurice Frechet, René Roy, and Maurice Allais, and could not understand why everybody was so nice to him. He admired André Piatier's efforts to build up statistical research against great odds and, in the evening, listened to Alfred Schutz express pessimism about his professional future – over dinner at La Tour d'Argent.

From there he went to Zurich, staying at Herman Weyl's room at the Dolder Waldhaus. In Basel, he later recalled, he met with one Furlan, a mathematician, who spoke against the "propaganda economists: Röpke, Hayek etc.".[25] Then, on June 29, for the first time in almost ten years, he arrived in Vienna, rejoining his family in their new apartment. "All are

---

[24] Diary, June 15, 1947, OMDU. Although after arriving in the United States he continued to keep his diary in German, on this visit back to Europe, Morgenstern switched temporarily to his still-imperfect English. Perhaps it was to increase the distance between him and those about whom he was writing, thus ensuring that the diary was a form of complete retreat from the world.

[25] Diary, Aug. 23, 1947, OMDU.

intellectually excellent, not a bit infected by the Nazi pest: on the contrary. Strange to be in Vienna after almost 10 years, among my own furniture but in an unknown apartment and with people who have had totally different experiences & problems".[26] He returned to the Institute, which, although now luxuriously appointed, showed "no signs of pure scientific research".[27] The following day, at a garden party at the U.S. Embassy, he spoke with Mrs. Dulles, a certain Prof. Cohn from Harvard, and others: "They all insist that they know still too little about certain Austrians e.g. Mayer (with whom they want no dealings), Drexler and others. They would like to hear more from me about all Austrian problems & persons".[28] The day after, he took coffee with Schams, his former Institute assistant. Schams had been "Pg" – that is, Nazi, but had been cleared and was now able to teach at the Technische Hochschule. "I think I can forgive him his Pg", wrote Morgenstern, but "it is different from Kamitz". Schams spoke of Mayer, who had apparently "applied voluntarily to be sent to a Nazi Schuling lager! He also asked Sch[ams] for help to become a Pg! (When the Russians arrived, Grassberger told me, Mayer, instead of going to the Univ., went to offer his 'services' to K. Gruber, who was know as leader of the Tyrol Resistance Movement. Shortly afterwards G. became Priv[at] Doc[ent] (!). There is nobody who has a good word to say for Mayer".[29] One Neider, of Gerold & Co., reported Mayer's actions in the dismissal of Jews from the Economics Society, a few days after the Anschluss, "even before any laws were made, orders given, etc.".[30]

A few days after visiting Kurt Gödel's family, he met at the Café Bastei with his old teacher, Mayer himself. The erstwhile critic of marginalist economics he found to be "[e]vasive, depressed, reduced".[31] Mayer astonished Morgenstern by asking whether Edgeworth was dead – he had died in 1926 – and asking him to resume the editorship of the *Zeitschrift für Nationalökonomie*. "Nothing doing. Also he would propose me in 2–3 years for the Univ.! as if this were something. He was – naturally – persecuted by the Nazis, etc. etc. Most disgusting. Then some talk about the theory of games which is totally unknown to him . . . He has a very bad conscience. It shows again that a certain amount of character is inseparable from

[26]   Diary, July 3, 1947, OMDU.
[27]   Diary, July 4, 1947, OMDU.
[28]   Diary, July 5, 1947, OMDU.
[29]   Diary, July 6, 1947, OMDU.
[30]   Diary, July 6, 1947, OMDU.
[31]   Diary, July 11, 1947, OMDU. Gödel's family, incidentally, did not seem to understand quite who Gödel was. "K. never really told them, in Vienna he is largely unknown" (*ibid*, July 10).

science . . . What a disappointment to see this man who has been a teacher for me & to whom I once looked up, although he made it soon & increasingly difficult" (*ibid*). For Morgenstern, it was a poignant moment, the confrontation with Mayer's treachery providing a moment of closure. Genetic-causal theory lay buried forever beneath the events of the mid-century.

The dreadful conditions and the despondency left an impression on him: "Every thing is lacking, food is bad & little, no real progress, & especially no hope . . . It is a sad story and even now I get weary. Here is perhaps one of those econ. situations where there may be no solutions (while economists naively think that every econ. problem has a solution [& that they even know it, of course!])" (*ibid*). At the same time, he wondered whether the Rockefeller Foundation could be persuaded to help bring in new people, visiting faculty. On July 16, he addressed the Austrian-American-British discussion club with a talk on the international spread of business cycles. Mayer, Marget, and Dulles were amongst those present. Some worried about his nihilism with respect to theory, his "negation of a 'system'".[32]

Morgenstern's abiding feeling in Vienna, however, was one of alienation. "The city makes me sad. Standing there, I enjoyed what I saw, but it gave me no *pang*. For that my interests & sympathies lie elsewhere & I cannot forget what has happened here e.g. to the Jews, how people plundered their neighbors etc." (*ibid*). He had uncovered further information about his former Institute assistant, "head and shoulders" Kamitz, who, like Schams, had been retained in his post by the Germans under Wagemann, while Wald and others had been let go. Even prior to 1938, it was alleged, Kamitz had been reporting statistical and economic information to "all sorts of Nazi offices": "Not surprising that Kamitz, soon after the Anschluss, could say to my sister that I had quite a few black points with the Gestapo. It is clear where they came from. I am glad I had them. But what these fellows did was ordinary plain treason & I don't want ever to have anything to do with them if I can possibly help it. How could they be good scholars?" (*ibid*).

After parting with his parents and sister, Hannichen, he headed to Berlin, where he encountered a people even more beaten down than the Austrians. He was distressed by the bomb damage to Munich, but as the flight path continued over Dachau and Nuremberg, any feelings of pity were soon extinguished. "It is impossible to say *what* one should feel in view of tragedies of these dimensions".[33]

From there he traveled to Copenhagen, where he briefly saw a Danish woman he had met previously in Princeton. The discovery that she had

---

[32] Diary, July 17, 1947, OMDU.
[33] Diary, July 30, 1947, OMDU.

since gotten married threw him into a melancholy state: "Now, sitting on the terrace of the Hotel Regina-Waldrand, slight rain falling, but much light, I feel my loneliness more than ever. I had a glimpse of another life. I think I shall crave company of a lovely woman more than ever. Only, in the background looms the work & its routine, with the absorption it always produces. But this time I do not want to succumb to it again".[34] He promised that, once back at Princeton, before it was too late, he would have a "most serious talk" with himself. But that only reminded him that he missed "Johnny" and Gödel greatly, and left him wondering about the lack of letters from the United States.

He left for London, confirmed in the opinion that he no longer belonged to continental Europe. He was not surprised that even the French artists, who had spent the war "disdainfully" in the United States, were now returning there again, having seen the state of postwar France. "I did not have to be convinced; I knew what my reaction would be . . . I *do* enjoy many things in Europe, but mostly because I have that wonderful feeling of not belonging here any more & being part of a really *free* country. Living in a totally different atmosphere".[35] At Cambridge, he spent the day with Piero Sraffa, affable, promising the collected works of Ricardo soon, "as always". On the whole, however, Cambridge lacked vigour, with Keynes, and now Gerald Shove, being sorely missed. At Oxford, he saw Roy Harrod again, was struck by how he had aged, and found him to be, like other Englishmen, "fundamentally quite touchy". Within a day or two, he returned to Princeton, promptly joining Klari von Neumann and others for cocktails. Thus began American life again. The trip had done him a lot of good, he felt, and had confirmed many opinions and impressions, but he was glad to be back in the States: "How interesting to see how much more I belong here, than to Europe".[36]

Silent Duels and Hidden Targets

When Shapley arrived at RAND, the institution was divided into the "engineers" and the "Williams people".[37] The former comprised the electronics and aeronautical engineers, the physicists working on nuclear topics. The

---

[34] Diary, Aug. 23, 1947, OMDU.
[35] Diary, Aug. 29, 1947, OMDU.
[36] Diary, Sept. 17, 1947, OMDU.
[37] "So it was, sort of, the 'engineers' and the John Williams people. He was probing into the social sciences, and the engineers thought their mission was: don't just plan the next generation of planes, but plan the faster ones after those, the continuation of that. So there was a lot of tension in RAND between the two" (Shapley interview with author).

Williams people, who soon consituted almost half the RAND staff, were a motley, interdisciplinary group. They included mathematicians such as Shapley, Helmer, and Bellman working on game theory and linear and dynamic programming; Speier's sociologists and political scientists[38]; experimental psychologists around John Kennedy; and the economists, who pushed for the application of economic criteria to the choice of weapon systems.[39] In this unique environment, under the liberal stewardship of Williams, there emerged an interdisciplinary institutional culture.[40]

The seminar in 1948 in which Shapley had made himself known to von Neumann was one of several that collectively gave rise to a RAND Report, "Mathematical Theory of Zero-Sum Two-Person Games with a Finite Number or a Continuum of Strategies", co-authored by Dresher, Helmer, Shapley, and others. It laid out the place of game theory under the new regime:

In the study of problems of national security and warfare, the competition of opposing forces is similar to a game of strategy between opposing players. Analogous to a game of strategy, national security depends not merely on our own actions but on the actions of our opponents as well. We have a system in which our objective is to maximize some function, such as defense or war power, of which we do not control all the variables. Simultaneously, the enemy aims to minimize this same function and he likewise does not control all the variables (Bohnenblust et al [1948], p. 1).

[38] Clearly, therefore, when I speak of social science at RAND, I am referring to the gamut of approaches, not just Speier's Social Science Division.

[39] There were tensions between the engineers and the economists. Unlike at the aircraft companies, such as Douglas and North American, the RAND physicists and engineers were not there to make machines or devices. They were there to talk about them, to consider the use of existing and potential technologies. They tended to be concerned with continuously improving technical performance, building bigger and faster bombers, for example – an attitude favoured, more or less, by an Air Force attached to technological superiority and very conscious of interservice rivalry – whereas the economists tended to ask disruptive questions about overall strategic purpose and the opportunity cost of particular choices. There were also tensions between the engineers and the mathematicians. The former were used to a normal working routine with normal hours. They were quite unprepared for the fact that the mathematicians were able to wander in when it suited them, and pursue their own interests without any immediate benefit to the military mission.

[40] "John Williams was the mastermind of this. He was not himself a mathematician, but he could understand... You could work when you wanted, work on what you wanted... It was just a pleasure having him there. An ideal academic administrator. As soon as he left, we had all sorts of things to justify". (Shapley interview with author). Again and again, RAND veterans point to Williams. "His attitude was: we have hired the best people we can find. We should leave them alone and never even critique their output because they're smarter than we are... We don't understand what they're doing; we're not competent to critique their output... The extreme concept of a scholar. That was John Williams' attitude". (Quade, NASM interview).

The report considered two-person, zero-sum games, showing several new results, and mentioning such tactical problems as the area defence problem: how to divide a defending force between two threatened areas, and how to divide the bombing force for the offence. As of 1948, there appeared dozens of game-theoretic analyses of tactics by Shapley, Bellman, Dresher, Girschick, and Helmer, extending the wartime work of Morse and Flood to situations involving air war and nuclear bombs. An early study by Paxson dealt with "Games of Tactics".[41] Amongst Richard Bellman's first papers were "A Bomber–Fighter Duel", "Application of Theory of Games to Identification of Friend and Foe", and "On Games Involving Bluffing". Olaf Helmer modelled conflict between heterogeneous forces, and Melvin Dresher became the military modeller par excellence with studies on "Local Defense of Targets of Equal Value", "Optimal Tactics in a Multistrike Air Campaign", "Optimal Timing in Missile Launching: A Game-Theoretic Analysis", and "A Game Theory Analysis of Tactical Air War".[42] His 1961 book, *Games of Strategy: Theory and Applications*, was a classified manual for the instruction of Air Force personnel, already in use for more than a decade when it was published.

Amongst the vast output of papers by Lloyd Shapley are a few dealing with such topics as "A Hidden Target Model" and "The Silent Duel, One Bullet Versus Two, Equal Accuracy". The hidden target model concerned the problem of an atomic bomb arriving in one of two identical bombers, one of which was protected, the other exposed. Player 1 has to choose the bomber in which to place the bomb, and Player 2 an order of attack on the bombers. Assuming known probabilities of success against each target, Shapley constructed a simple model casting the problem as a two-person, zero-sum game. In the simplest case of two consecutive attacks, each by a different fighter, and in each of which one of the bombers is targeted, Shapley showed how the optimal sequence of target choice shifted discretely from pure to mixed strategies, depending on bomber vulnerability. Herbert Scarf, who spent two years at RAND from 1954 and later became a prominent general equilibrium theorist, began his career working with Shapley on differential games with survival payoffs, and games with partial information, in which each player learns of his opponent's move only with a time lag.[43]

---

[41] Paxson (1948). For an extensive list of the unclassified material, see RAND Corporation (1989).

[42] See also Dresher (1950), (1951), and Caywood and Thomas (1955).

[43] For example, see Scarf (1955), Scarf and Shapley (1956). Scarf came to Princeton in 1951 for a doctorate in mathematics, which dealt with diffusion processes on manifolds. While

At the Air War College at Maxwell Air Force Base, Colonel Oliver Haywood completed a student thesis, soon to be a RAND report, on the "Military Doctrine of Decision and the von Neumann Theory of Games", which sought to show how military commanders should incorporate mixed strategies into their decisions.[44] He illustrated this by considering the situation at the Avranches gap in France in August 1944, when General Omar Bradley faced three choices for the disposition of his army reserve, and the German commander, General Günther von Kluge, could either attack or withdraw (Figure 13.1).

As it happened, Bradley based his decision on the attack capabilities of the Germans and sat tight for one day. Von Kluge opted to retreat eastwards.[45] Haywood showed how Bradley, had he randomised between his strategies 2 and 3, could have achieved an expected payoff higher than that yielded by conservatively choosing pure strategy 3 (Figure 13.2).

Work of this kind at RAND was encouraged by von Neumann. In the late 1940s, he made frequent visits to Santa Monica, giving talks, spending most of his time with the physicists and providing mathematical help to those who queued up to see him. To others, he provided advice by letter.[46]

there, he knew about game theory only indirectly through the work of fellow students John Nash, Shapley, and Martin Shubik. In 1954, he went to RAND, keen to do some work in applied mathematics. Given the work there on games and linear programming, RAND, at that time, was the place to go. "Many people were doing game theory, and I'd found out about the relationship between game theory and linear programming, and fell in love with linear programming and it's never really left me in a way" (Leonard interview, December 5, 1991, Cowles Commission, New Haven). Scarf stayed for two years, before moving on to Stanford. It was there, in proximity to Arrow, Hurwicz, and Uzawa, that he developed an interest in general equilibrium theory, its relationship to cooperative game-theory ideas such as the core, and issues concerning computability of general equilibria. Interestingly, Scarf revealed himself to be less than enthusiastic about noncooperative game theory, regarding it as a methodology to which economics had become hostage and which had "wasted the energies of a generation". "I cannot see questions rising to the level of the entire economy being studied by noncooperative game theory or some of the refinements of noncooperative game theory. I don't hear a major question whose resolution in a century would shed information on a serious practical policy issue. I don't see it! Now maybe that's just somebody getting older and caught in his old ways and unable to accept the excitement of a new field, that's perfectly possible" (*ibid*).

[44] See Haywood (1949), which was then issued in 1951 as a RAND report, *RM-528*, and finally published in revised form as "Military Decision and Game Theory" in the *Journal of the Operations Research Society of America* (Haywood [1954]).

[45] In fact, von Kluge was prevented from carrying out this decision by Hitler who ordered him to attack and close the Avranches Gap. There followed a two-day battle that led to the remnants of the German army being almost surrounded but managing to retreat, upon which von Kluge committed suicide.

[46] For example, in May 1948, Paxson wrote to "John", requesting his opinion "in the Harley St. sense . . . on a problem in serial correlation theory which is of fundamental

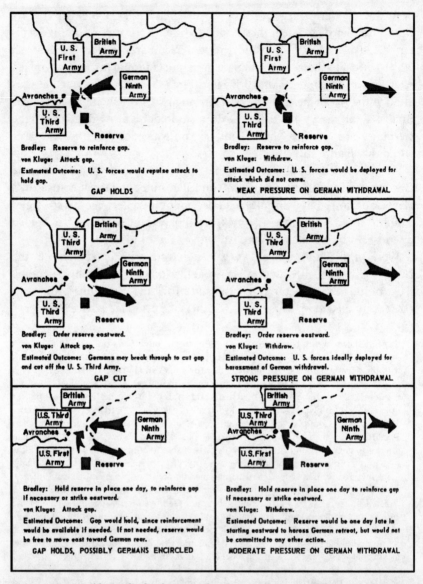

Figure 13.1. *Possible Battles for the Avranches-Gap Situation.* Six different engagements of forces may result from the interaction of Bradley's three strategies with von Kluge's two strategies. *Credit:* Reprinted by permission from Haywood, O. G., "Military Decision and Game Theory", *Journal of the Operations Research Society of America*, Vol. 2, No. 4, 1954. Copyright 1954, the Institute for Operations Research and the Management Sciences, 7240 Parkway Drive, Suite 300, Hanover, Maryland 21076.

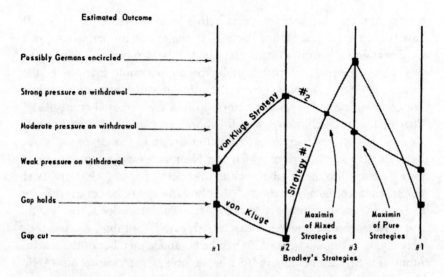

A commander normally decides to implement one of his alternative strategies. Thus, Bradley can select Strategy #1, #2, or #3, shown in the diagram by vertical lines. These are called 'pure strategies' by von Neumann. With von Neumann's 'mixed strategies', Bradley is free to select any intermediate position between his strategies #1 and #2, or between #2 and #3, or #3 and #1. Rather than assuring the 'maximin of pure strategies', he can obtain an expectation of the 'maximin of mixed strategies'.

Figure 13.2. *Mixed Strategies for the Avranches-Gap Situation. Credit:* Reprinted by permission from Haywood, O. G., "Military Decision and Game Theory", *Journal of the Operations Research Society of America*, Vol. 2, No. 4, 1954. Copyright 1954, the Institute for Operations Research and the Management Sciences, 7240 Parkway Drive, Suite 300, Hanover, Maryland 21076.

Within a few years, however, von Neumann was making himself scarce at RAND. He remained encouraging of game theory and took part in several conferences at Princeton, but his own interests had shifted towards the design and construction of computers. It was this related work in the theory of automata, and the development of nuclear weapons that would

importance in fire control and hunting studies". The problem stemmed from a 1946 paper by Cunningham and Hynd in the Supplement to the *Journal of the Royal Statistical Society* dealing with the chance that M shots will all miss a target when the rounds are correlated in a stationary way with all preceding rounds. Paxson and his team were unhappy with the series analysis in the paper and sought help with the term exp {-Lpq}, which they had tried to interpret using double LaPlace transforms, but which, in turn, had led to a tricky multiple integration situation. Von Neumann duly replied, suggesting two different approximations for the term depending on sample size. Paxson to von Neumann, May 17, 1948, VNLC, Container 25, File RAND Corp.; von Neumann to Paxson, May 21, 1948, *loc cit.*

occupy him for his remaining years.[47] In a letter to Weaver, he wrote: "I have spent a good deal of time lately on trying to find numerical methods for determining 'optimum strategies' for two-person games. I would like to get such methods which are usable on an electronic machine of the variety which we are planning, and I think that the procedures that I can contemplate will work for games up to a few hundred strategies".[48] That year, at the Ballistics Research Laboratory at the Aberdeen Proving Ground and at the Operations Evaluation Group of the Office for Naval Research (ONR), to both of which von Neumann was advisor, solutions were developed to the bomber–fighter duel developed at RAND and later published in confidential reports.[49] By the early 1950s, however, with the increase in his consulting activities to a range of institutions, both public and private, von Neumann's contacts with RAND had thinned, and they were trying to lure him back.[50] When he became a member of the federal Atomic Energy Commission in 1952, he was obliged to terminate his RAND consultancy.

Work of the Shapley and Haywood kind was welcomed at RAND, and when it came to his experimental work on games and learning, Merrill Flood pointed to Haywood's work as an early example of how complex "real-life" decisions could be effectively reduced to choices of strategy.[51] However, this burst of game-theoretic analysis of tactics in the late 1940s and early 1950s, although intensive, was short-lived. Although formally impeccable, little of it ultimately found its way into the larger systems analyses conducted at RAND, the studies most influential upon the Air Force client. Radar engineer Ed Barlow coordinated the construction of a radar

---

[47] See Aspray (1990) for a comprehensive treatment of von Neumann's work on computing. For a portrayal of von Neumann as a "cyborg scientist" responsible for bringing "machine rationality" into economics, see Mirowski (2002). Von Neumann's postwar work on games included his (1947b), (1948), (1950), (1953) and (1954b).

[48] vN to Weaver, Mar. 1, 1948, VNLC, Container 32, File: Corresp. with Warren Weaver.

[49] See Ballistics Research Laboratories, Aberdeen Proving Ground, *Results of Computations of the Fighter-Bomber Duel*, 1949 (Confidential); *Solution of the Fighter–Bomber Duel by the Theory of Games*, 1949 (Confidential); and Operations Evaluation Group, ONR, *A Fighter Bomber Duel: Optimal Firing Schedules of Both Fighter and Bomber as Determined by the Theory of Games*, 1949 (Confidential); all cited in Haywood, Col. Oliver G. (1951), "Military Doctrine of Decision and the von Neumann Theory of Games", RAND report *RM-528*, Feb. 2.

[50] By 1951, RAND statistician Alex Mood was writing to him, quietly doubling his consultancy fee, "[s]ince we have not had much luck luring you out here during the past few years". Mood to vN, Oct. 1, 1951, VNLC, *loc cit.* Mathematician Robert Specht, who came to RAND in 1949, would later recall: "Never laid eyes on von Neumann" (NASM interview, June 26, 1989, p. 9).

[51] See Flood's (1954b) "Game-Learning Theory and Some Decision-Making Experiments".

system adequate to warn of intervention by Soviet bombers. "Linear programming and game theory were largely theoretical developments", he said. "They were tools that were developing, and we felt – I felt – that these are going to be important for some class of problems someday, but they don't do much for now in systems analysis. That was my feeling. And if you try to look at what problems people tried to use them for, for very, very few in the systems analysis, did they use those tools . . . they were not major players in the scene".[52]

Again and again, one encounters variants of this judgement, whether in the writings of Albert Wohlstetter, who, using methods even less technically sophisticated than Barlow, conducted RAND's third and most influential systems analysis, the 1953 Basing Study; or Charles Hitch, whose stewardship of the cost–benefit analysis of defense procurement saw economics reign triumphant at RAND.[53] Repeatedly, one encounters the opinion that, as far as systems analysis was concerned, the value of game theory at RAND was not in providing numerical solutions to particular strategic problems, but in providing qualitative insight, in stressing the importance of reciprocation, opposing intentions, the credibility of threats, and the like. Game theory played a role in systems analysis, therefore, but not as a generator of numerical solutions to well-specified problems.

Notwithstanding its limited value as a guide in precise decision-making, game theory remained important, however, at RAND. Whether directly or indirectly, it became an important point of reference in a range of experimental activities that emerged in RAND social science. These include the experiments in psychology and group dynamics by John Kennedy and colleagues at the Systems Research Laboratory; the experiments in game

---

[52] Barlow, NASM interview. Indeed, in retrospect, Barlow insisted that a highly mathematical approach to systems analysis was of limited value. The Air Defence Study was highly quantitative, using Monte Carlo studies and incorporating such probability distributions as that of a bomber's surviving an attack by an intercepting fighter. And this, in turn, might be a function of a set of other probabilistic events. Barlow recalled that "if you put in what you think are realistic estimates of your uncertainty about each of these things, some of your conclusions wash out entirely. You can't support them because the uncertainties are enough to just wash them out" (*ibid*).

[53] Albert Wohlstetter (1964a) took a "temperate view of the present uses of both game theory and operational gaming" (p. 130). Game theory was helpful conceptually, but "as far as its theorems are concerned, it is still a long way from direct application to any complex problem of policy" (*ibid*). See Wohstetter 1964a, "Analysis and Design of Conflict Systems", in Hitch and McKean, *Economics of Defense in the Nuclear Age*, pp. 103–48; 1964b, "Sins and Games in America", in Shubik, M. (ed.) (1964), *Game Theory and Related Approaches to Social Behavior*. See also Brewer and Shubik (1979), which draws on work by Berkovitz and Dresher.

theory itself by Flood, Nash, and others; and the wargames of Helmer, Shapley, and others.

Psychology at RAND: Individual Discipline

At Princeton, as of 1948, Morgenstern resumed the role of research entre-preneur similar to that played in pre-war Vienna. From ONR directors Mina Rees and then Joachim Weyl, he obtained funding for several proj-ects, including extensive work on inventory control and military and indus-trial supply systems. He was closely involved with George Dantzig, Albert Tucker, and Harold Kuhn, who worked on the development of linear pro-gramming and other methods of optimal resource allocation and inventory control. This activity provided a "nursery ground" for Allen Newell, Martin Shubik, R. W. Shephard, T. W. Whitin, Gerald Thompson, and, later, Robert Aumann. As he had in Vienna, Morgenstern drew young researchers such as these into his projects, framing what he considered to be the impor-tant conceptual issues, stimulating those with the training to undertake the mathematical work. Beginning in 1950, he was closely involved with an annual logistics conference held at the George Washington University, and was founding editor of the *Naval Research Logistics Quarterly* in 1954. Much like von Neumann, he enjoyed the contact with the Rear Admirals and Navy brass.

Unable to attend the RAND summer conference of 1948 because of another European trip, he wrote to Charlie Hitch on return, expressing an interest in their work in game theory.[54] Within a few months, Olaf Helmer was sending him papers on game theory. Morgenstern insisted that he also wanted to be kept abreast of RAND's work on panics, interactions, and crowd psychology. He had long been sensitive to such questions: from civil-war Vienna in the mid-1930s, he had written to Frank Knight about the "quiet threat to psychological equilibrium" and, on his postwar return to Europe, he had written in his diary about the bombing of Munich and the psychological despair of the defeated Germans.

The RAND work on psychology and war had been begun earlier in 1948 by Yale psychologist, Irving Janis, and was given a certain impetus by the news in September 1949 that the Russians had successfully completed nu-clear tests.[55] Although atomic bombing was unlikely to produce severe and

---

[54]   OM to Hitch, Nov. 19, 1948, OMDU, Box 14, Folder RAND.
[55]   For example, see Janis's (1949) "Are the Cominform countries using hypnotic techniques to elicit confessions in public trials?" or his 1952 book on *Air War and Emotional Stress*. On the latter subject, see also George (1952).

permanent psychiatric disorders, Janis warned, it could well induce temporary excessive anxiety and affect the individual's ability to work. Effective civil defense might require the use of psychiatric first-aid techniques to be administered by nonprofessionals. Atomic bombing might also cause widespread panic and fright, which could lead to irrational, maladaptive behaviour and demoralization. Drawing on Janis' work, George (1952) wrote that the citizen "may become defeatist, his loyalty to his group may weaken and he may be less willing as a result to work for the achievement of his group's aims".[56] That simple phrase, a reminder that the atomic bomb threatened not only the rationality of the citizen but the cohesion of the social group, unwittingly captures the tectonic shift taking place at RAND at the time. With the threat of general eclipse, there was a clear interest in questions of group functioning and organisational cohesion.

That pattern is captured well in the work of the person who was, perhaps, RAND's most important psychologist in the early years, John Kennedy. During the war, as Technical Aide at the OSRD's Applied Psychology Panel, he had been active in that Cambridge milieu described earlier, at the intersection of psychology and engineering, close to Philip Morse and Sid Stevens. He then became professor at Tufts College, and founder of its Institute of Applied Experimental Psychology, where, under contract to the ONR's Special Devices Center, he continued his wartime work on human performance in man–machine systems: how best to ensure performance of, say, the pilot controlling the bomber or the radar operator in surveillance.[57] In 1951, he came to RAND to head the Systems Research Laboratory. Initially located in Hans Speier's Social Science Division, Kennedy's unit was soon transferred to Williams' section, alongside the game theorists and mathematicians in Military Worth.

Kennedy was brought to RAND because they valued the sort of work he had been doing at Tufts, an important part of which stemmed from intelligence reports that had leaked from Germany towards the end of the war, concerning the activities of the physiologist Kornmüller.[58] The latter, it was claimed, had built a portable, lightweight device to monitor brain-wave activity and warn of declining alertness in military personnel. When the brain's alpha frequency slowed down sufficiently, an alarm sounded, reawakening the pilot and bringing him back to the task at hand. After the war, the ONR asked Kennedy to investigate the practicality of such an alertness indicator, and he had worked for several years trying to develop a good device.

[56] Op cit, pp. 11–12.
[57] See Kennedy (ed.) (1949), and also Viteles (1945).
[58] See Kennedy (1952).

Early investigations showed that alpha rhythm was a less-than-reliable indicator. Subjects, even when wide awake, showed great variance in alpha activity, which seemed to be independent of their state of alertness. Kennedy soon discarded the idea, and shifted to using muscle action potentials. Surface electrodes were placed on the body to monitor the 'tonus' of the muscle. This looked more promising, especially when sponge rubber electrodes, soaked in saline solution, were strapped to the forehead. Muscle spike (electrical) activity seemed to be correlated to the speed of response to unanticipated demands during tasks of monotonous vigilance. With the help of the Electrodyne Co. of Boston, Kennedy and his team were able to construct an automatic electronic monitor that sounded the alarm when the subject's muscle tension, and thus alertness, fell below a critical level.[59] However, it turned out that the effectiveness of the machine depended very much on the individual involved. For normal subjects, the test situation had to be very monotonous for muscle spike oscillation to occur. Others, such as a sleep-deprived airport meteorologist, showed terrific oscillation such that neither the red light nor a loud buzzer were sufficient to wake him, and the experimenter had to intervene. Kennedy spoke also about the practical problems of using such a device. They tended to require operation by a specialised technician. People did not like to wear headbands, nor to be strapped to electrical devices. In fact, said Kennedy, it was not clear that the device was preferable to the old-fashioned jaw-spike, strapped under the chin, or even the stimulant drug Benzedrine.

For a while, after his arrival at RAND, Kennedy continued to work on human engineering of this type. Collaborating with colleagues William Emmons and Robert Hennessy, he investigated the claim in an 1949 article in *Science* magazine that stimulation of the visual area of the cortex, by a bar of light moving slowly across the visual field, gave rise to electrical activity due to a change of state of the brain cells.[60] The authors of the study, Köhler and Held, claimed to have recorded this DC (direct current) change through an electrode placed over the occipital region, and they called the phenomenon "The Cortical Correlate of Pattern Vision". At RAND, Kennedy's group replicated the experiment.

In a room, shielded so as to prevent electrical and light interference, the subject sat before a six-foot screen, instructed to fix his view on a mark in

---

[59] The subject was placed before a screen on which he operated a joy-stick, steering towards an imaginary target. When muscle tension dropped below the threshold, a red warning light appeared in the periphery of the screen, and the pilot had to react as quickly as possible by depressing a foot pedal. The reaction times here were taken as an indicator of the subject's readiness and were measured against the readings of the alertness indicator.

[60] See Köhler and Held (1949), and Emmons, Hennessy, and Kennedy (1953).

the middle. A projected horizontal bar of light was slowly passed up and down the screen, through an eighteen-degree arc, crossing the midpoint. The subject was attached to a detector by means of three electrodes placed on the head. The electrodes were refined by fusing chlorided silver wire into glass tubes filled with saline solution, and using a cotton stopper to contact the open end of the tube to the scalp. From the subject's head, the cables ran outside the shielded room to a balancing and calibrating unit which recorded the DC signals emitted.

Eleven subjects were given a total of 638 trials and, as in the Köhler-Held experiment, a positive deflection of 15 microvolts or more was regarded as a "success". The average rate of success per subject in the RAND experiment was only 26.5 percent, and significant DC deflections were observed even in subjects without the stimulus. Kennedy concluded that the phenomenon of DC change on the scalp was not related to visual stimulation. In fact, it even seemed to be caused by corneo-retinal potential due to eye movement. The Köhler-Held claim that pattern vision had a cortical correlate, said Kennedy, was premature.

Merrill Flood: From Individual to Group Learning

Whereas Kennedy's subjects surveyed illuminated screens, those of Merrill Flood monitored lamps. As we have seen, Flood had been one of the first, at wartime Princeton, to use a combination of economic analysis and game theory for the development of a rational approach to strategic bombing. Following a postwar stint as Deputy Director of Research at the Department of the Army, he came to Santa Monica. There, in June 1949, he began engaging in informal experiments on game theory, stimulated by his attempt to buy a car from Herman Kahn. The experiments were not written up until three years later, when the experimental culture was in full blossom.[61] Citing the work of Shapley, Nash, and Raiffa, Flood insisted that game theory

---

[61] With many RAND researchers visiting for short periods from outside, there was a lively trade in second-hand cars. Flood's experiments appeared in a 1952 RAND memo, "Some Experimental Games" (*RM-789–1*). Flood wrote, "Although this is the first time that I have written about these scattered experiments, I have used them in many meetings to illustrate one or another point about game theory. So much interest has been shown in them that I am encouraged to report them here, in spite of their limited extent and uncertain conclusions" (1952b, p. 4). The actual experiments reported involved RAND summer visitors, who were indicated by initials but whose identities will be immediately obvious to anyone familiar with the period. For example, HK is no doubt Herman Kahn; MF, Flood himself; RB, Raoul Bott; JW, John Williams; and AA, Armen Alchian. The report was later published, in 1958, under the same title in the five-year-old journal *Management Science* (see Flood [1958]).

was mathematically rigorous and of great value, but of questionable validity insofar as it had not been shown to stand up to experimental test. Even two-person zero-sum theory, the most reliable part, said nothing about how a player would actually play in even simple games. In "Matching Pennies", it predicted the use of randomization but said nothing about how a player might use observations of his opponent's past play, something which even a child could be expected to do. In the case of the car negotiations, as it happened, Kahn chose not to sell his automobile, and Flood found himself in the market two months later for one being sold by Albert Tucker, who was returning East after summer in California. In both the Kahn and Tucker experiments, Flood examines the usefulness of game-theoretic predictions regarding how the profit between the negotiators will be divided. Suffice it here to say that, when buying an Oldsmobile from Tucker, Flood found it impossible to settle on any game-theoretic solution that could predict the outcome of the bargaining stage: "A theory of haggling is needed for the bargaining problem if it is to be treated as a dynamic process in game-theoretic fashion" (1952b, p. 14). This encounter with haggling subjects marked the beginning of Flood's dissatisfaction with static game theory. At RAND, game theory was no less important in its ability to provoke dissatisfaction and resistance.

He ran another interesting experiment in January 1950, designed to test the Nash equilibrium, which had just become known to the community at RAND.[62] He and Dresher roped Williams and Alchian into a two-person, nonconstant sum game, to see whether they tended to "play Nash", or veer towards the von Neumann–Morgenstern solution, the split-the-difference principle, or something else. The game they faced was the following:

		JW	
		1 (Defect)	2 (Cooperate)
AA	1 (Cooperate)	$-1, 2$	$1/2, 1$
	2 (Defect)	$0, 1/2$	$1, -1$

[62] In fact, it was so early that neither Alchian nor Williams knew about Nash's work. Flood had learned of this from his advisor, Albert Tucker, who had spent the summer at Stanford, and visited RAND. It was at Stanford, while explaining the Nash equilibrium to the psychology department, that Tucker devised the Prisoner's Dilemma story, to accompany the curious game matrix shown him by Flood and Dresher at RAND.

The Nash equilibrium is the pure strategy pair (2, 1), which is also a dominant pair. Over the course of 100 plays of the experiment, it would have given Alchian and Williams total payoffs of 0 and $0.50, respectively. Even though the game was noncooperative and thus not a "game" in the strict sense of von Neumann–Morgenstern, had the game permitted side-payments, the von Neumann–Morgenstern solution would have been (1, 2), with Williams making a side-payment to Alchian no greater than $0.50 and no less than − $0.50. Splitting the difference would also lead to (1, 2), with no side-payments. In fact, in a bid to avail of the higher payoffs, the players played (1, 2) sixty percent of the time, and the Nash equilibrium only fourteen percent. Over the course of the 100 plays of the game, there was a tendency for the players to begin with the (inefficient) Nash equilibrium and then shift towards the more efficient split-the-difference choice.

Flood distributed the results of the experiment to various colleagues, including Nash, who, as ever, was keen to defend his solution concept. The fact that the game was to be played 100 times, said Nash, ruled it out as a test of the single game. He pointed out that the strategies "A plays 1 till B plays 1, then 2 ever after" and "B plays 2 till A plays 2, then 1 ever after" were very nearly a Nash equilibrium in this cycle – and actually were a Nash equilibrium in an infinite cycle with interest on utility or in a cycle with an indeterminate stop point. If Flood wanted to test the theory, said Nash, then he should remove the interaction between plays by rotating players and not making the game history known. True to form, he concluded with a jab at Alchian and Williams, noting that they were very inefficient in obtaining the rewards: "One would have thought them more rational" (p. 24a). Flood replied that Nash's comments were useful but did nothing to alter his conclusions about the results. And just as Williams had recorded the "protocols" of that Social Science meeting in New York a few years previously, Flood recorded his subjects' utterances in this experiment, complete with Williams' laughing remarks about Alchian's "instability".[63]

It was this general dissatisfaction with the empirical validity of solutions such as the Nash equilibrium, and the inability of game theory to say anything about how players learnt in the course of play, that pushed Flood in the direction of other, behaviourist, analyses of social interaction. Initially, he continued his exploration of the individual subject, drawing on a model of stochastic learning that Bush and Mosteller had developed to explain the results of the experiments on rats in B. F. Skinner's 1938 *The Behavior of*

[63] See Appendix to Flood paper.

Organisms. In their 1951 paper, Bush and Mosteller had constructed a mathematical "fusion model", which seemed to describe the way in which rats learn to respond to the probability of reward in simple choice experiments.[64] For simple experiments in which the rat had to choose between two rooms or two press-bars, each with different (unknown) probabilities of reward or punishment, the model captured the observed convergence in rats towards the option yielding the higher probability of reward. In "On Game-Learning Theory and Some Decision-Making Experiments", Flood took the "fusion model" and applied Monte Carlo techniques to produce a pattern of play for a simple two-choice game, in which he could vary the parameters governing rewards and probabilities. The rat's decision, said Flood, was similar to the human's choice of game-theoretic strategy, insofar as biological organisms in general reduced their available choices to a small feasible range.[65] The decision paths thus simulated he called those of the "Stat-rat". He then took several simple games and had human subjects play them against the Stat-rat, and also compared humans and the Stat-rat playing the same game. Throughout, Flood talks of the "organism", be it the rat or the human, the human subjects being particularly suitable for experimentation because "the conditions of the experiment can be explained to them easily, and because their choices are made quite rapidly" (p. 149). The results of the games between humans and the Stat-rat were skimpy and inconclusive, but tended towards the conclusion that humans with a knowledge of game theory and statistics were able to defeat the rat-generated play.

Flood pushed ahead with another behaviourist experiment, "Environmental Non-Stationarity in a Sequential Decision-Making Experiment", drawing on recent experimental work in psychology by Bill Estes of the University of Indiana, another RAND Summer Seminar participant.[66] Estes had conducted experiments in which the human subject (student) had to sit before a console equipped with a lamp and two switches. In each round, the subject had to choose a switch: if the "correct" one was chosen, the light went on. It was unknown to the subject that the mechanism had been preprogrammed so that the light would respond to each switch with a certain probability. Estes developed a linear operator model – a variant of the Bush–Mosteller learning model with parameter restrictions – which seemed to capture the learning behaviour of the subjects. This model,

---

[64] See Bush and Mosteller (1951).

[65] It was here that Flood noted that the reduction of strategic choices had, to date, been made most effectively in the application of game-theory to military decision, citing the 1949 report by Oliver Haywood discussed earlier.

[66] Flood (1954a). See Estes, W. K. (1950) and (1954).

and its associated experimental behaviour, was criticised by game theorists because it did not predict complete convergence (that is, a pure strategy) to the switch with the higher probability of success. Increasingly disenchanted with game theory by now, Flood sought to counter these criticisms by suggesting that mixed play might be rational if the subject believed there to be a pattern – that is, nonstationarity, in the reward sequence. His experiment was designed to discern whether subjects who were told the sequence was stationary played differently from those who were not so told (and were thus free to suspect nonstationarity). The results obtained by Flood regarding the effect of knowledge of stationarity were vague and inconclusive.

In line with the pattern at RAND, the emphasis in Flood's work soon shifted from the rationality of the individual to the behaviour and functioning of the social group. This was helped by his encounter with R. F. Bales of Harvard's Laboratory of Social Relations, whom he got to know through Fred Mosteller. Bales spent the summer of 1951 at RAND, along with A. S. Householder of the Oak Ridge National Laboratory, and Flood spent time at the Harvard Laboratory, where he ran a series of "preference experiments". Observing their Harvard subjects through one-way glass in their laboratory, Bales and Householder had developed a formal, mathematical model of group decision, the Interactor: a social "robot", complete with Super-ego and Rational (social) Mind. Flood took this, *n*-person game theory and an extension to the group of the stochastic model of individual learning, to create a synthesis he termed his Model G, which purported to describe the process underlying the interaction of individuals as they reached a collective decision regarding how to select an item from a range of objects or divide rewards amongst group members.[67]

In Flood's work, as in others', it was only a short step from modelling the group of subjects to disciplining them. Thus, while Irving Janis wondered whether the Comecon countries were using hypnosis to elicit public confessions from dissidents and Kennedy pondered the effects of the jaw strap and Benzedrine, Flood considered the possible extensions of his own experiments: "'Stooges' may be used to give the experimenter stronger control over the subjects. Drugs, hypnosis, and surgery could eventually be employed for similar purposes. It may be instructive to make some trials with rats, or pigeons, as well as with normal and abnormal human subjects" ("A Preference Experiment" P-256, p. 2). Flood's work was part of an array

---

[67] See Flood (1952c). See also R. F. Bales (1951), *Interaction Process Analysis*, Cambridge: Addison Wesley Press.

of interventionist experimental practices conducted at RAND and when he left in 1952 for Columbia University, it was to direct its Institute for Research in Management and Industrial Production.[68]

Von Neumann and Nash at Princeton

If von Neumann was less and less present at RAND, he was available at Princeton, where Albert Tucker ran seminars and organised small conferences under the aegis of his Logistics Research Project. Here, von Neumann continued to exert an influence and, in particular, to offer a view of n-person game theory that emphasised not the computational possibilities featured in some of his papers but rather the central idea of the 1944 book – namely that the theory captured the possibilities of social organisation. The main problem continued to be the lack of a general proof of the existence of a von Neumann–Morgenstern stable set solution for all n-person games. A 1953 conference featured a small number of papers and extensive discussion amongst the forty or so participants, which included Kuhn, Tucker, Nash, Shapley, Jacob Marschak of the Cowles Commission, and von Neumann.[69] In the closing discussion, von Neumann said that the inability to settle the existence question seemed to be related to the lack of general properties of individual solutions and the difficulty in describing the entire set of solutions. He felt that useful heuristic information concerning the existence problem for general n-person games might be had by studying more closely

---

[68]   At a conference on game theory at Princeton in March 1953, there was general discussion involving von Neumann and Dalkey, Nash, Thompson, Nering, and Flood. They spoke about an extensive game that had been considered at the Michigan summer seminar at Santa Monica the previous summer in which, like "So-long Sucker", double-crossing and embezzlement were profitable. Flood noted that such games could be used to train people in bargaining situations. After three years at Columbia, Flood went in 1956 to the University of Michigan, where he was professor of industrial engineering for the remainder of his career.

[69]   See Harold Kuhn (ed.) (1953), "Report of an Informal Conference on the Theory of N-Person Games", Logistics Research Project, Department of Mathematics, Princeton University, March 20–21. I am grateful to Harold Kuhn for providing me with a copy of this report and the one mentioned later, edited by Philip Wolfe. Amongst the remaining participants were Julian Bigelow (Institute for Advanced Study), D. W. Blackett (Princeton), N. Dalkey (RAND), M. Flood (RAND), D. B. Gillies (Princeton), G. K. Kalisch (Univ. Minnesota), J. B. Kruskal (Princeton), H. Raiffa (Columbia), M. Shubik (Princeton), R. M. Thrall (Univ. Michigan), G. Thompson (Princeton), and W. Vickrey (Columbia). The six papers of the first day included Kalisch on "Generalized Quota Games", D. B. Gillies on "Location of Solutions", and Shapley on "Solutions Containing Arbitrary Closed Components". The three the day after included Shapley on "A n-Person Auction", and Shubik on "Non-cooperative Games and Economic Theory".

two special classes of games – qualified majority games and tower games.[70] He also felt that discriminatory solutions, in which Player *i* received a fixed amount in all imputations of the solution, were especially important, and he conjectured that "preferred" solutions might be defined in terms of the set of these fixed payments. The remaining discussions ranged widely. Nash reported on the experiments conducted at RAND. Thompson emphasised the psychological aspects of extensive form games such as "So-Long Sucker!" There was also an interesting discussion in which Marschak bemoaned the lack of a theory for a two-person bargaining situation resembling what would later become the ultimatum game. Finally, von Neumann suggested that it might be useful to experimentally investigate the effects of modifying the cooperative game by restricting communication between the players.[71]

Two years later, at a similar conference, von Neumann was again an active participant.[72] Most of the first day was devoted to games of the kind used to model pursuit situations, such as that of a battleship by a bomber. For example, Herbert Scarf presented a theorem on sequential games, showing that optimal strategies existed for each player, given perfect recall, monotonically increasing information about the opponent's moves, and continuity of the payoff function. In characteristically lucid prose, John Nash presented a verbal account of how to tackle pursuit games with an information lag. He introduced the notion of a fictitious "agent", one for each player – it was reminiscent of von Neumann's fictitious $(n + 1)$th player – who would know the information unavailable to that player, and who would be instructed by the player to do whatever randomisation was appropriate. The transformed game became one in which there was perfect information as far as the actions of the players were concerned.[73] The following day, they

---

[70] *Ibid*, p. 23–25.

[71] Note that von Neumann is still dealing here with characteristic function – that is, coalitional, games, albeit in modified form. In particular, this was not an endorsement of Nash's noncooperative game, which was predicated on zero communication.

[72] Wolfe, Philip 1955, "Report of an Informal Conference on Recent Developments in the Theory of Games", Logistics Research Project, Department of Mathematics, Princeton University, Jan. 31–Feb. 1. In addition to Kuhn, Tucker, Nash, Shapley, Shubik, Morgenstern, and von Neumann, those present included R. Aumann, then at Princeton's Analytical Research Group (a military consulting unit), M. Beckman (Cowles Commission), G. Dantzig (RAND), M. Dresher (RAND), M. Flood (Columbia), R. D. Luce (Center for Advanced Study in the Behavioral Sciences, Stanford), J. Milnor (Princeton), R. Radner (Cowles) and H. Scarf (RAND). Some fifteen papers were presented.

[73] It was this, or something close to it, that Scarf remembered years later: "Shapley and I got involved in something called the Bomber Battleship Duel . . . trying to get the optimal strategies, and then came to the conclusion at one point, I remember very vividly, that a

moved to n-person games, including Shapley on "Markets as cooperative games" and, in a session chaired by Morgenstern, Shubik on "Game theory models of economic situations".

The closing session saw von Neumann preside over Nash's presentation of his "Opinions on the future development of theory of n-person games". Nash said the main issue to be faced was the great multiplicity of solutions to games, whether cooperative or noncooperative. He called for further study of the embedding of cooperative game in the noncooperative game with built-in coalition rules. Restrictions could be placed on coalition form-ation. For example, coalitions "could be formed at random, each player being payed upon entry the amount by which he increases the coalition's worth; or the formation of coalitions could be made irreversible" (Wolfe (ed.) (1955), p. 24).[74] Experimental studies, he said, suggested that players often implicitly assumed such restrictions. Another way of filtering out "acceptable" solutions might be to impose prior bounds on what could be regarded as reasonable payments. In response, von Neumann defended the great variety of solutions, referring, as we have already seen in Chapter 10, to the great variety of observed stable social structures.

Von Neumann then proposed a new approach to the cooperative game, which would require some theory of the rules of games. Beginning with the noncooperative game, one could introduce "admissible extensions" cov-ering, for example, communication, negotiation, and side-payments. The "stronger" the extension, the more cooperative the game. The aim would then be to find a "maximal" extension – that is, a set of rules such that the noncooperative solutions for the game did not change with further exten-sions. "If this were possible, it would be the ideal way in which to dispose of the completely cooperative game" (*ibid*, p. 25).[75] Martin Beckmann of

---

lot depended on the lag in information that either one of the agents has. For the one and two move lags, there was a very simple solution to the problem . . . but for the three move lag, we came to the conclusion once – we never published this – that the optimal strategy was no longer Markovian: the whole history mattered in some way. [These] were examples of games with some information problem in the sense that even though players moved simultaneously, they only knew the other player's move with a certain time lag. And John Nash happened to be passing through at that moment and he gave us a very clever idea of how to write down functional equations for the optimal strategies for a general game with time lags in the information. It was really very ingenious what he thought of – it was some enlargement of the state space" (Interview with author, *op cit*).

[74] This, of course, was close to the idea underlying the Shapley Value.

[75] To this, Steven Vajda asked whether the reverse might be possible, namely to examine the effect on the cooperative game of restrictions on communication or of ignorance concerning possibilites of cooperation. Von Neumann replied that this was more difficult as it was easier to build limited opportunities for cooperation into a game than it was to

the Cowles Commission made the general observation that although game theory had been useful in the areas of strategy, tactics, and politics, if it did not offer something soon, it was likely to be dismissed by scientists in search of applications of mathematics to other areas of social science. Von Neumann replied that "the construction of some mathematical model for any social science problem presented no conceptual difficulties". Indeed, "[a]ny mathematical model free of contradiction could have its counterpart in a (possibly quite artificial, but stable) real-life situation" (p. 27).

Nash's Game Experiments at RAND

The experiments referred to by Nash were those he conducted at RAND. While Flood tried to move away from strict game theory and develop models of group behaviour that captured learning and interaction, Nash and others persisted with experiments to test game-theoretic solution concepts for cooperative games.[76] A cooperative game, with side-payments, such as that featured in von Neumann and Morgenstern, was played several times, with varying numbers of players and repetitions. Subjects were rotated to discourage permanent coalitions, and were instructed to play in a "spirit of calmly aggressive selfishness". Towards the other players, they were to adopt an "attitude of polite reserve and have no favorites" (Kalisch et al (1952), p. 5). Before each game, players were presented with the characteristic function describing the payoffs to different coalitions (see Plate 25, Chapter 11).

The experiments revealed a tendency for coalitions of two to form, and only then, if at all, increase in size, and for players to split the proceeds of a coalition evenly between members. There was also a tendency for players privileged by the game not to extract the entire benefits in principle available to them. As with Flood's account of haggling, personality differences were very important in Nash's experiments, with loquaciousness and aggressiveness helping in securing coalition membership. In the five-person game, the players' positions at the table seemed to matter, with facing pairs likely to enter in coalition. "In spite of an effort to instill a completely selfish and competitive attitude in the players, they frequently took a fairly cooperative attitude", reported Nash (p. 17). The assumption that players'

incorporate measures excluding cooperation. To von Neumann's new suggestion, Shapley responded with the observation that, when sufficiently strong cooperative moves were added to a game, every possible outcome became a Nash equilibrium. Von Neumann replied that this might be overcome by restricting the class of permissible extensions.

76 See Kalisch et al (1952), "Some Experimental n-Person Games", *RAND RM-948*, Aug. 25, later published in Thrall, Coombs, and Davis (eds.) (1954).

utility was linear in the payoffs was made, but the tendency for some players to favour randomisation in coalition formation seemed to undermine that, suggesting convexity of the utility function.

Comparing the outcomes with theoretical solutions, the authors found the results to agree reasonably well with the Shapley Value. It was difficult to judge the von Neumann–Morgenstern solution, they said, for the reasons highlighted earlier by Herbert Simon and other early critics: it wasn't quite clear what the theory asserted. Only in a minority of games did the dominance conditions for the observed imputations seem to hold. Another set of trials was devoted to a negotiation model, or cooperative game with a formalized negotiation procedure. In the three-person variant, the players were blindfolded and indicated to the umpire their offers – that is, with whom they were willing to form a coalition and what share of the proceeds they wished to claim for themselves. The observed results were preliminary but promising, the authors said. Yet another set involved the introduction of a "stooge" who was instructed to play in a particular, competitive, way. The result was a tendency for aggressiveness to spread amongst the set of players.

Social Discipline: The Systems Research Laboratory

To accomodate new recruits, while they awaited security clearance, RAND had what was called a "contamination ward". Mathematician Robert Belzer remembered his time in decontamination in 1948:

[I]t was a long time that I was in what was called the "contamination ward". We tried to work on some problems and things like that, but it was mostly a matter of waiting until we did get a clearance . . . it did take several months. There were a number of fellows in there with me at the same time. I mean there must have been – oh, I don't know – 15 or 20 people . . . It wasn't as difficult for me as it was for some fellows that were in there, because a lot of the work on game theory of course was unclassified, because it was theoretical and so on, but nevertheless you were limited. You know, you couldn't move around and you couldn't go to other parts of the building where these people were located, except for a short period of time into your administrative offices or things like that. Because you had problems with a lot of the walls in aerospace companies; they do not go to the ceiling and things that are going on in one office can be heard in another. [77]

---

[77] NASM; Belzer interview, p. 9. Belzer came to RAND in 1948, after attaining a master's degree in mathematics at Stanford, and teaching for two years at the University of Santa Clara. He recalls being recruited by a Douglas Aircraft recruiter: "All they could tell you about it was, it was very interesting. They had mathematicians down there, a lot of big names, and they paid good salaries. And it was top secret . . . Something that was big

Considerations such as this, as well as the tensions involved in housing an academic research unit in a private company, made it inevitable that RAND would have to find its own premises.[78] In 1951, when the RAND Corporation, as it had become, was planning the new quarters in which it would consolidate its activities, John Williams intervened in the architectural deliberations. He wanted to ensure the greatest communication possible amongst the researchers so as to encourage interdisciplinary cooperation. If, to take one extreme, the offices were all placed along a single length of corridor, then the distance dividing the most separate pair of occupants would be at its greatest. By folding that corridor into a square, that distance would be reduced. By breaking that square down into subsquares, a matrix of patios separated by corridors, it would be reduced even further, increasing the possibility of hallway encounters and interdisciplinary exchange. Williams – the same person who had rolled wartime pennies through an array of coins to simulate the operation of a balloon barrage – prevailed upon the architects so that when the RAND building was finally completed in 1953, with its rectilinear pattern of corridors and patios, it looked something like a chessboard from above (see Plate 22). Architecture and simple decision theory came together to yield a space in which physical encounters would affect creativity – not unlike Nash's experiments in which table position affected negotiation outcomes. When the researchers moved into the new building, Williams, rather than grouping his mathematicians together in one area, distributed half of them in offices throughout the system of corridors, "for mutual education" (Specht, NASM interview, p. 11).[79]

on the agenda down there, thanks to John Williams, was work on the theory of games and economic behaviour . . . A lot of mathematicians thought that the theory of games and economic behaviour was a very exciting area mathematically to work in, and that it was going to, you know, promise significant payoffs. Also in the military field . . . " (p. 6). Mathematician Ed Quade took a year's leave from the University of Florida in 1948 to spend some time at RAND. His first shock was when he walked in the door of the building and was greeted with "Good morning, Dr. Quade" by an unknown security guard. "[T]hey had obtained pictures . . . which you sent in on the application, and those guards studied the pictures . . . So they knew you" (Quade, NASM interview, p. 8).

78 For Williams' brilliantly comic description of these tensions, see the material quoted in Specht (1960).

79 Again and again, often in the form of something vaguely sensed, RAND veterans allude to the effect of architectural design and the organisation of space. "I think I can remember a vague feeling that if you're down the hall from somebody, you see him often. If you're in another building, something different has really happened, and it is harder, and you only meet at conferences and things. And that being together within a few tens of feet is important. You could almost tell what the other man was doing if his door was open, you know, and you saw his visitors, and he saw your visitors, and you just were a lot closer that way" (Barlow, NASM interview). "'We want you to walk around the halls and talk

In that new building, Williams had a special area constructed for psychologist Kennedy's experiments. The Systems Research Laboratory (SRL) opened its doors in January 1952.[80] A fortnight later, Kennedy was at the University of Pittsburgh, giving a talk in the lecture series "Psychology in the World Emergency". In a presentation on "The Uses and Limitations of Mathematical Models, Game Theory, and Systems Analysis in Planning and Problem Solution", he elliptically described the SRL without giving away anything about its true activities.[81]

"Why study systems rather than components?" he asked (Kennedy [1952], p. 98). Consider the relation, he said, between the size of a control dial and the performance of the pilot accomplishing a particular military mission. The human engineer who focuses on the component aspect alone will show that the speed and accuracy of dial reading improve with dial size. But the aircraft designer, interested in dial-reading as merely one part of the system constituted by the plane and the pilot, is concerned with the performance of the system as a whole: "He wants to be told that either this instrument component is critical to system performance so that he should maximize its space, or that its particular size, within wide tolerances, is not critical to the successful performance of the military task for which the plane is designed" (p. 99). Psychological research, said Kennedy, had a role to play in ensuring the reliability of the human parts of man–machine systems, even if the boundaries of the system were somewhat arbitrary. Was it the bomber, the fleet, the entire attack system?

Kennedy portrayed his laboratory experimentation as a response to the inability of formal, mathematical approaches to cope with the intricacies of human interaction. In many such models, he said, the basic information provided to the mathematician was inadequate. For example, the response of the human operator in a target-tracking system had been described by

to economists, and you don't know what to talk about but you've got to learn what to talk about, see. You haven't done it this way before but you've got to learn. You've got to talk and commingle this way, see' . . . [I]n planning this building, John Williams says 'Look,' – and he persuaded RAND and the architects – 'let's make this building such that you have hallways that maximize communication between people that are in the halls, passing, if nothing more'. If everybody's got to go around this way, everybody's got to meet, so here's another cell over there, and everybody goes around that way, and finally you can made a figure eight. This building, with patios in the middle and all those little checkerboard things, was deliberately designed – and even the architect may not know it – it's to maximize communications among disciplines" (Rumph, NASM interview, p. 35). Robert Belzer and Robert Specht also both referred to the same phenomenon. (See Belzer and Specht NASM interviews).

[80] Alongside Kennedy at the SRL were Allen Newell, W. C. Biel, and R. L. Chapman.
[81] See Kennedy (1952b).

a second-order linear differential equation, involving the absolute error between actual and desired position, the rate of change of the error, and the integral of past error. The model thus described the human gunner as a continuous servomechanism, all of which, said Kennedy, ignored significant factors such as human learning, task interest, and rapport with the experimenter.[82] The servomechanical description did not embed the human subject in a setting that was sufficiently complex. The other problem was the embryonic state of the mathematics that had been developed to deal with complexity, be it the theory of stochastic processes or the theory of games. Kennedy briefly described Bales' two-person interactor model, passing no particular judgement, and, turning to game theory, used the solution to the two-person game, Le Her, to show how the actions of individual players in complex interaction could be completely described. Game theory was a promising attack, he said, but it had important weaknesses: it ignored those large segments of human behaviour that were not rational; it required very detailed description of all strategic choices, and it was weak in describing anything other than zero-sum, purely competitive, situations. Kennedy said that if these mathematical treatments of social interaction had showed limitations, it was because they lacked the appropriate kind of information. For that, it was necessary to get into the laboratory. By looking at how people actually behaved in an experimental setting, the SRL mathematician would discover "by actual experience whether or not his assumptions about behavior are credible and consistent with reality" (p. 114).

If Kennedy was deliberately cryptic about the SRL in his public talk in Pittsburgh, his colleague, Robert Chapman, had been clearer in an internal RAND Memo distributed a few weeks previously.[83] Like Householder's Social Relations laboratory at Harvard, the SRL at RAND had a central observation deck from which the experimenter watched his subjects.

The subjects were a team of military personnel, engaged in the coordination of an air defence centre during simulated air attack, over the period of 200 hours devoted to the experiment. Their function was to defend a geographical area of about 100,000 square miles, represented on a map that dominated the room, in the face of some 10,000 flights over the area. They acted on information about incoming planes, received either as paper messages or via an input presentation device that produced imitation radar scopes.

---

[82] Here Kennedy cites recent work by Ellson, Taylor, Tustin, and Hick.

[83] See Chapman (1952). See also Chapman and Kennedy (1955) for further discussion of the SRL and group-learning under stress.

The job of the military personnel under examination was to coordinate with each other so as to spot and monitor incoming aircraft, some of which were hostile, others friendly. Results took the form of not only tracking, successfully or inadequately, but also the communicative means by which the individuals coordinated themselves as a group. Contact between them was through a custom-built telephone net to which handsets, headsets, and switch boxes were attached. The experimenters on the observation mezzanine were plugged into the same system so that all information flows could be recorded and monitored. Microphones placed in the laboratory allowed normal conversation to be monitored as well.

The emphasis was on pushing the group further and further. Learning depended on motivation and sustained performance depended on continual reinforcement. A distinction was made here between effective performance by a single individual and that of the group. Kennedy and his colleagues were interested in the latter. The information processing unit was not the individual brain but the social organism, of which the individual was merely a well-motivated cog: "Performance will be reinforced with knowledge of group performance. Organized group conferences will be held to promote group interaction and to accelerate the learning process. When peformance levels off, experimentation will begin. The substance of the research program is here in performance. Performance and more performance" (Chapman [1952], p. 12).

With the SRL, the work of Kennedy and his team had shifted from human engineering – the pilot's alpha waves and pattern vision, the alertness of the individual subject – to a curious form of social engineering: scrutinising groups, recording and analysing their protocols, making them behave better. The relevant flow of information was no longer that within the individual's nerve net, but that over the set of communication channels that constituted the group. Recordings, time studies, micromotion studies, "motion picture photographs", watching affective reactions – the modes of surveillance were many. Five years previously, Williams, Weaver, and Collbohm had monitored the social scientists in New York, analysing them later to see who might be best suited to pursue social science in a nation about to embark on a war. Now, in 1952, the SRL spoke to a better-defined set of anxieties – those stemming from external nuclear attack. If one could smooth out the informational wrinkles and optimise the group's ability to process information, then no planes would slip through and safety would be ensured.

Just as there was a strong managerial undercurrent in Flood's and Householder's work on social organisation, the SRL, too, soon presented extended possibilities for governance. For example, in mid-1953, the same Chapman addressed a professional association of personnel managers on the relationship between systems research and their discipline. Remaining opaque regarding the exact nature of the laboratory's activities, he described how the group experiments helped identify the importance of tacit knowledge in making groups work. RAND systems research was finding its way into corporate personnel management.

Nor did it end there. In the mid-1950s, Kennedy became involved in a collaboration with Yale political scientist, Harold Lasswell; Cornell anthropologist, Allan Holmberg; and Yale economist, C. E. Lindblom, in the Andean highlands of north-central Peru. There, at Hacienda Vicos, an ancient indigenous community stood in the way of a French-financed hydroelectric project being promoted by the Peruvian government. That government turned to the Department of Anthropology at Cornell University for aid in removing the obstacle, propelling the Indians into modernity through forced cultural and technical change. The community had been rented by the government to the Cornell anthropologists, as a "laboratory for the study of cultural change". Fresh from three years immersion in the SRL in Santa Monica, Kennedy saw the Peruvian Hacienda as an interacting complex system with its own problems of internal communication. Modernising the community was a management problem. Kennedy's contribution to the project was to develop a project "map room" complete with "contextual map". It was the equivalent of the SRL's bombing invasion map at RAND, this time plotting target and actual values of the economic, social, and attitudinal variables to be changed amongst the Indians. At Santa Monica, the SRL had been dominated by a map of the geographical area to be defended, with the tracks of the incoming flights labelled and carefully monitored, the whole serving as the group memory, imposing order, inducing learning. Here, in the Andean highlands, in a Hacienda-turned-control-centre, another map, with tracks and targets of a different kind, served a very similar purpose – disciplining the group as they, in turn, imposed order on the Indians working in the Hacienda outside.

Unlike the previous work on individual psychology, the SRL work was a thoroughly empirical affair with little or no theoretical substrate. It was, as they said, a "flesh and blood model". Some attempts were made to give the project some of the surface rigour associated with other areas of RAND social science. For example, speaking at the Western Psychological Association conference at Berkeley, Chapman tried to draw together some of the patterns inductively emerging from four years of SRL experiments to create a theory of organizational behaviour. His talk of organizational stability and collapse, and of operating practices being "assaulted first by one force and then another", showed how his theoretical vocabulary, like his laboratory practice, was shaped by the prevailing anxieties.[84]

Chapman's was a rather vague attempt to capture in theoretical terms, and thus make more scientifically respectable, an area the practical success of which at RAND had outstripped that of many others. His less-than-rigorous theoretical speculations were unlikely to satisfy the game theorists, and they didn't, but, at the same time, the work was so prominent that it could not be ignored. If, four years previously, Kennedy had alluded to the promise for organisation theory of stochastic learning models and game theory, the SRL in the interim had gone from one rich experimental success to another, largely sweeping aside the earlier concern for reaching out to formal theories. By the mid-1950s, the SRL was a veritable hub of activity. Perhaps this activity spoke to a perceived need to "do something", and offered something that the Air Force could immediately understand. Perhaps the spectacle, the hue and cry, of the simulated interception of nuclear bombers laid greater claim to the attention of administrators than did the stylised niceties of mathematical models of bomber–fighter duels. At any rate, with the pulsing activity of Kennedy's group, RAND's institutional antennae became even more sensitive to discussion of system, organisation, and stability under threat.[85]

Epistemology's New Terrain

Throughout 1949, Morgenstern continued to insist on being kept informed of what was going on at RAND, and was particularly interested in Paul Samuelson's RAND papers on linear programming.[86] At a conference on that subject in Chicago in mid-1949, Morgenstern finally met many of the RAND people – Williams, Armen Alchian, Ed Paxson, and others –

---

[84] See Chapman (1956). For an attempt to construct a rough theory of organizational development, see Weiner (1954). For an attempt to mould the training of future psychologists to correspond to the ideal embodied in RAND's research, see Harman (1955).

[85] In 1957, when the SRL became the Systems Development Corporation, Kennedy went to Princeton, appearing to have cut links with the Santa Monica group. He stayed at Princeton until 1966, continuing experimental work at the intersection of psychology and game theory, and collaborating with Oskar Morgenstern, amongst others. From there, he went to the Institute for Educational Development. He died in the 1980s. For information on Kennedy, I am grateful to Professor Carl Sherrick of Princeton's Subcutaneous Communications Laboratory.

[86] See Samuelson, Paul (1949), "Market mechanisms and maximization", *RAND P-69*, March 28. Also by Samuelson, RAND reports *LPC-804*, *RM-107*, *RM-179*, *RM-210* and *RM-270*.

and, soon, Williams had him sent the various professional service papers necessary to his becoming a consultant. By 1950, he was in contact with Merrill Flood, discussing the similarity of RAND's interest in logistics with the subject of his own project, which was concerned with problems arising from the administration of the large naval supply base at Bayonne, New Jersey.

Flood became Morgenstern's main contact at RAND, inviting him to spend two months at a RAND summer session on logistics, which required "Secret" – the lowest – clearance, and where Morgenstern presented his "Notes on the Formulation of a Study of Logistics" (*RAND LOGS 67*). In the following months, from Flood and Hitch, he received batches of game theory papers – including a copy of John Nash's Princeton thesis, Nash having spent the summer at RAND as an intern – and further work by Samuelson on game theory and programming. Morgenstern, in turn, kept Flood informed of work on inventory control by his student Tom Whitin, under the ONR project at Princeton. In 1951, Hitch sent Morgenstern a copy of Arrow's RAND report and book, *Social Choice and Individual Values*, which prompted Morgenstern to suggest that Menger's 1934 book on social organisation might be of interest to Arrow. He spent the period from September 1951 to February 1952 in Europe, where he appears to have been involved in some classified work for the Navy, probably in connection with the ONR logistics work. From the office of the Naval Attaché at the American Embassy in Rome, he sent material to RAND's Hitch, who replied in cryptic letters, requesting any further information "on those subjects which you may learn".[87] This drew Morgenstern into a more intimate relationship with RAND, and he was soon attributed Top Security clearance.[88]

He also had contact with Allen Newell who, in mid-1950, having spent a few months as his assistant on the Navy project, abandoned a doctorate in

---

[87] See letter, Hitch to OM, Dec. 20, 1951, *loc cit*. In a further letter to Williams later, Morgenstern refers to his being about to prepare "another report for Charles Hitch", which he could not write while abroad, "because of difficulties of transmitting classified material". OM to Williams, Feb. 6, 1952, *loc cit*. See also letter, Hitch to OM, Feb. 29, 1952, *loc cit*. Morgenstern wanted a third report, which went to Hitch in March 1952, to be classified because he wanted to "protect [his] sources of information and because of the general delicacy of some of the matters mentioned", OM to Hitch, Mar. 10, 1952, *loc cit*.

[88] In late 1952, Morgenstern was procuring for Williams "all the information about pursuit among the Poles and Czechs I could get" (OM to Williams, Sept. 24, 1952, *loc cit*). This included Steinhaus's "Definitions Necessary for the Theory of Games and Pursuit", apparently secured by Harold Kuhn through Stan Ulam. See letter, OM to Helmer, Oct. 7, 1952, *loc cit*.

mathematics and went to RAND.[89] Once there, in addition to completing work on the logistics of supply and transport begun at Princeton, Newell became involved in the theory of organisation and in the work of the SRL.[90] Olaf Helmer also worked on this topic and was in communication with Morgenstern.[91] Morgenstern spent five months at RAND in mid-1951, which allowed him to attend the meetings of the Econometric Society, being held there, and become involved in Flood's work on the theory of organisation, giving a talk on "Communication and Organisation".[92]

Morgenstern's most important paper during this period was his "Prolegomena to a Theory of Organisation" (*RAND RM-734*), which was distributed to the RAND external readership. Also mentioned in the correspondence is a top secret paper "Command and Organisation" (*RAND 269-A*). In anticipation of Morgenstern's summer visit in 1953, Helmer, who was to be absent then, wrote to him about his work with Norman Dalkey and Fred Thompson on modelling the economy under stress. They were trying to develop a model of the Russian economy under air attack and, through war-gaming, simulate its behaviour against the recuperation process.[93] At RAND, where he met Arrow, Morgenstern became involved in this and became very interested in questions of the stability and defence postures under attack:[94] "In general, the procedure should be to discover breaking points under pressure, similar to the physical techniques of determining such for material and shapes. All this could be tied up closely with the problems of target analysis, the construction of a national redoubt, etc."[95]

Behind the locked door of the physics department, Herman Kahn and the nuclear physicists considered how the molecular structures of various

---

[89] Newell finished writing up the work for the Bayonne project, which was related to the logistics of rail networks, while at RAND. See OM to Newell, June 27, 1950, and Newell to OM, Aug. 17, 1950, OMDU, Box 14, RAND file.

[90] See Kruskal and Newell (1950); Newell (1951); and Newell and Kruskal (1951).

[91] See Helmer, "Research in Organisation Theory", *RAND D-868*.

[92] See Letter, Marian Centers, Mathematics Division, RAND to OM, Aug. 21, 1951, *loc cit*. The preparations for this visit involved Morgenstern obtaining Air Force security clearance – Navy clearance for the ONR being neither the same thing nor sufficient – and instructions on how to store classified materials. See the 1951 correspondence between OM and RAND's security officer R. J. Dieudonné in OMDU, Box 14, RAND file.

[93] Letter, Helmer to OM, June 17 1953, *loc cit*.

[94] He wrote to Hitch in late 1953, requesting an Office of Defense Mobilization study on the dispersion of U.S. industry outside cities as a defence strategy. Hitch was unable to trace the report, but remarked that he personally thought the dispersal strategy to be quite infeasible, and sent him *RAND RM-1103*, "A Comparison of Some Alternative Defense Programs", (Secret); OM to Hitch Oct. 27, 1953; Hitch to OM, Nov. 6, Nov. 18, 1953, *loc cit*.

[95] Morgenstern (1954b), p. 19.

building materials would react under the extreme temperatures of nuclear explosion. Down the corridor, Kennedy and Chapman remarked on the structural cohesion of their team of subjects as they coped with the stress of all-out attack by Soviet bombers. Morgenstern mused about the stability of economic systems subject to bombardment. One person who was struck by this shared language of strains and structures in the material and social domains was Olaf Helmer. This occurred as he watched Williams, Specht, and others cluster around the war map, arguing over how to respond to the hypothetical arrival of bombers on the coast of New England.

Helmer had been one of the first mathematicians brought to Santa Monica by Williams. He was, perhaps, the most philosophically inclined of an already reflective group centered upon Shapley, Dresher, Mood, and McKinsey. His early work followed in the mould of Flood's wartime analysis of military worth: game-theoretical models of conflict situations. One of RAND's very first papers, for example, was Helmer's 1947 "Combat between Heterogeneous Forces", a generalization of Lanchester's conflict model, in which the rate of attrition of one side's various specialised allocations – for example, tanks, medium bombers, anti-aircraft guns, etc. – depended on the enemy's allocations of same.[96] By 1949, he was dealing, appropriately that year, with "Local Defense of Targets of Equal Value".[97] Within a few years, however, on the foot of the experimental work on games and the new developments surrounding Kennedy's laboratory, Helmer's writings began to suggest that he had reached something of a threshold: "I am . . . asserting (a) that both game theory and organization theory are in real trouble today, (b) that organization theory can be viewed as a very natural extension of game theory as far as applications are concerned, and (c) that by giving proper recognition to this intimate relationship between the two fields they are both likely to overcome their present difficulties" (1957, p. 1).

Without naming Chapman, Morgenstern, or the related work of Kruskal and Herb Simon, Helmer dismissed organisation theory as being without substance. There had been much talk about it in the previous decade, he said, but there had been "no serious and successful attempt to build up an adequate conceptual framework . . . The notions which people working in this general area have with regard to this concept seem to be quite uncertain and frequently at variance with one another" (pp. 1–2). As for game theory, his remarks, coming from one who had spent a decade modelling tactics

---

96 See Helmer (1947).
97 Belzer, Dresher, and Helmer, "Local Defense of Targets of Equal Value", RAND *RM-319*, November.

and conflict, are arresting. Its basic concepts, he said, were inadequate "for grasping the realities of conflict situations among people" (p. 2). In economic applications, it was absurd to assume that player utilities were linear in the money payoffs. In military applications, the situation was even worse. There, in order to make game-theoretic treatment possible, modellers had to contrive a completely fictitious measurable utility, and make untenable assumptions about linearity and, to ensure the zero-sum property, about uniformity in the perception of outcomes.[98] In non–zero-sum variants, the von Neumann–Morgenstern theory not only presumed cooperation but also required strictures on the transferability of utility, all of which reduced its applicability to military engagements. As for the equilibrium point – the Nash equilibrium – although it made no strictures on commensurability of utilities, it did presuppose complete noncooperation, which, said Helmer, made it "rarely applicable directly to a real-life situation" (p. 6). There were many games, he said, in which the gains to a modicum of cooperation were so great that the Nash solution was unreasonable, yet game theory had nothing to say about how players would exercise their cooperative options. This depended on extra-game considerations such as players' attitudes towards fellow players and their observation of past patterns, to discuss which one had to go beyond game theory proper. What Helmer proposed was a research programme combining game theory and the study of organisations. A worthwhile theory of organisation would not consist in vague generalities of Chapman's kind, it would be an extension of game theory; it would put flesh on the basic strategic model by explaining how individuals with particular dispositions towards fellow players, towards risk, and with regard to learning, would interact in decision-making situations under given constraints.

There were two ways forward, said Helmer. One was to use high-speed computing methods, Monte Carlo fashion, to simulate the effects of different input parameters. Although this sounds remarkably similar to the models developed earlier by Flood and Householder, which sought to parameterise attitudinal features such as willingness to cooperate and capacity to learn in order to construct a "social robot", Helmer makes no reference to

[98] He cited the Blotto game, for example, which dealt with how two opposing commanders should distribute their battalions amongst different battlefields. The payoff was usually assumed to be the difference in the number of surviving battalions plus so many credit points for each position held. Even if this was an accurate reflection of military utility to one side, he said, there was little reason to suspect that it was linear with respect to probability combinations of outcomes or that one side's utility should be assessed by the other using the same formula. See *loc cit*, p. 5.

them and elaborates no further, neither here nor in his later work. What he had in mind was something less mechanical, an approach to organisation that had emerged from his engagement with games as actually played by people.

At RAND, a favourite lunchtime occupation was *Kriegspiel* (war game), a form of blind chess, at which Lloyd Shapley was the resident expert. Each player moved without being able to see his opponent's pieces, forming an impression of the other's game when the umpire indicated that a proposed move was impossible, or that a piece had been lost. It was a war game against a partly visible enemy. More realistic simulations of war, played not on chessboards but on twin maps in separate rooms, with an umpire passing between the two teams, were an important activity at RAND in the early days. A map of the world was spread out, tokens representing the armaments of the Blue and Red teams were distributed, and the umpire began the game with, say, the announcement that the French had left Vietnam. "Then they talk about it for a while and discuss . . . very practical attitudes about what range of choices are available to the United States, and what would be the impact politically if they did this or that, and therefore the Blue Team finally says: 'I'll send 100,000 men to the area . . . Okay, Red Team, what are you going to do?'" (Barlow, NASM interview, p. 124). Plate 26 (Chapter 11) shows a wargame being played in the early 1950s by Philip Morse, by then director of Brookhaven Laboratories, Caltech's Lee Dubridge, and others. Watching all from the near left-hand corner is John Williams.

One such operational game was the Strategic War Planning game, SWAP, developed by Shapley and Helmer. The opposing teams were, again, located in separate rooms. The game began with a five-year procurement phase, during which forces were built through a combination of prescribed addi- tions and discretionary choices, the latter depending partly on intelligence concerning the actions of the other side, filtered back and forward by an umpire. With procurement completed, the dice were rolled and the players found themselves back in one of those five years, ready for war. Here, for a day, with moves taking place hourly, players made decisions concerning air strikes, takeoffs, refuelling in the air, penetration of enemy borders, alloca- tion of bombers to targets, and so on. "Assessment of the war outcome is in terms of losses in aircraft, air bases and missile launching sites, fighter bases and local-defense installations, urban destruction, and mortalities (both direct and fallout)" (Helmer [1960], p. 7). In the close-up of New England shown in Plate 27, the token in hexagon D4 indicates that the cell of Russian penetrating bombers has been detected and will be tracked as it proceeds across the map.

Given the growing disenchantment with formal, mathematical game theory as a guide to strategy, and given the prominence of the SRL's resolutely nontheoretical efforts, there is a sense in which operational games such as these may be regarded as the game theorists' response to the Air Force demand for "mission-oriented" work.[99] Some of these wargames were, in Helmer's terms, quite rigid: the strategies were chosen at the outset and, given the rules and the parameters, the consequences were calculated by computer. There was little room for interpretation and judgement after the button had been pushed, so to speak. Helmer's interest, however, was in flexible games such as his and Shapley's SWAP game, or map exercises of the kind involving Morse and Williams discussed earlier. These depended crucially on the exercise of both player judgement and umpire discretion – for example, in filtering intelligence or ruling out certain moves as unrealistic.[100]

Such wargames were, in many respects, the antithesis of the formal modelling in which Helmer had been involved since his arrival at RAND. In the bomber–fighter duels and allocation problems, as we have seen, the task was to construct the model and derive the solution, showing how it depended on the parameters of the game. Here, in these flexible wargames, something quite different occurred. They were open-ended, developmental. You could not say at the outset how the game was going to end, nor did you know what you might learn along the way. It was like a game of chess, played by teams. When assessing the situation, evaluating the enemy's stance, or deciding on strategies and responses, it became clear that human processes of discussion and debate actually mattered for the achievement of group consensus. At key moments, players looked to figures of authority, to those who brought a certain expertise to the table. At other times, tempers flared and pieces were scattered, in another reminder that rationality was contextual.[101]

---

[99] Dalkey interview, UCLA, Feb. 2, 1990.

[100] See Dalkey (1964a) and (1964b). For example, Dalkey constructed rigid wargames, casting the allocation of nuclear forces to countervalue (civilian) and counterforce (military) targets as a noncooperative game with a Nash equilibrium. With hindsight, Dalkey was quite sceptical of the value of these simulations, saying that they were much too detailed and, given that they were one-time simulations of partly random events, should not have been taken too seriously (Dalkey interview, UCLA, Feb. 2, 1990).

[101] Thomas Schelling was another figure whose early work was an extended critique of game theory. His 1960 *Strategy of Conflict*, which discusses, amongst other things, the relationship between actual gaming and game theory, was written while Schelling visited RAND in 1957–58.

Through such games, with their emphasis on tacit expert knowledge, Helmer saw the need for the development of a new epistemology, in which social science was likened to medicine, architecture or the physics of extreme temperatures.[102] Faced with his colleagues clustered around the map, trying to decide on the likely effects of a particular bombing strategy, he began to see social science differently.[103] It was not the case that social science was inexact in reflection of its subject matter, while the natural sciences were exact, relying on clear, formal, hypothetico-deductive epistemological procedures. Helmer said that although certain subfields of physics were, indeed, very exact, in several areas of science, such as engineering, aerodynamics, or the physics of extreme temperatures (that is, in nuclear explosion), one found exact procedures intermingled with unsystematised expertise, the use of rules of thumb, educated approximations, and tacit knowledge. This was especially the case in areas of application to the complexities of the real world, such as architecture and medicine. Here, satisfactory explanations and predictions – for example, concerning a building's ability to withstand collapse or the state of a patient's health given certain symptoms – relied not only on exact deduction but also on the exercise of expert judgement, poorly defined and intuitive, but genuinely felt, accumulated through long experience in similar situations. This recognition was a radical departure for one steeped in Carnapian philosophy.

Helmer argued that social science shared many of these features. The role of expert judgement needed to be recognised and explored. Thus, together with Norman Dalkey, he developed what became known as the "Delphi Technique". Faced with a range of opinions over, say, the likely effects of a particular bombing strategy, the opinion and reasons of each expert subject were individually canvassed. These were then redistributed to all subjects so that they might reevaluate their decisions, and the procedure was continued until a consensus or an irreconcilable but clearly defined stalemate emerged. To Helmer, the method allowed for the use of intrinsic expertise, the importance of which had been highlighted in the wargames, but without the complicating psychological factors that arose in discussions around the map.[104]

---

[102] See Helmer and Rescher (1960), and Helmer (1966).

[103] See Dalkey and Helmer (1962), their paper on the use of Delphi techniques to gauge the hypothetical effects of atomic bombing on industrial targets.

[104] For example, specious persuasion, the unwillingness to abandon publicly expressed opinions, the bandwagon effect of majority opinion. See Helmer and Rescher, *op cit*, p. 33.

Kennedy's odyssey took him from engineering the alertness indicator, to the maps and group dynamics of the SRL, to the extension of these methods to management practices and even the cultural transformation of Peruvian Indians. Helmer passed from the early formal models of tactics to the pragmatic looseness of wargames, out of which grew the Delphi Method as a means of making predictions about the unknown future.[105] As the nuclear severities of the early 1950s yielded to the urban and racial problems of the 1960s, Helmer, the mathematician-turned-social philosopher, sought to apply the method to gauging the future of American society.[106] In his 1966 book, *Social Technology* – dedicated, incidentally, to the memory of his protector John Williams – he explicitly compared the Delphi Method to Kennedy's Contextual Map at Hacienda Vicos, seeing it as a form of social engineering, a means of harnessing expert opinion to achieve consensus and social stability.[107]

Conclusion

RAND's early support of mathematical research in game theory in the late 1940s and early 1950s was based on the belief that there would be direct applications to the future war with the Soviet Union. However, that early belief gradually gave way to the view that the theory could, at best,

---

[105] In 1957, RAND's president, Frank Collbohm, faced with a tightened budget, undertook some cuts in strange places, for reasons that remain mysterious. One victim was the SRL, and that year saw Kennedy leave RAND for Princeton's psychology department. Control of the SRL was seized by some of its former staff, excluding Kennedy, who took it into the private sector as the Systems Development Corporation, consulting to the Air Force to train staff through computer-simulated defense systems, such as SAGE. For an internal history of the SDC, see Baum (1981). See also Edwards (1997). Baum's account seems to suggest that Kennedy broke completely with the SDC.

[106] Collbohm's second intended victim was Olaf Helmer, for reasons that are, again, unclear. (If it is true, as has been suggested, that Collbohm's opposition to Herman Kahn's work on "Civil Defense" was due to what he perceived as the vaguely "pinko" connotation of the word "civil", then the matter is clarified a little.) Furious, and threatening to leave RAND, Williams entered into battle with Collbohm to protect Helmer. He won: Helmer was retained and Williams took a year's sabbatical. This he spent as consultant to the National Council of Teachers of Mathematics on the place of mathematics in American society. The first part of his report he released as a RAND report, "Some Attributes of the Changing Society" (RAND, *RM-2285-RC*, Nov. 10, 1958). Improved mathematical education, said Williams, would be crucial in addressing the Communist threat. The planet was likened to a padded cell, three-quarters flooded, with an uneven floor, and competition for the dry ground.

[107] Helmer left RAND in 1968 and, within two years, had formed his own Institute for the Future. In 1973, he took a chair in "Futures Research" at the University of Southern California.

provide, not direct quantitative guides regarding when to pull the trigger or how to design a bomber convoy, but rather general conceptual guidance, a thought framework. As time went on, hopes were pinned less on the technical apparatus of game-theoretic operations research and more on the conceptual benefits of reasoning in a strategic way. Little of the formal game-theoretic work found its way into the larger studies conducted at RAND, the studies most important in terms of influence with the Air Force client. As we have already seen, this perception of game theory's limited value was expressed clearly by some of RAND's most prominent members, including Ed Barlow, Charles Hitch, and Albert Wohlstetter.

At the same time, game theory nonetheless became a central point of reference in an array of social scientific practices at RAND. Whether in Flood's and Nash's experimental work on games of bargaining and coalition formation; or as part of the background for Kennedy's laboratory experiments and the ensuing organization science; or, again, as point of departure for Helmer in his odyssey from operations research to Delphi Method for social engineering, game theory was a touchstone for many researchers and it pervaded RAND's activities in a complex and diffuse manner.

It is difficult to escape the impression that, for all the professed rationalism of the times, these collective experimental activities spoke to other, deeper needs. Whether it was the SRL, with its subjects absorbed in tracking bombers while being surveyed themselves from the mezzanine by psychologists and military officers; or RAND's mathematicians clustered about the map in simulated war; or students bargaining over a game of division behind the one-way glass of the laboratory; it is difficult to escape the impression that these collective activities bear characteristics of ritual and therapy: collective meditations, so to speak, at a time of anxiety and strain.

Conclusion

That the image with which we finished our story is so similar to the one with which we opened it provides a certain closure to this account. We began with figures clustered over the chess tables of Germany and the Austro-Hungarian Empire. In this cultural setting, the gameplay of the chessmasters inspired literary treatments, provoked psychological investigations, and provided a point of departure for a nascent mathematics of strategy. This was broached in the writings of Emanuel Lasker and became an object of mathematical treatment at the hands of Borel and von Neumann. Thirty years later, on the West Coast of the United States, the gameplayers at RAND are once again clustered around the chessboard – playing *Kriegspiel*, gathered over the map playing wargames. Although the world has changed and the context is different, *Homo Ludens* is still trying to have the "game" reveal its secrets about the world.

Similarity of those images notwithstanding, the historical path connecting them departed quickly from the chessboard, showing the creation of game theory to have been a rich, multifaceted story, defying any easy classification. Only in part can it be described as history of economics, for it also owes much to debates on the nature and significance of modern mathematics; the great events in the political and social history of the interwar period; and the biographical accidents and psychological proclivities of our main protagonists. Indeed, so deeply was it rooted in the events, debates, and tragedies of the period that the creation of game theory may be taken as a parable for the first half of the twenthieth century.

In 1928, the young von Neumann cut to the heart of a problem then being discussed, in various fashions, by several mathematicians. In the restricted universe in which two game players confronted each other – that of the chess club, the café, or the casino – there existed an equilibrium

344

solution that transcended all psychologising. Ten years later, as an expatriate Hungarian Jew confronted with the desecration of his world, von Neumann returned to game theory, this time extending it to embrace the phenomena of social equilibrium and change. This episode, as we have seen, is a story of mathematical creativity, laden with psychological considerations that could be confronted only indirectly: the role of willful creation versus discovery; the therapeutic effect of mathematical work; the utopian element in social theorizing.

Through the contribution of Oskar Morgenstern – another exile coming to terms with his new situation in a manner very different from von Neumann's – the influence of interwar Viennese discussions of economics and mathematics was felt at Princeton. Those debates present layer upon layer of richness and complexity, and one is quickly led from Morgenstern and Hayek's discussions of foresight and equilibrium to Karl Menger's mathematics of social compatibility. That, in turn, is connected to Austrian politics and to debates on the foundations of mathematics. The degree of interlinkage and intricacy is suggestive of a highly ramified graph or net – much beloved, as it happens, of certain Viennese and Budapest mathematicians.

If the social instability that preceded World War II was important to von Neumann's effort to construct a new mathematics of coalitions, the effect of the war itself was to see the ambition of his new theory curtailed. The idea of capturing social configurations and change was put aside by the demands of history, and a minor part of the mathematics put to work in the domain of operations research. The theory's relevant object became the placement of planes and submarines, so to speak, not the relative standing of social groups. This was a symbolic moment, in my opinion, with this shift in mathematical focus echoing the great transition from the now-lost *Mitteleuropa* to the new American age of power and technique.

Von Neumann stood at that cusp and went with the transition. The loss of his universe – Hilbert's Göttingen and Ortvay's Budapest – at the hands of the Nazis and their followers unleashed something in him. As an advisor, he was important to the war effort, Hiroshima included, and he contributed to the postwar momentum that led to the establishment of the RAND Corporation. There, now outgrowing its founder, game theory was sustained, initially as a technique promising precision and control in military engagement, and then as a machinery of thought, so to speak, helping the qualitative analysis of strategic problems. Throughout, the new analytics of bargaining and coalition-formation helped constitute a new culture of laboratory experimentation and game-play.

It was in this nexus of military and academic research that the future central figures of game theory made their appearance – John Nash, Lloyd Shapley, Martin Shubik, John Harsanyi, and Robert Aumann – to be succeeded by an exponentially growing progeny inhabiting a new discipline. The names of von Neumann and Morgenstern are known to them all, but few are aware of the complex drama underlying the creation of the theory of games.

Bibliography

Abrahams, Gerald. (1974), *Not Only Chess*, London: George Allen & Unwin

d'Abro, A. (1939), *Decline of Mechanism*, New York: Dover

Adelman, Irma (1987), "Fellner, William John (1905–1983)", in *The New Palgrave: A Dictionary of Economics*, Volume 2, edited by John Eatwell, Murray Milgate, and Peter Newman, New York: Stockton Press, 1987, p. 301

Albers, Donald J. and G. L. Anderson (eds.) (1985), *Mathematical People*, Boston: Birkhauser

Alexanderson, G. L., et al (1987), "Obituary of George Pólya", *Bulletin of the London Mathematical Society*, Vol. 19, pp. 559–608

Allen, Robert Loring (1990), *Opening Doors: The life and work of Joseph Schumpeter*, Transactions

Alt, Franz (1936), "Über die Messbarkeit des Nutzens", *Zeitschrift für Nationalökonomie*, Vol. 7, pp. 161–69

———— (1997), Interview with Seymour Kass, Bert Schweitzer, Abe Sklar, and Mrs. Annice Alt, May 17, New York

Arány, Daniel (1924), "Note sur 'Le troisième problème de jeu'", *Acta Scientiarum Mathematicarum* (Acta Universitatis Szegediensis), Vol. 2, pp. 39–42

———— (1927), "Verallgemeinerung des problems der Spieldauer für de fall von drei Spielern", *Mathematikai és Physikai Lapok*, Vol. 34, pp. 96–105 (in Hungarian). Reviewer: D. König, Budapest

———— (1928), "Sur la Généralisation du Problème de la Durée du Jeu pour Trois Joueurs", *International Congress of Mathematicians*, Bologna, pp. 73–75

———— (1929a), "Considerations sur le problème de la durée du jeu", *Tohoku Mathematical Journal*, Vol. 30, pp. 157–81

———— (1929b), "Note sur le 'Seconde problème de la durée de jeu dans le cas de trois joueurs'", *Association Francaise pour l'Avancement des Sciences*, Vol. 53, pp. 33–35

———— (1933a), "Le problème des parcours", *Tohoku Mathematical Journal*, Vol. 37, pp. 17–22

———— (1933b), "Le problème des parcours", *Assocation Française pour l'Avancement des Sciences*, pp. 20–23

Armatte, Michel (1997), "Les Mathématiques sauraient-elles nous sortir de la crise économique? X-Crise au fondement de la technocratie", Actes du Colloque Mathématiques sociales et expertise, Besançon, 30–31 octobre

Arrow, K., D. Blackwell, and M. Girschick (1949), "Bayes and Minimax Solutions of Sequential Decision Problems", *Econometrica*, Vol. 17, pp. 213–44

Asch, Mitchell (1995) *Gestalt Psychology in German Culture, 1890–1967*, Cambridge and New York: Cambridge University Press

Aspray, William (1990), *John von Neumann and the Origins of Modern Computing*, Cambridge, Massachusetts: MIT Press

Aspray, William, et al (1989), "Discussion: John von Neumann – A Case Study of Scientific Creativity", *Annals of the History of Computing*, Vol. 11, No. 3, pp. 165–69

Auman, R. (1985), "What is Game Theory Trying to Accomplish?" in Arrow and Honkapohja (eds.), *Frontiers in Economics*, Oxford: Basil Blackwell

—— (1989), "Game Theory" in Eatwell, et al (eds.) *The New Palgrave Dictionary of Economics*, New York: W. W. Norton, pp. 460–82

—— (1989), *Lectures on Game Theory*, Boulder, Colorado: Westview Press

—— (1991), Letter to R. Leonard, November 30

Bales, R. F., M. M. Flood, and A. S. Householder (1952), "Some Group Interaction Models", *RAND RM-953*, October 10

Barber, William J. (1981), "The United States: Economists in a Pluralistic Polity", *History of Political Economy*, Vol. 13, No. 3, pp. 513–47

Bassett, Gilbert W. Jr. (1987), "The St. Petersburg Paradox and Bounded Utility", *History of Political Economy*, Vol. 19, No. 4, pp. 517–23

Batterson, Steve (2006), *Pursuit of Genius: Flexner, Einstein and the Early Faculty at the Institute for Advanced Study*, Wellesley, Massachusetts: A. K. Peters

Baum, Claude (1981), *The System Builders: The Story of SDC*, Santa Monica, California: Systems Development Corporation

Baumol, W. J. and S. Goldfeld (eds.) (1968), *Precursors in Mathematical Economics*, LSE Series Reprints of Scarce Works on Political Economy, No. 19, London: LSE

Bauschinger, Sigrid (1999), "The Berlin Moderns: Else Lasker-Schüler and Café Culture", in Emily D. Bilski (ed.) (1999), *Berlin Metropolis: Jews and the New Culture, 1890–1918*, Berkeley: University of California Press, pp. 58–83

Baxter, James Phinney (1946), *Scientists Against Time*, Boston: Little Brown

Bellman, Richard, et al (1949), "Application of Theory of Games to Identification of Friend and Foe", *RAND RM-197-PR*, July

—— and D. Blackwell (1949a), "A Bomber–Fighter Duel", *RAND RM-165-PR*, June

—— and D. Blackwell (1949b), "Some Two-Person Games Involving Bluffing", *RAND P-84*, May

—— and D. Blackwell (1950), "On Games Involving Bluffing", *RAND P-168*, August

Benacerraf, P. and H. Putnam (eds.) (1991) (1964), *Philosophy of Mathematics, Selected Readings*, 2nd. ed., Cambridge and New York: Cambridge University Press

Berchtold, Jacques (ed.) (1998), *Echiquiers d'encre. Le jeu d'échecs et les lettres (XIXe-XXe siècles)*, prologue de George Steiner, Genève: Droz

Bergson, Henri (1902), "L'Effort intellectuel", *Rev. Phil. de la France et de l'étranger*, Vol. 13, pp. 1–27

Berkley, George E. (1988), *Vienna and Its Jews: The Tragedy of Success, 1880s–1980s*, Cambridge, Massachusetts: Abt Books; Lanham, MD: Madison Books

Bertrand, J. (1883), "Théorie mathématique de la richesse sociale", *Bulletin des Sciences Mathématiques et Astronomiques*, première partie, pp. 293–303

Beveridge, W. H. (1921), "Weather and Harvest Cycles", *Economic Journal*, Vol. 31, pp. 429–52

—— (1922), "Wheat Prices and Rainfall in Western Europe", *Journal of the Royal Statistical Society*, Vol. 85, pp. 412–78

Biel, W. C., et al (1957), "The Systems Research Laboratory's Air Defense Experiments", *RAND P-1202*, October 23

Binet, Alfred (1894), *Psychologie des grands calculateurs et des jouers d'échecs*, Paris, Genève: Slatkin, republished in 1981 with an introduction by François Le Lionnais

—— (1969), *The Experimental Psychology of Alfred Binet: Selected Papers*. Edited by Robert H. Pollack and Margaret W. Brenner; translated by Frances K. Zetland and Claire Ellis. New York: Springer Publishing Co. (republished 1995)

—— (2004 [1906]) *La Graphologie. Les Révélations de l'écriture d'après un contrôle scientifique*, Introduction de Serge Nicolas. Paris: l'Harmattan

Birkhoff, G. (1940), *Lattice Theory*, Providence, Rhode Island: American Mathematical Society

—— (1958), "Von Neumann and Lattice Theory", *Bulletin of the American Mathematical Society*, Vol. 64, No. 3, pt. 2, 50–56

Blackwell, D. and M. A. Girschick (1954), *Theory of Games and Statistical Decisions*, New York: Wiley

Böhm, David (1952), "A Suggested Interpretation of the Quantum Theory in Terms of 'Hidden' Variables, I and II", *Physical Review*, Vol. 85, No. 2, pp. 166–93

von Böhm-Bawerk, Eugen (1896), *Karl Marx and the Close of His System*, reprinted as "Unresolved Contradiction in the Marxian Economic System", in *Shorter Classics of Eugen von Böhm-Bawerk*, Vol. I, South Holland, IL: Libertarian Press, 1961, pp. 201–301

—— (1914), *Macht oder Okonomisches Gesetz*, Wien: Manzsche k. u. k. Hof-Verlags und Universitats-Buchhandlung, pp. 205–71

Bohnenblust, H. F., et al (1948), "Mathematical Theory of Zero-Sum Two-Person Games with a Finite Number or a Continuum of Strategies", Santa Monica, California: The RAND Corporation (no RAND document number), September 3

Bohnenblust, H. F., L. S. Shapley, and S. Sherman (1949), "Reconnaissance in Game Theory", *RAND RM-208*, August 12.

Borch, K (1973), "The Place of Uncertainty in the Theories of the Austrian School", in J. R. Hicks and W. Weber (eds.), *Carl Menger and the Austrian School of Economics*, Oxford: Clarendon, pp. 61–74

Borel, Émile (1908), "Le calcul des probabilités et la méthode des majorités", *L'Année psychologique*, Vol. 14, pp. 125–51, reprinted in (1972) *Oeuvres de Émile Borel*, 4 vols., Vol. 2, Paris: Centre National de Recherche Scientifique, pp. 1005–31

—— (1914), *Le Hasard*, Paris: Alcan

—— (1921), "La théorie du jeu et les équations intégrales à noyau symétrique", *Comptes Rendus, Académie des Sciences*, Vol. 173, pp. 1304–08. Translated in Maurice Fréchet (1953), "Emile Borel, Initiator of the Theory of Psychological Games and its Application", *Econometrica*, Vol. 21, pp. 95–127

—— (1923), "Sur les jeux où interviennent l'hasard et l'habileté des joueurs", *Association Française pour l'Avancement des Sciences*, Vol. 177, pp. 79–85

—— (1924a), "Sur les jeux où l'hasard se combine avec l'habileté des joueurs", *Comptes Rendus de l'Académie des Sciences*, Vol. 178, pp. 24–25

———— (1924b), "Sur les jeux où interviennent l'hasard et l'habileté des joueurs", *Eléments de la théorie des probabilités*, Paris: Librairie Scientifique Hermann, pp. 204–24

———— (1926), "Un Théoreme sur les systèmes de formes linéaires à detérminant symétrique gauche", *Comptes Rendus, Académie des Sciences*, Vol. 183, pp. 925–27, avec erratum p. 996

———— (1927), "Sur le système de formes linéaires à déterminant symétrique gauche et la théorie générale du jeu", in "Algèbre et Calcul des Probabilités", *Comptes Rendus, Académie des Sciences*, Vol. 184, pp. 52–53

———— (1936), "Quelques remarques sur l'application du calcul des probabilités aux jeux de hasard", *Congrès international de mathématiciens*, Oslo, Tome 2, pp. 187–90, reprinted in (1972) *Oeuvres de Émile Borel*, 4 vols., Vol. 2, Paris: Centre National de Recherche Scientifique

———— et al (1938), *Traité du Calcul des Probabilités et de ses Applications*, Tome IV, Fasc. 2, Applications aux Jeux de Hasard, Paris: Gauthier-Villars

———— (1965), (1950), *Elements of the Theory of Probability*, Englewood Cliffs: Prentice-Hall. Translated by John E. Freund

———— and Paul Painlevé, 1910, *L'Aviation*, Paris: Alcan

Born, Max 1978, *My Life*, New York: Scribner's

Botz, Gerhard (1987), "The Jews of Vienna from the Anschluss to the Holocaust" in Oxaal, et al (eds.) (1987), pp. 185–204

Braham, Randolph L. (1981), *The Politics of Genocide: the Holocaust in Hungary*, Vol. I, New York: Columbia University Press

Brewer, Garry and M. Shubik (1979), *The War Game*, Cambridge, Massachusetts; Harvard University Press

Brodie, Bernard (1949), "Strategy as a Science", *World Politics*, Vol. 1, July, pp. 467–88

Broos, Kees (1979), *Symbolen voor onderwijs en statistiek, Symbols for Education and Statistics. 1928–1965 Vienna-Moscow-The Hague*, The Hague: Mart Spruijt

Brouwer, L. E. J. (1905a), *Leven, Kunst en Mystiek*, Delft: J. Waltman

———— (1905b), "Over moraal, Propria Cures", Jg 16, No. 10, p. 16, translated and quoted in Van Stigt, "L. E. J. Brouwer, The Signific Interlude", in A. S. Troelstra and D. Van Dalen (eds.) (1982), *The L. E. J. Brouwer Centenary Symposium*, proceedings of the conference held in Noordwijkerhout, 8–13 June 1981 (Amsterdam: North Holland), pp. 505–12

Brown, G. W. and J. von Neumann (1950), "Solutions of Games by Differential Equations", in Kuhn and Tucker (eds.) (1950), *Contributions to the Theory of Games I.* Annals of Mathematics Studies 24. Princeton University Press, Princeton, pp. 73–79

Buckley, Wm. F. Jr. and L. Brent Bozell (1954), *McCarthy and his Enemies*, Chicago: Henry Regnery

Bulmer, Martin and Joan Bulmer (1981), "Philanthropy and Social Science in the 1920's: Beardsley Ruml and the Laura Spelman Rockefeller Memorial, 1922–29", *Minerva*, XIX, Autumn, p. 385ff

Burns, Eve (1929), "Statistics and Economic Forecasting", *Journal of the American Statistical Association*, Vol. 24, pp. 152–63

Bush R.R. and C.F. Mosteller (1951), "A Mathematical Model for Simple Learning", *Psychological Review*, Vol. 58, pp. 313–23

Caldwell, Bruce (1988), "Hayek's Transformation", *History of Political Economy*, Vol. 20, No. 4, Winter, pp. 513–41

—— (2004), *Hayek's Challenge: An Intellectual Biography of F. Hayek*, Chicago: University of Chicago Press

Carnap, Rudolf (1928), *Der Logische Aufbau der Welt*, Berlin: Weltkreis-Verlag. Translated as *The Logical Structure of the World and Pseudoproblems of Philosophy*, London: Routledge & Kegan Paul, 1967

—— (1928), *Der Logische Aufbau der Welt*, Leipzig: Felix Meiner Verlag. Translated as *The Logical Structure of the World: Pseudoproblems in Philosophy*. Berkeley: University of California Press, 1967

—— (1934), *Logische Syntax der Sprache*. Translated as *The Logical Syntax of Language*, New York: Humanities, 1937

Carsten, Francis L. (1977), *Fascist Movements in Austria: From Schönerer to Hitler*, London and Beverley Hills: SAGE

Cartwright, N., et al (1996), *Otto Neurath: Philosophy between Science and Politics*, Cambridge and New York: Cambridge University Press

Cassel, Gustav (1918), *Theoretische Sozialökonomie*. Translated as *A Theory of Social Economy*, 1924, New York: Harcourt Brace, xiv, 654pp.

Caywood, T. E. and C. J. Thomas (1955), "Applications of Game Theory in Fighter Versus Bomber Combat", *Journal of the Operations Research Society of America*, Vol. 3, pp. 402–11

Chamberlin, E. H. (1933), *The Theory of Monopolistic Competition*, Cambridge, Massachusetts: Harvard University Press

Champernowne, D. (1945), "A Note on J. v. Neumann's Article on 'A Model of Economic Equilibrium'", *Review of Economic Studies*, Vol. 13, pp. 10–18

Chapman, Robert L. (1952), "The Systems Research Laboratory and Its Program", *RAND RM-890*, Jan. 7

—— (1953), "Systems Research and Personnel Management", *RAND P-443*, Oct. 23

—— (1956), "A Theory of Organizational Behavior Deriving from Systems Research Laboratory Studies", *RAND P-802*, Mar. 12

—— and John L. Kennedy (1955), "The Background and Implications of the Systems Research Laboratory Studies", *RAND, P-740*, Sept. 21

Chevalley, Claude (1945), Review of *Theory of Games and Economic Behaviour*, in *View, The Modern Magazine*

Christie, Lee. S., R. Duncan Luce, and Josiah Macy, Jr. (1952), "Communication and Learing in Task-Oriented Groups", M.I.T. Research Laboratory of Electronics, Technical Report No. 231 (reprinted as *RAND, RM-1163*, Dec. 1, 1953)

Clare, G. (1982), *Last Waltz in Vienna: The Destruction of a Family, 1842–1942*, London: Pan

Collbohm, F., et al (1964), "John Davis Williams, 1909–1964, In Memoriam", The Memorial Service, Santa Monica Civic Auditorium, Santa Monica, California, Dec. 6, 1964

Collingwood, E. F. (1959), "Émile Borel", *Journal of the London Mathematical Society*, Vol. 34, pp. 488–512

Congdon, Lee (1991), *Exile and Social Thought: Hungarian Intellectuals in Germany and Austrian, 1919–1933*, Princeton: Princeton University Press

Coriat, Isador H. (1941), "The Unconscious Motives of Interest in Chess", *Psychoana-lytical Review*, Vol. 28, pp. 30–36

Cornides, T. (1983), "Karl Menger's Contribution to the Social Sciences", *Mathematical Social Sciences*, Vol. 6, pp. 1–11

Corry, Leo (1999), "Hilbert and Physics (1900–1915)", in J. Gray (ed.), pp. 145–88

———— (2000), "The Empiricist Roots of Hilbert's Axiomatic Approach", in V. F. Hendricks et al (eds.), pp. 35–54

———— (2004), *David Hilbert and the Axiomatization of Physics (1898–1918)*, Dordrecht: Kluwer

Courant, Richard and Kurt Friedrichs (1944), *Supersonic Flow and Shock Waves*, New York: Courant Institute of Mathematical Sciences, New York University

Courant, Richard and Kurt Friedrichs (1948), *Supersonic Flow and Shock Waves*, New York: Interscience Publishers

Cournot, Antoine Augustin (1838), *Recherches sur les principes mathématiques de la théorie des richesses*, Paris: Hachette. Translated as *Researches into the Mathematical Principles of the Theory of Wealth*, by Nathaniel T. Bacon, 1929, New York: Macmillan

Cowles Commission (1952), *Economic Theory and Measurement, A Twenty Year Research Report 1932–1952*, Chicago: Cowles Commission

Crain, Tom (1998), *Schlechter's Chess Games*, Yorklyn, Delaware: Caissa Editions

Craver, Earlene (1986a), "The emigration of the Austrian economists", *History of Political Economy*, Vol. 18, No. 1, pp. 1–32

———— (1986b), "Patronage and the Directions of Research in Economics: The Rocke-feller Foundation in Europe, 1924–1938", *Minerva*, XXIV, pp. 205–22

Crum, W. L. (1923), "Cycles of Rates on Commercial Paper", *Review of Economic Stat-istics*, Vol. 5, pp. 17–29

Cubeddu, Raimondo (1993), *The Philosophy of the Austrian School*, London: Routledge

Cunningham, Jacqueline L. (1997), "Alfred Binet and the Quest for Testing Higher Mental Functioning", in Wolfgang Bringmann, et al (eds.) *A Pictorial History of Psychology*, Chicago: Quintessence Club

Dalkey, Norman (1964a), "Solvable Nuclear War Models", RAND *RM-4009-PR*, April

———— (1964b), "Games and Simulations", RAND *P-2901*, April

Dalkey, Norman and Olaf Helmer (1962), "An Experimental Application of the Delphi Method to the Use of Experts", RAND *RM-727-1*-Abridged.

Dauben, Joseph (1990), *Georg Cantor*, Princeton: Princeton University Press

Dawkins, Richard (1976), *The Selfish Gene*, New York: Oxford University Press

Debreu, Gerard (1959), *Theory of Value*, New Haven and London: Yale University Press

———— (1991), "The Mathematization of Economic Theory", *American Economic Review*, Vol. 81, No. 1, pp. 1–17

———— (1992), Interview with R. Leonard, April 15, Berkeley, California

De Groot, Adriaan (1965), *Thought and Choice in Chess*, The Hague and Paris: Mouton & Co. (originally published as *Het Denken van de Schaker*, Amsterdam: North-Holland, 1946)

de Ville, P., and C. Ménard (1989), "An Insolent Founding Father", *European Economic Review*, Vol. 33, pp. 145–57

Digby, James (1989), "Operations Research and Systems Analysis at RAND, 1948–1967", *RAND N-2936-RC*, April

———— (1990), "Strategic Thought at RAND, 1948–1963", *RAND N-3096-RC*, June

Dimand, Mary Ann, and Robert Dimand 1996, *A History of Game Theory: From the Beginnings to 1945*, London: Routledge

Djakow, I. N., N. W. Petrowski, and P. A. Rudik, 1927, *Psychologie des Schachspiels*, Berlin and Leipzig: Walter de Gruyter & Co.

Dore, M., R. Goodwin, and S. Chakravarty (eds.) (1989), *John von Neumann and Modern Economics*, Oxford: Clarendon Press

Dresher, Melvin (1949), "Local Defense of Targets of Equal Value", *RAND RM-319-PR*

⸻ (1950), "Methods of Solution in Game Theory", *Econometrica*, Vol. 18, pp. 179–80

⸻ (1951), "Games of Strategy", *Mathematics Magazine*, Vol. 25, pp. 93–99

⸻ (1954), "Optimal Tactics in a Multistrike Air Campaign", *RAND RM-1335-PR*

⸻ (1959), "A Game Theory Analysis of Tactical Air War", *RAND P-1592*

⸻ (1961), "Optimal Timing in Missile Launching: A Game-Theoretic Analysis", *RAND RM-2723-PR*

⸻ (1961), *Games of Strategy*, Englewood Cliffs: Prentice-Hall

⸻ (1948), "Mathematical Theory of Zero-Sum Two-Person Games with a Finite Number or a Continuum of Strategies", RAND Corp., Sept. 3

Eatwell J., M. Milgate, and P. Newman (eds.) (1989), *The New Palgrave: Game Theory*, New York & London: Norton

Edgeworth, F. (1925), *Papers Relating to Political Economy*, I, London: Macmillan, pp. 111–42

Edwards, Paul (1997), *The Closed World: Computers and the Politics of Discourse in Cold War America*, Cambridge, Massachusetts: MIT Press

Einstein, A. and L. Infeld (1938), *The Evolution of Physics: The Growth of Ideas from Early Concepts to Relativity and Quanta*, New York: Simon & Schuster

Emmons, William H., Robert Hennessy, and John L. Kennedy (1953), "Experiments on 'the Cortical Correlate of Pattern Vision'", *RAND P-447*, Nov. 3.

Enke, Stephen (1965), "Using Costs to Select Weapons", *American Economic Review, Papers & Proceedings*, May

Enthoven, A. and K. Wayne Smith (1971), *How Much is Enough?* New York: Harper & Row

Estes, W. K. (1950), "Towards a Statistical Theory of Learning", *Psychological Review*, Vol. 57, pp. 94–107

⸻ (1954), "Individual Behavior in Uncertain Situations: An Interpretation in Terms of Statistical Association Theory", in R. M. Thrall, et al (eds.), pp. 127–37

Etkind, Alexander (1997), *Eros of the Impossible. The History of Psychoanalysis in Russia.* Translated by Noah and Maria Rubens. Boulder, Colorado: Westview Press

Evans, Griffith C. (1930), *Mathematical Introduction to Economics*, New York: McGraw-Hill

Feffer, Loren Butler (1998), "Oswald Veblen and the Capitalization of American Mathematics: Raising Money for Research, 1923–1928", *Isis*, Vol. 89, No. 3, pp. 474–97

Fellner, W. (1949), *Competition Among the Few*, New York: A. Knopf, pp. 328

Feyerabend, P. (1957–58), Review of von Neumann (1955), "Mathematical Foundations of Quantum Mechanics", *British Journal for the Philosophy of Science*, February, 32, pp. 343–47

Fienberg, S.E., et al (eds.) (1990), *A Statistical Model: Frederick Mosteller's Contributions to Statistics, Science, and Public Policy*, New York: Springer-Verlag

Fine, Reuben (1956), "Psychoanalytic Observations on Chess and Chess Masters", *Psychoanalysis, Monograph* 1, 4, No. 3

Fischman, M. and Emeric Lendjel (no date), "X-Crise et modèle des frères Guillaume" (mimeographed, Clersé-Université Lille I and Grese-Université Paris I).

Fisher, Franklin (1989), "Games Economists Play: A Noncooperative View", *RAND Journal of Economics*, Vol. 20, No. 1, pp. 113–24

Flanagan, John C., et al (1952), *Current Trends: Psychology in the World Emergency*, University of Pittsburgh Press

Fleming, Donald and Bernard Bailyn (eds.) (1969), *The Intellectual Migration*, Cambridge, Massachusetts: Harvard University Press

Fleming, J., and S. Strong, 1943, "Use of Chess in the Therapy of an Adolescent Boy", *Psychoanalytical Review*, Vol. 30, pp. 399–416

Flexner, Abraham (1908), *The American College*, New York: The Century Co.

Flood, Merrill M. (1944), "Aerial Bombing Tactics: General Considerations" (later published as *RAND RM-913*, Sept. 2, 1952)

—————— (1951a), "A Preference Experiment", *RAND P-256*, Nov. 13

—————— (1951b), "A Preference Experiment" (Series 2, Trial 1), *RAND P-258*, Dec. 5

—————— (1952a), "A Preference Experiment" (Series 2, Trials 2, 3, 4), *RAND P-263*, Jan. 25

—————— (1952b) "Some Experimental Games", *RAND RM-789-1*, June 20

—————— (1952c), "Testing Organisation Theories", *RAND P-312*, Nov. 1

—————— (1952d), "Testing Organization Theories", in Bales, R. F., M. Flood, and A. S. Householder (1952), "Some Group Interaction Models", RAND Report *RM-953*, Oct. 10

—————— (1952e), "On Stochastic Learning Theory", *RAND P-353*, Dec. 19

—————— (1954a), "Environmental Non-Stationarity in a Sequential Decision-Making Experiment", in Thrall, Coombs and Davis (eds.), pp. 287–300

—————— (1954b), "Game-Learning Theory and Some Decision-Making Experiments", in Thrall, Coombs and Davis (eds.), pp. 139–58

—————— (1958), "Some Experimental Games", *Management Science*, Vol. 5, pp. 5–26, (initially published as *RAND, RM-789–1*, revised June 20, 1952)

von Foerster, Heinz et al (eds.) (1950), *Cybernetics: Circular, Causal and Feedback Mechanisms in Biological and Social Systems. Transactions of the sixth conference March 24–25, 1949.* New York: Josiah Macy, Jr. Foundation

Fortun, M., and S. Schweber (1993), "Scientists and the Legacy of World War II: The Case of Operations Research (OR)", *Social Studies of Science*, Vol. 23, No. 4 (November), pp. 595–642

Fosdick, Raymond B. (1952), *The Story of the Rockefeller Foundation*, New York: Harper

Fossati, Eraldo (1957), *The Theory of General Static Equilibrium*, Oxford: Basil Blackwell, edited by G. L. S. Shackle

Fraenkel, A. (1928), *Einleitung in die Mengenlehre*, 3rd ed., Berlin: Springer

Frank, Tibor (2001), "Networking, Cohorting, Bonding: Michal Polanyi in Exile", *Polanyiana*, Vol. 10, pp. 108–26

Fréchet, Maurice (1953), "Émile Borel, Initiator of the Theory of Psychological Games and its Application", *Econometrica* Vol. 21, pp. 118–27

—————— (1955), *Les Mathématiques et le Concret*, Paris: Presses Universitaires

—————— (1965), "La Vie et l'Oeuvre d'Émile Borel", *L'Enseignement Mathématique*, Tome 1, Fasc. 1, pp. 1–97

Freeman, Harold (1968), "Wald, Abraham", in *International Encyclopaedia of the Social Sciences*, New York: Macmillan, Vol. 16, pp. 435–38

Friedman, J. (1977), *Oligopoly and the Theory of Games*, Amsterdam: North Holland

Frischauer, Willi (1938), *Twilight in Vienna*, London: Collins, translation by E. O. Lorimer

Frojimovic, Kinga, et al (1999), *Jewish Budapest, Monuments, Rites, History*, Budapest: Central European University Press

Gaither, H. Rowan, et al (1949), *Report of the Study Committee for the Ford Foundation on Policy and Program*, Detroit, Michigan: Ford Foundation

Galison, P. (1990), "Aufbau/Bauhaus: Logical Positivism and Architectural Modernism", *Critical Inquiry*, Vol. 16, pp. 709–52

—————— (1994), "The Ontology of the Enemy: Norbert Wiener and the Cybernetic Vision", *Critical Inquiry* Vol. 21, Autumn, pp. 228–66

Gallai, Tibor (1936), "Dénes König: A Biographical Sketch", in Dénes König, *Theorie der endlichen und unendlichen Graphen.* (Leipzig). Translated by Richard McCoart as *Theory of Finite and Infinite Graphs*, Boston: Birkhäuser, 1986, pp. 423–26

Gauthier, David (1986), *Morals by Agreement*, Oxford: Clarendon Press

George, Alexander L. (1952), "Emotional Stress and Air War" (a lecture given at the Air War College Air University, Nov. 28, 1951), *RAND P-302*, May 27

Gerschenkron, Alexander (1977), *An Economic Spurt that Failed*, Princeton: Princeton University Press

Gigerenzer, Gerd, et al (1989), *The Empire of Chance: How Probability Changed Science and Everyday Life*, Cambridge: Cambridge University Press

Giocoli, Nicola (2003), *Modeling Rational Agents: From Interwar Economics to Early Modern Game Theory*, Cheltenham: Edward Elgar

Glavinic, Thomas (1999), *Carl Haffner's Love of the Draw*, London: Harvill

Glimm, James, John Impagliazzo, and Isadore Singer (eds.) (1990), *The Legacy of John von Neumann*, Proceedings of Symposia in Pure Mathematics, Vol. 50, Providence, Rhode Island: American Mathematical Society

Goldenweiser, E. A. (1947), "The Economist and the State", *American Economic Review*, XXXVII, No. 1, pp. 1–12

Goldman, Warren (1994), *Carl Schlechter: Life and Times of the Austrian Chess Wizard*, Yorklyn, Delaware: Caissa Editions

Goldstein, J. R. (1961), "RAND: The History, Operations and Goals of a Nonprofit Corporation", *RAND P-2236–1*, April

Goldstine, Herman (1972), *The Computer from Pascal to von Neumann*, Princeton: Princeton University Press

Goldstine, Herman H. and Eugene P. Wigner (1957), "Scientific Work of J. von Neumann", *Science*, Vol. 125, No. 3250, pp. 683–84

Gray, Jeremy J. (ed.) (1999), *The Symbolic Universe: Geometry and Physics, 1890–1930*, Oxford University Press

Gray, Jeremy J. and Karen Hunger-Parshall (eds.) (2000), *Episodes in the History of Modern Algebra (1800–1950)*, Providence, RI: American Mathematical Society and London: London Mathematical Society

Groves, Leslie R. (1962), *Now It Can Be Told: The Story of the Manhattan Project*, New York: Harper

Gruber, Helmut (1991), *Red Vienna, Experiment in Working Class Culture, 1919–1934*, New York and Oxford: Oxford University Press.

Guerbstman, Alexander (1925), *Psichoanaliz sacmatnoj igri*, Moscow [The Psychoanalysis of Chess. An Interpretative Essay]

Guilbaud, G. (1949), "La Théorie des Jeux", *Economie Appliquée*, Vol. II, No. 1, pp. 275–319

———— (1952), "Les Problèmes de Partage", *Economie Appliquée*, Vol. V, No.1, pp. 93–137

———— (1954), "Leçons sur les éléments principaux de la théorie mathématique des jeux", in G. Guilbaud, P. Massé, et R. Hénon (eds.), *Stratégies et Décisions Économiques Études Théoriques et Applications aux Entreprises*, Cours et Conférences de Recherches 1951–1953, Centre d'Économétrie, C.N.R.S.

———— (1955), "La Théorie des Jeux", *Revue d'Economie Politique*, Tome LXV, pp. 153–88

Guillaume, G., and E. Guillaume (1932), *L'Economique rationelle*, Paris: Gauthier-Villars

Gumbel, E. J. (1945), Review of *Theory of Games and Economic Behavior*, in *The Annals of the American Academy of Political and Social Science*, Vol. 239, No. 1, pp. 209–210

Haag, J. (1976/7), "Othmar Spann and the Quest for a 'True State'", *Annual Austrian History Yearbook*, Vol. 12/13, pp. 227–50

Haberler, Gottfried (1984), "William Fellner In Memoriam", in Fellner W., *Essays in Contemporary Economic Problems, Disinflation*, Washington and London: American Enterprise Institute, pp. 1–6

Hacohen, Malachi (2001), *Karl Popper – The Formative Years, 1902–1945: Politics and Philosophy in Interwar Vienna.* Cambridge: Cambridge University Press

Hadamard, Jacques (1966), *The Mathematician's Mind: the Psychology of Invention in the Mathematical Field*, Princeton: Princeton University Press, originally (1945), *The Psychology of Invention in the Mathematical Field*, Princeton University Press

Hahn, Hans (1929), "Empirismus, Mathematik, Logik", *Forschungen und Fortschritte*, Vol. 5. Translated as "Empiricism, Mathematics and Logic" in Hahn (1980), pp. 39–42

———— (1930), *Überflüssige Wesenheiten (Occams Rasiermesser)* (pamphlet published by A. Wolf, Vienna). Translated as "Superfluous Entities, or Occam's Razor" in Hahn (1980), pp. 1–19

———— (1930–31), "Die Bedeutung der wissenschaftlichen Weltaufassung, insbesondere für Mathematik und Physik", *Erkenntnis*, Vol. 1. Translated as "The Significance of the Scientific World View, Especially for Mathematics and Physics" (presented August 1929) in Hahn (1980), pp. 20–30

———— (1931–32), "Diskussion zur Grundlegung der Mathematik", *Erkenntnis*, Vol. 2. Translated as "Discussion about the Foundations of Mathematics" (presented September 1930) in Hahn (1980), pp. 31–38

———— (1933a), "Logik, Mathematik und Naturkennen", *Einheitwissenschaft, Heft*, Vol. 2, Vienna: Gerold. Translated as "Logic, Mathematics, and Knowledge of Nature" in McGuinness (ed.) (1987), pp. 24–45

———— (1933b), "Die Krise der Anschauung" in *Krise und Neuaufbau in den exakten Wissenschaften. Fünf Wiener Vorträge*, Leipzig and Vienna: F. Deuticke. Translated as "The Crisis in Intuition" in Hahn (1980), pp. 73–102

———— (1934), "Gibt es Unendliches?" in *Alte Probleme – Neue Lösungen in den exakten Wissenschaften. Fünf Wiener Vorträge*, Leipzig and Vienna: F. Deuticke. Translated as "Does the Infinite Exist?" in Hahn (1980), pp. 103–31

————— (1980), *Empiricism, Logic and Mathematics*, Dordrecht: Kluwer

Halberstam, David (1993), *The Fifties*, New York: Fawcett Columbine

Haller, Rudolf (1991), "The First Vienna Circle", in Uebel (ed.) (1991), pp. 95–108

Halmos, Paul (1973), "The Legend of John von Neumann", *American Mathematical Monthly*, Vol. 80, pp. 382–94

Halperin, Israel (1984), Interview. The Princeton Mathematics Community in the 1930s. Transcript Number 18 (PMC18). www.princeton.edu/~mudd/finding_aids/mathoral/pm02.htm

————— (1990), "The Extraordinary Inspiration of John von Neumann", in *Proceedings of Symposia in Pure Mathematics*, Vol. 50: *The Legacy of John von Neumann*, edited by James Glimm, John Impagliazzo, and Isadore Singer, Providence, Rhode Island: American Mathematical Society, pp. 15–17

————— (1993), Communication with R. Leonard

Hannak, Jacques (1959), *Emanuel Lasker: The Life of a Chess Master*, New York: Simon and Schuster, an English translation by Heinrich Fraenkel of the 1942 biography in German, with a Preface by Albert Einstein

Harman, Harry H. (1955), "The Psychologist in Interdisciplinary Research", *RAND P-708*, July 25

Harrod, Roy (1959), *The Prof: A Personal Memoir of Lord Cherwell*, London: Macmillan

Hausdorff, F. (1927), *Mengenlehre*, 2nd ed., Berlin and Leipzig: Walter de Gruyter

Hayek, Friedrich (1933), *Monetary Theory and the Trade Cycle. London: Cape*. Translation of (1929) *Geldtheorie und Konjunkturtheorie (Beitrage zur Konjunkturforschung, herausgegeben vom Österreichisches Institut für Konjunkturforschung, No. 1)*. Vienna 1929, by N. Kaldor and H. M. Croome

————— (1935), *Prices and Production*, 2nd ed., revised and enlarged, New York: Augustus M. Kelly

————— (1937), "Economics and Knowledge", *Economica*, Vol. 4 (New Series), pp. 33–54

————— (1942–43), "Scientism and the Study of Society", *Economica*, New Series, Vol. 9, No. 35 (Aug., 1942), pp. 267–291; and Part II, *Economica*, New Series, Vol. 10, No. 37 (Feb., 1943), pp. 34–63

————— (1942–44), "Scientism and the Study of Society", *Economica* (New Series), Vol. IX, No. 35, August 1942, pp. 267–91; Vol. X, No. 37, February 1943, pp. 34–63; Vol. XI, No. 41, February 1944, pp. 27–39

————— [1944](1976), *The Road to Serfdom*, Chicago: University of Chicago Press

————— (1992), *The Collected Works of F.A. Hayek*, Vol. 4, "The Fortunes of Liberalism, Essays on Austrian Economics and the Ideal of Freedom", edited by Peter G. Klein, Chicago: University of Chicago Press

————— (1994), *Hayek on Hayek: An Autobiographical Dialogue*, edited by Stephen Kresge and Leif Wenar, London: Routledge

Haythorn, William W. (1957), "Simulation in RAND's Logistics Systems Laboratory", *RAND P-1075*, Apr. 30.

Haywood, Col. Oliver G. (1949), "Military Doctrine of Decision and the von Neumann Theory of Games", *Air University Quarterly*, Vol. 4, pp. 17–30

————— (1951), "Military Doctrine and the von Neumann Theory of Games", *RAND RM-528*, February

———— (1954), "Military Decision and Game Theory", *Journal of the Operations Research Society of America*, Vol. 2, November, No. 4, pp. 365–85

Heims, Steve J. (1980), *John von Neumann and Norbert Wiener*, Cambridge, Massachusetts: MIT Press

Helmer, Olaf (1947), "Combat Between Heterogeneous Forces", *RAND RM-6*, May 5

———— (1957), "The Game-Theoretical Approach to Organization Theory", *RAND P-1026*, Feb. 19

———— (1960), "Strategic Gaming", *RAND P-1902*, Feb. 10

———— Bernice Brown, and Theodore Gordon (1966), *Social Technology*. New York: Basic Books

———— and Nicholas Rescher (1960), "On the Epistemology of the Inexact Sciences", *RAND R-353*, February, no date.

Hendricks, V. F., et al (eds.) (2000), *Proof Theory: History and Philosophical Significance*, Dordrecht: Kluwer

Herken, Gregg (1985), *Counsels of War*, New York: Knopf

Hersh, Reuben and Vera John-Steiner (1993), "A Visit to Hungarian Mathematics", *The Mathematical Intelligencer*, Vol. 15, No. 2, pp. 13–26

Herzog, Arthur (1963), *The War-Peace Establishment*, New York: Harper & Row

Heywood, R. B. (ed.) (1947), *The Works of the Mind*, Chicago: University of Chicago Press

Hicks, John (1933), "Gleichgewicht und Konjunktur", *Zeitschrift für Nationalökonomie* Vol. 4, p. 445

———— (1939), *Value and Capital*, Oxford: Clarendon Press

Hicks, J. R. and W. Weber (eds.) (1973), *Carl Menger and the Austrian School of Economics*, Oxford: Clarendon

Hilbert, David (1899), *Grundlagen der Geometrie*, Leipzig: Teubner

———— (1918), "Axiomatisches Denken", *Mathematische Annalen*, Vol. 78, p. 405ff

———— (1926), "Über das Unendliche", *Mathematische Annalen*, Vol. 95, pp. 161–90. Translated as "On the infinite", by E. Putnam and G. J. Massey in Benacerraf and Putnam, (eds.) (1991), pp. 183–201

Hilbert, John S. (2001), "Emanuel Lasker: The Challenge for a Biographer", on Chess-Cafe.com, May 15th, www.chesscafe.com/text/skittles154.pdf

Hitch, Charles (1953), "Suboptimization in Operations Problems", *Journal of the Operations Research Society of America*, Vol. 1, May, No.3, pp. 87–99

———— (1963), "Plans, Programs and Budgets in the Department of Defense", *Journal of the Operations Research Society of America*, Vol. II, pp. 1–17

———— and Roland McKean (1960), *Economics of Defense in the Nuclear Age*, Cambridge, Massachusetts: Harvard University Press

Hoffman, Dassie (2000), "Sandor Ferenczi and the Humanistic Psychologists" (mimeographed, New York: Saybrook Graduate School)

Hofmann, Paul (1988), *The Viennese*, New York: Anchor Books

Holmes Wolf, Theta (1973), *Alfred Binet*, Chicago: University of Chicago Press

Hounshell, David (1997), "The Cold War, RAND, and the generation of knowledge, 1946–1962", *Historical Studies in the Physical and Biological Sciences*, Vol. 27, pp. 237–67.

Huizinga, J. (1950) [1933], *Homo Ludens: A Study of the Play Element in Culture*. New York: Roy Publishers

Hurwicz, L. (1945), "The Theory of Economic Behavior", *American Economics Review*, Vol. 35, pp. 909–25

Ingrao, Bruna and Giorgio Israel (1991), *The Invisible Hand*, Cambridge, Massachusetts: MIT Press

Innocenti, Alessandro and Carlo Zappia (2005), "Thought- and performed experiments in Hayek and Morgenstern", in Fontaine and Leonard (eds.), *The Experiment in the History of Economics*, London: Routledge, pp. 71 – 97

Isaacs, R. P. (1951), "Games of Pursuit", *RAND P-257*, November

––––––– (1955), "The Problem of Aiming and Evasion", *RAND P-642*, March

Israel, Giorgio, and Ana Milan Gasca (2009), *The World as a Mathematical Game: John von Neumann and Twentieth Century Science*, Basel: Birkhäuser

Janik, Allan, and Stephen Toulmin (1973), *Wittgenstein's Vienna*, New York: Touchstone

Janis, Irving (1949), "Are the Cominform countries using hypnotic techniques to elicit confessions in public trials?", *RAND RM-161*, Apr. 25

––––––– (1952), *Air War and Emotional Stress*, New York: McGraw-Hill

Jardini, David (1997), *Out of the Blue Yonder: the RAND Corporation's Diversification into Social Welfare Research* (unpublished PhD thesis, Carnegie-Mellon University)

Johnston, William M. (1983 [1972]), *The Austrian Mind, An Intellectual and Social History, 1848–1938*, Berkeley and Los Angeles: University of California Press

Jones, Ernest (1931), "The Problem of Paul Morphy. A Contribution to the Psychoanalysis of Chess", *International Journal of Psychoanalysis*, Vol. 12, pp. 1–23, reprinted in Jones, Ernest (1964), *Essays in Applied Psychoanalysis*, Vol. 1, New York: International Universities Press, pp. 165–96.

Kac, Mark (1985), *Enigmas of Chance*, New York: Harper & Row

Kádár, Gábor and Zoltán Vági (2004a), "Rationality or Irrationality? The Annihilation of Hungarian Jews", *The Hungarian Quarterly*, Vol. XLV, No. 174, pp. 32–54

––––––– (2004b), *Self-Financing Genocide: The Gold Train, the Becher Case and the Wealth of Hungarian Jews*, Budapest and New York: Central European University Press

Kahn, Herman (1962), *Thinking about the Unthinkable*, New York: Avon

––––––– and Irwin Mann (1957), "Game Theory", *RAND P-1166*, July 30.

Kalisch, G., et al (1952), "Some Experimental n-Person Games", *RAND RM-948*, Aug. 25. Reprinted in Thrall, Coombs, and Davis (eds.) (1954), pp. 301–327

Kalmár, Lazsló (1928/29), "Zur Theorie der abstrakten Spiele" (translated as "On the Theory of Abstract Games"), *Acta Litterarum ac Scientiarum, Regiae Universitatis Hungaricae Francisco-Josephinae*, Sectio: Scientiarum Mathematicarum, Szeged. Vol. IV, pp. 65–85

Kant, Emmanuel (1911, orig. 1781), *Critique of Pure Reason*, New York: Macmillan

––––––– (1969, orig. 1785), *Foundations of the Metaphysics of Morals*, Indianapolis: Bobbs-Merril

Kaplan, Fred (1983), *The Wizards of Armageddon*, New York: Simon & Schuster

Kaplanski, I. (1945), "A contribution to von Neumann's theory of games", *Annals of Mathematics*, Vol. 46, pp. 474–79

Katz, Barry (1989), *Foreign Intelligence*, Cambridge, Massachusetts: Harvard University Press

Katzburg, Nathaniel (1981), *Hungary and the Jews: Policy and Legislation, 1920–1943*, Ramat-Gan: Bar-Ilan University Press.

Kaufmann, Felix (1936), *Methodenlehre der Sozialwissenschaften*, Vienna: Julius Springer

——— (1944), *Methodology of the Social Sciences*, London and New York: Oxford University Press

Kaysen, C. (1946), "A Revolution in Economic Theory?" *Review of Economic Studies*, Vol. 14, pp. 1–15

Kecskeméti, Paul (1935), "Ethics and the 'Single Theory'", *Social Research*, Vol. 2, pp. 210–21

Kennedy, John L. (ed.) (1949), *Handbook of Human Engineering Data for Design Engineers*, Special Devices Center Technical Report, SDC 199-1-1

——— (1952a), "Some Practical Problems of the Alertness Indicator", *RAND*, February 29

——— (1952b), "The Uses and Limitations of Mathematical Models, Game Theory, and Systems Analysis in Planning and Problem Solution", in Flanagan, John C., et al, *Current Trends in Psychology in the World Emergency*, Pittsburgh, Pennsylvania: University of Pittsburgh Press, pp. 97–116.

——— (1955), "The Contextual Map", *RAND RM-1575*, Oct. 24.

Kertész, Imré (1992), *Fateless*. Translated by Christopher C. Wilson and Katharina M. Wilson, Evanston, Illinois: Northwestern University Press

——— (1997), *Kaddish for a Child not Born*. Translated by C. C. Wilson and K. M. Wilson, Evanston, Illinois: Hydra Books

Keynes, J. M. (1921), *A Treatise on Probability*, London: Macmillan

Kindleberger, Charles (1978), "World War II Strategy", *Encounter*, Vol. 51, pp. 39–42

——— (1980), "The Life of an Economist", *Banca Nazionale de Lavoro*, Vol. 134, September, pp. 231–45

Kirzner, Israel (ed.) (1994), *Classics in Austrian Economics: A Sampling in the History of a Tradition*, London: William Pickering

Kjeldsen, T. H. (2001), "John von Neumann's Conception of the Minimax Theorem: A Journey Through Different Mathematical Contexts", *Archive of the History of the Exact Sciences*, Vol. 56, pp. 39–68

Klaus, Georg (1965), "Emanuel Lasker – ein philosophischer Vorläufer der Spieltheorie", *Deutsche Zeitschrift für Philosophie*, Vol. 13, pp. 976–988

Klausinger, Hansjörg (2008), "Policy Advice by Austrian Economists: The Case of Austria in the 1930s", *Advances in Austrian Economics*, Vol. 11, pp. 25–53

Klein, Judy (2000), "Economics for a Client: The Case of Statistical Quality Control and Sequential Analysis," *Toward a History of Applied Economics*, edited by Roger Backhouse and Jeff Biddle, Annual Supplement to the *History of Political Economy*, Vol. 32, Durham, North Carolina: Duke University Press, 2000, pp. 27–69.

Kline, Morris (1980), *Mathematics: The Loss of Certainty*, Oxford and New York: Oxford University Press

Köhler, W., and R. Held (1949), "The Correlate of Pattern Vision", *Science*, Vol. 110, pp. 414–19

König, Dénes (1927), "Über eine Schlussweise aus dem Endlichen ins Unendliche" (translated as "On a Method of Conclusion from the Finite to the Infinite"), *Acta Litterarum ac Scientiarum, Regiae Universitatis Hungaricae Francisco-Josephinae*, Sectio: Scientiarum Mathematicarum, Szeged. Vol. III, pp. 121–30

——— (1936), *Theorie der endlichen und unendlichen Graphen*. Leipzig: Teubner. Translated as *Theory of Infinite and Infinite Graphs* (1986) by Richard McCoart. Boston: Birkhäuser.

Koopmans, Tjalling C. (ed.) (1951), *Activity Analysis of Production and Allocation*, Cowles Commission Monograph 13, New York: Wiley

Kraft, Victor (1953), *The Vienna Circle*, New York: Philosophical Library

Kramer, Edna (1982), *The Nature and Growth of Modern Mathematics*, Princeton: Princeton University Press

Kreps, David (1990), *Game Theory and Economic Modelling*, Oxford: Oxford University Press

Kruskal, J. B. Jr. and Allen Newell (1950), "A Model for Organization Theory", *RAND LOGS-103*

Kruskal, William and Judith M. Tanur (eds.) (1978), *International Encyclopedia of Statistics*, New York: Free Press

Kuhn, Harold (1952), *Lectures on the Theory of Games*, issued as a report of the Logistics Research Project, Office of Naval Research, Princeton University

———— (1991), Interview with R. Leonard, Dec. 11, Princeton

———— (1992), Personal communication with R. Leonard, Sept. 30

Kuhn, H. and A. W. Tucker (eds.) (1950), *Contributions to the Theory of Games I*, Annals of Mathematics Studies 24, Princeton: Princeton University Press

———— (1958), "John von Neumann's Work in the Theory of Games and Mathematical Economics", *Bulletin of the American Mathematical Society*, Vol. 64, No. 3, Part 2, May, pp. 100–22

Kuratowski, Kazimierz (Casimir) (1945), "A Half Century of Polish Mathematics", *Fundamenta Mathematicae*, Vol. XXXIII, pp. v–ix

Kürschàk, József (1963), *Hungarian problem book: Based on the Eötvös competitions, 1894-[1928]*. Revised and edited by G. Hajós, G. Neukomm, and J. Surányi, translated by Elvira Rapaport, New York: Random House.

Kurz, H. and N. Salvadori 1993, *European Journal for the History of Economic Thought*, Vol. 1, No. 1, pp. 129–60

Lasker, Emanuel (1965) [1896], *Common Sense in Chess*, New York: Dover

———— (2001, orig. 1907), *Kampf*, New York: Lasker's Publishing Co.; reprinted in 2001 by Berlin-Brandenburg: Potsdam, with foreword by Lothar Schmidt

———— (1976), *Lasker's Manual of Chess*, New York: Dover (orig., *Lehrbuch des Schachspiels*, 1926. First English translation, 1927)

———— (1941), *The Community of the Future*, New York: M. J. Bernin

Latour, B. (1987), *Science in Action*, Cambridge, Massachusetts: Harvard University Press

Laufenburger, Henry (1934), "Review of Morgenstern (1934) *Die Grenzen der Wirtschaftspolitik*", *Revue d'Economie Politique*, Vol. 48, No. 3, pp. 1084–85

Lax, Peter (1990), "Remembering John von Neumann", in Glimm, et al (eds.), pp. 5–7.

Leonard, R. (1991), "War as a 'Simple Economic Problem'", *History of Political Economy* Vol. 23, Special Issue: Economics and National Security, pp. 261–83

———— (1992), "Creating a Context for Game Theory", *History of Political Economy*, Vol. 24, Special Issue: "Toward a History of Game Theory", pp. 29–76

———— (1994), "Reading Cournot, Reading Nash: The Creation and Stabilization of the Nash Equilibrium", *Economic Journal*, Vol. 104, No. 424, May, pp. 492–511

———— (1995), "From Parlor Games to Social Science: von Neumann, Morgenstern, and the Creation of Game Theory, 1928–1944", *Journal of Economic Literature*, Vol. XXXIII, June, pp. 730–61

———— (1998), "Ethics and the Excluded Middle: Karl Menger and Social Science in Interwar Vienna", *Isis*, Vol. 89, pp. 1–26

———— (2004), "Structures sous tension: Théorie des jeux et psychologie sociale à la RAND", in Amy Dahan and Dominique Pestre (eds.), *Les Sciences pour La Guerre*, Paris: Ecole des Hautes Etudes en Sciences Sociales, pp. 83–127

Lewin, Kurt (1936), *Principles of Topological Psychology*. Translated by F. Heider and G. Heider, New York and London: McGraw-Hill

Litschel, Rudolf Walter (1974), *1934- Das Jahr der Irrungen*, Linz: Oberosterreichischer Landesverlag, No. 36

Loomis, Lynn H. (1946), "On a Theorem of Von Neumann", in *Proceedings of the National Academy of Science*, Vol. 32, Aug. 15, No. 8, pp. 213–15

Lorch, Edgar R. (1993), "Szeged in 1934", Reuben Hersh (ed.), *American Mathematical Monthly*, Vol. 100, pp. 219–30

Lotka, A. J. (1924), *Elements of Physical Biology*, later republished (1956) as *Elements of Mathematical Biology*, New York: Dover

Lucas, W. F. (1969), "The proof that a game may not have a solution", *Transactions of the American Mathematical Society*, Vol. 137, pp. 219–29

Luce, R. Duncan and H. Raiffa (1957), *Games & Decisions*, New York: Wiley

Lukács, György (1983), *Record of a Life*, London: Verso

Lukacs, John (1988), *Budapest, 1900*, New York: Grove Weidenfeld

———— (1998), *A Thread of Years*, New Haven, Connecticut: Yale University Press

Lutfalla, G. (ed.) (1938), *Recherches sur les principes mathématiques de la théorie des richesses*, Paris: Rivière, 255 pp.

Macdonald, Dwight (1957), *The Ford Foundation*, New York: Reynal Press

Macdougall, G. D. A. (1951), "The Prime Minister's Statistical Section" in D. N. Chester (ed.), *Lessons of the British War Economy*, Cambridge: Cambridge University Press, pp. 58–68

MacLane, Saunders (1989), "The Applied Mathematics Group at Columbia in World War II", in Peter Duren, et al (eds.), *A Century of Mathematics in America*, Part III, Vol. 3, Providence, Rhode Island: American Mathematical Society, pp. 495–515

Macrae, Norman (1992), *John von Neumann: The Scientific Genius Who Pioneered the Modern Computer, Game Theory, Nuclear Deterrence, and Much More*, New York: Pantheon Books

Mahr, Alexander (1956), "Hans Mayer – Leben und Werk", *Zeitschrift für National-ökonomie*, Bd. 16, pp. 3–16

Marget, A. W. (1929), "Morgenstern on the Methodology of Economic Forecasting", *Journal of Political Economy*, Vol. 37, pp. 312–39

Marrow, Alfred (1969), *The Practical Theorist: The Life and Works of Kurt Lewin*, New York: Basic Books

Marschak, J. (1946), "Von Neumann and Morgenstern's New Approach to Static Economics", *Journal of Political Economy*, Vol. 54, pp. 97–115

Marshall, Andrew W. (1958), "Experimentation by Simulation and Monte Carlo", *RAND P-1174*, Jan. 28.

Marshall, James (1988), "Fellner, William J." in R. Sobel and B. S. Katz (eds.), *Biographical Directory of the Council of Economic Advisors*, Westport, CT: Greenwood Press

Maurensig, Paolo (1998), *The Lüneberg Variation*, New York: Henry Holt (orig. *La variante di Lüneberg*, Milan: Adelphi, 1993)

Mayberry, J. P., J. F. Nash, and M. Shubik (1953), "A comparison of treatments of the duopoly situation", *Econometrica*, Vol. 21, pp. 141–54

Mayer, Hans (1911), "*Eine neue Grundlegung der theoretischen Nationalökonomie*", *Zeitschrift für Volkswirtschaft, Sozialpolitik und Verwaltung*, Vol. 20, pp. 181–209

———— (1921–22), "Untersuchung zu dem Grundgesetz der wirtschaftlichen Wertrechnung", *Zeitschrift für Volkswirtschaftslehre und Sozialpolitik*, Vol. 2, pp. 1–23

———— (1928), "Zurechnung", *Handwörterbuch der Staatwissenschaften*, Vol. VIII, 4th ed., pp. 1206–28. Translated and reprinted as "Imputation", in Kirzner (ed.) (1994), pp. 19–53

———— (1932), "Der Erkenntniswert der funktionellen Preistheorien", in Mayer (ed.) *Die Wirtschaftstheorie der Gegenwart*, Vienna, Vol. 2, pp. 14–239b. Translated and reprinted as "The Cognitive Value of Functional Theories of Price, Critical and Positive Investigations Concerning the Price Problem", in Kirzner (ed.) (1994), pp. 55–168

McCagg, William O. Jr, (1972), *Jewish Nobles and Geniuses in Modern Hungary*, Boulder, Colorado: East European Quarterly

McCulloch, Warren and Walter Pitts (1943), "A Logical Calculus of the Ideas Immanent in Nervous Activity", *Bulletin of Mathematical Biophysics*, Vol. 5, pp. 115–33

McDonald, John (1950), *Strategy in Poker, Business and War*, New York: Norton

McGlothlin, W. H. (1958), "The Simulation Laboratory as a Developmental Tool", *RAND P-1454*, Aug. 7

McGuinness, B. F. (ed.) (1979), *Ludwig Wittgenstein and the Vienna Circle: Conversations Recorded by Friedrich Waismann*, Oxford: Blackwell

McKinsey, J. C. C. (1952a), "Some Notions and Problems of Game Theory", *Bulletin of the American Mathematical Society*, Vol. 58, No. 6, pp. 591–611

———— (1952b), *Introduction to the Theory of Games*, New York: McGraw-Hill

Mehrtens, Herbert (1987), "Ludwig Bieberbach and "Deutsche Mathematik", in Esther Phillips (ed.), *Studies in the History of Mathematics*, Washington, D.C.: The Mathematical Assocation of America, pp. 195–241

Ménard, C. (1978), *La Formation d'une Rationalité économique: A. A. Cournot*, Paris: Flammarion

Menger, Carl (1871), *Grundsätze der Volkswirtschaftslehre*. Translated by J. Dinwall and B. Hoselitz (1981), as *Principles of Economics*, New York: NYU Press

———— (1923, [1871]) *Grundsätze der Volkswirtschaftslehre*, 2nd ed., edited by Karl Menger, Wien: Hölder-Pichlen-Tempsky

Menger, Karl (1923), "Über die Dimensionalität von Punktmengen, I. Teil", *Monatshefte für Mathematik und Physik*, Vol. 33, pp. 148–60

———— (ed.) (1928–1936), *Ergebnisse eines Mathematischen Kolloquiums*, seven volumes, Leipzig and Vienna: Deuticke

———— (1930), "Der Intuitionismus", *Blätter für Deutsche Philosophie*, Vol. 4, pp. 311–25. Translated by R. Kowalski as "On Intuitionism" in Menger (1979), pp. 46–58

———— (1933), "Die neue Logik", in *Krise und Neuaufbau in den Exakten Wissenschaften. Fünf Wiener Worträge*, Leipzig and Vienna: Deuticke, pp. 94–122. Translated by H. B. Gottlieb and J. K. Senior as "The New Logic" in *Philosophy of Science* 4, pp. 299–336. Reprinted in Menger (1979), pp. 17–45, with prefatory notes on "Logical Tolerance in the Vienna Circle" pp. 11–16

———— (1934a), "Bernoullische Wertlehre und Petersburger Spiel", *Ergebnisse eines mathematischen Kolloquiums*, Vol. 6, pp. 26–27

———— (1934b), "Ein Satz über endliche Mengen mit Anwendungen auf die formale Ethik", *ibid*, pp. 23–26

———— (1934c), "Das Unsicherheitsmoment in der Wertlehre. Betrachtungen in Anschluss an das sogenannte Petersburger Spiel", *Zeitschrift für Nationalökonomie* Vol. 5, pp. 459–85. Translated in Menger (1979) as "The Role of Uncertainty in Economics", pp. 259–78

———— (1934d), *Moral, Wille und Weltgestaltung. Grundlegung zur Logik der Sitten*, Vienna: Julius Springer. Translated (1974) as *Morality, Decision and Social Organization: Towards a Logic of Ethics*, Dordrecht: Reidel

———— (1935), "Hans Hahn", *Fundamenta Mathematicae*, Vol. 24, pp. 317–20

———— (1936a), "Einige neuere Fortschritte in der exakten Behandlung sozialwissenschaftlicher Probleme", in *Neuere fortschritte in den exakten Wissenschaften. Fünf Wiener Vorträge, Dritter Zyklus*, Leipzig and Wien: F. Deuticke, pp. 103–32

———— (1936b), "Bemerkungen zu den Ertragsgesetzen", *Zeitschrift für Nationalökonomie*, Vol. 7, pp. 25–26, and "Weitere Bemerkungen zu den Ertragsgesetzen", *ibid*, pp. 388–97. Translated as "The Logic of the Laws of Return. A Study in Meta-Economics", in Morgenstern (ed.) (1954), pp. 419–81. Revision of translation as "Remarks on the Law of Diminishing Returns. A Study in Meta-Economics" in Menger (1979), pp. 279–302

———— (1937), "An Exact Theory of Social Relations and Groups," in *Report of Third Annual Research Conference on Economics and Statistics*, Cowles Commission for Research in Economics, Colorado Springs, Colorado, pp. 71–73

———— (1938), "An Exact Theory of Social Groups and Relations", *American Journal of Sociology* Vol. 43, pp. 790–98

———— (1952), "The Formative Years of Abraham Wald and his Work in Geometry", *Annals of Mathematical Statistics*, Vol. 23, pp. 14–20

———— (1955), *Calculus. A Modern Approach*, Boston: Ginn, pp. xviii and 354

———— (1956), "Why Johnny Hates Math", *The Mathematics Teacher*, Vol. 49, pp. 578–84

———— (1973), "Austrian Marginalism and Mathematical Economics", in Hicks and Weber (eds.), pp. 38–60

———— (1979), *Selected Papers in Logic and Foundations, Didactics, Economics*, Dordrecht: Reidel

———— (undated), "My Memories of the Early Days of I.I.T." (unpublished manuscript, 4 pp., Karl Menger Papers, Duke University Library)

———— (1982), "Memories of Moritz Schlick", in Eugene T. Gadol (ed.) (1982), *Rationality and Science. A Memorial Volume for Moritz Schlick in Celebration of the Centennial of His Birth*. Wien and New York: Springer-Verlag, pp. 83–103.

———— (1994), *Reminiscences of the Vienna Circle and the Mathematical Colloquium*, edited by Louise Golland, Brian McGuinness, and Abe Sklar, Dordrecht: Kluwer

———— (1998), *Ergebnisse eines Mathematischen Kolloquiums*, [1928–1927], edited by E. Dierker and K. Sigmund, Wien and New York: Springer

Mensch, A. (ed.) (1966), *Theory of Games, Techniques and Applications*, New York: American Elsevier

Miller, L., et al (1989), "Operations Research and Policy Analysis at RAND, 1968–1988", *RAND N-2937-RC*, April

Miller, Martin A. (1998), *Freud and the Bolsheviks: Psychoanalysis in Imperial Russia and the Soviet Union*, New Haven, Connecticut: Yale University Press

Milnor, J. W. (1955), "On Games of Survival", *RAND P-622*, Jan. 11

Mirowski, Philip (1989), *More Heat Than Light*, New York: Cambridge University Press

―――― (1991), "When Games Grow Deadly Serious", *History of Political Economy*, Vol, 23, Special Issue: Economics and National Security, A History of their Interaction, pp. 227–60

―――― (1992), "What Were von Neumann and Morgenstern Trying to Accomplish?" *History of Political Economy*, Vol. 24, Special Issue: "Toward a History of Game Theory", pp. 113–47

―――― (2002), *Machine Dreams: Economics Becomes a Cyborg Science*, New York and Cambridge: Cambridge University Press

von Mises, L. (1920), "Economic Calculation in the Socialist Commonwealth", in *Collectivist Economic Planning*, edited by F. A. Hayek. New York: Augustus M. Kelley

―――― (1922), *Socialism: An Economic and Social Analysis*, Indianapolis: Liberty Fund

―――― (1949), *Human Action: A Treatise on Economics*, Chicago: Henry Regnery

―――― (1960 [1933]), *Epistemological Problems of Economics*, Princeton: Van Nostrand

―――― (1971 [1912]), *The Theory of Money and Credit*. (Translated by H. E. Batson). Irvington, New York: Foundation for Economic Education.

―――― (1978), *Notes and Recollections*, Amsterdam: South Holland

―――― (1980), *Planning for Freedom*, South Holland, Illinois: Libertarian Press

von Mises, R. (1951) [1939], *Positivism*, Cambridge, Massachusetts: Harvard University Press

Mitchell, W. C. (1913), *Business Cycles and their Causes*, Berkeley: University of California Press

―――― (1927), *Business Cycles: The Problem and its Setting*, New York: National Bureau of Economic Research

Mongin, Philippe (2003), "L'axiomatization et les théories économiques", *Revue économique*, Vol. 54, pp. 99–138

Monk, R. (1990), *Ludwig Wittgenstein. The Duty of Genius*, New York: Penguin

Moore, H. L. (1914), *Economic Cycles – Their Law and Cause*, New York: Macmillan

―――― (1923), *Generating Economic Cycles*, New York: Macmillan

―――― (1925), "A Moving Equilibrium of Demand and Supply", *Quarterly Journal of Economics*, Vol. 39, pp. 357–71

Morgan, M (1990), *The History of Econometric Ideas*, Cambridge and New York: Cambridge University Press

Morgenstern, Oskar (1926), "Bemerkungen zu Cassels Preistheorie" (unpublished manuscript), OMDU, Box 21, Writings and Speeches, Alphabetical "Bemerkungen zu Cassels Preistheorie", 1925–1926

―――― (1927a), "Francis Y. Edgeworth", *Zeitschrift für Volkswirtschaft und Sozialpolitik*, Vol. 5, No. 10–12, pp. 646–52, translated in Schotter (ed.) 1976, pp. 477–80

―――― (1927b), "Friedrich von Wieser, 1851–1925", *American Economic Review*, Vol. 17, No. 4, December, pp. 669–74, translated in Schotter (ed.) 1976, pp. 481–85

―――― (1928), *Wirtschaftprognose: Eine Untersuchung ihrer Voraussetzungen und Möglichkeiten*, Vienna: Julius Springer, pp. iv and 129

―――― (1929), "Allyn Abbot Young", *Zeitschrift für Nationalökonomie*, Vol. 1, May 1929, pp. 143–45, translated in Schotter (ed.) 1976, pp. 487–88

———— (1931), "Mathematical Economics", in Edwin R.A. Seligman and Alvin Johnson (eds.), *Encyclopaedia of the Social Sciences*, Vol. 5, pp. 364–68, New York: Macmillan

———— (1934a), *Die Grenzen der Wirtschaftspolitik*, Vienna: Julius Springer, 136 pp. Translated by Vera Smith and revised as *The Limits of Economics*, London: Hodge, 1937, pp. v and 151

———— (1934b), "Das Zeitmoment in der Wertlehre", *Zeitschrift fur Nationalökonomie*, Vol. 5, No. 4, pp. 433–58. Translated as "The Time Moment in Economic Theory" in Schotter (ed.) (1976), pp. 151–67

———— (1935a), "Vollkommene Voraussicht und wirtschaftliches Gleichgewicht", *Zeitschrift für Nationalökonomie*, Vol. 6, No. 3, pp. 337–57. Translated by Frank Knight (mimeographed, University of Chicago). Reprinted in Schotter (ed.) (1976), pp. 169–83

———— (1935b), "Report on the Activities of the Austrian Institute for Trade Cycle Research 1931–1935", Feb. 13, AIRAC, Folder 37, Austrian Center for Trade Cycle Research, Vienna 1935–1936

———— (1936), "Logistics and the Social Sciences", *Zeitschrift fur Nationalökonomie*, Vol. 7, No. 1, pp. 1–24. Translated in Schotter (ed.) (1976), pp. 389–404

———— (1937), *The Limits of Economics*, London: W. Hodge. Translation of Morgenstern (1934) by Vera Smith

———— (1941a), "Professor Hicks on Value and Capital", *Journal of Political Economy*, Vol. 49, No. 3, pp. 361–93

———— (1941b), "Quantitative Implications of Maxims of Behavior" (unpublished manuscript, Princeton University)

———— (1951a), "Joseph A. Schumpeter, 1883–1950", *Economic Journal*, Vol. 61, No. 241, March, pp. 197–202, reprinted in Schotter (ed.) (1976), pp. 489–92.

———— (1951b), "Abraham Wald, 1902–1950", *Econometrica*, Vol. 19, No. 4, October, pp. 361–67

———— (1951c), "Notes on the Formulation of a Study of Logistics", *RAND LOGS 67*, May 28

———— (ed.) (1954), *Economic Activity Analysis*, New York: Wiley

———— (1954a), "Consistency Problems in the Military Supply System", *RAND RM-1296*, July 14

———— (1954b), "The Compressibility of Organizations and Economic Systems", *RAND RM-1325*, Aug. 17

———— (1958), "Obituary. John von Neumann, 1903–57", *The Economic Journal*, Vol. 68, March, pp. 170–174. Reprinted in Schotter (ed.), (1976), pp. 499–503

———— (1968), "Schlesinger, Karl", in *International Encyclopedia of the Social Sciences*, Vol. 14, pp. 51–52

———— (1976), "The Collaboration between Oskar Morgenstern and John von Neumann on the Theory of Games", *Journal of Economic Literature*, Vol. 14, pp. 805–16

———— , Oskar Morgenstern Papers, Rare Book, Manuscript and Special Collections Library, Duke University (OMDU)

Morris, Charles (1984), *A Time of Passion*, New York: Harper & Row

Morris, Edie and Leon Harkleroad (1990), "Rózsa Péter: Recursive Function Theory's Founding Mother", *The Mathematical Intelligencer*, Vol. 12, pp. 59–64

Morse, Marston and William L. Hart (1941), "Mathematics in the Defense Program", *American Mathematical Monthly*, Vol. 48, pp. 293–302

Morse, Philip M. (1948), "Mathematical Problems in Operations Research", *Bulletin of the American Mathematical Society*, Vol. 54, pp. 602–21

—— (1977), *In at the Beginnings*, Cambridge, Massachusetts: MIT Press

—— and George E. Kimball (1951), *Methods of Operations Research*, New York: Technology Press and Wiley (originally in classified form as 1946, same title, OEG Report 54)

Mosteller, Frederick et al (1961), *Probability with Statistical Applications*. Reading, Massachusetts: Addison-Wesley

—— (1968), "Wilks, S.S.", in *International Encyclopaedia of the Social Sciences*, Vol. 16, edited by David L. Sills. New York: Macmillan and Free Press, pp. 550–53

—— and Robert R.Bush (1955), *Stochastic Models for Learning*, New York: Wiley

Muller, Adam (1809), *Die Elemente der Staatskunst*, Berlin: J. D. Sander

Müller, Karl H. (1991), "Neurath's Theory of Pictorial-Statistical Representation" in Uebel (ed.) (1991), pp. 223–250

Nabokov, Vladimir (1964), *The Defense*, New York: G.P. Putnam's Sons (orig. *Zashchita Luzhina*, Berlin: Slovo, 1930)

Nagy, Dénes, Péter Horváth, and Ferenc Nagy (1989), "The von Neumann–Ortvay Connection", in J.R. Brink and C.R. Haden (eds.) 1989, *The Computer and the Brain: Perspectives on Human and Artificial Intelligence*, Amsterdam, New York, Oxford, Tokyo: North Holland, pp. 227–39

Nagy, Ferenc (1987), *Neumann János és a "Magyar Titok", A Dokumentumok Tükrében* (John von Neumann and the "Hungarian Secret"), Budapest: Országos Müszaki Információs Központ és Könyvtár

Nasar, Sylvia (1998), *A Beautiful Mind: A Biography of John Forbes Nash, Jr.*, New York: Simon & Schuster

Nash, J. F. Jr. (1950a), "The Bargaining Problem" *Econometrica*, Vol. 18, pp. 155–62

—— (1950b), "Equilibrium Points in N-Person Games", *Proceedings of the National Academy of Science*, Vol. 36, pp. 48–49

—— (1950c), *Non-cooperative Games*, Ph.D dissertation, Princeton University

—— (1950d) with L. Shapley, "A Simple Three-person Poker Game", in Kuhn and Tucker (eds.) *Contributions to the Theory of Games*, pp. 105–16

—— (1951a), "Non-cooperative games", *Annals of Mathematics*, Vol. 54, pp. 286–95

—— (1951b), "N-Person Games, an Example and a Proof", *RAND RM-615*, June 4

—— (1954), "Continuous Iteration Method for Solution of Differential Games", *RAND RM-1326*, Aug. 18

—— (1954b), "Parallel Control", *RAND RM-1361*, Aug. 27

—— (1991), Interview with R. Leonard, Dec. 11, Princeton

—— (1993), Letter to R. Leonard, Feb. 22

Neumann, Olaf (2000), "Divisibility Theories in the Early History of Commutative Algebra and the Foundations of Algebraic Geometry", in Gray and Hunger-Parshall (eds.), pp. 73–104

von Neumann, John (1925), "Eine Axiomatisierung der Mengenlehre", *Journal für die Reine und Angewandte Mathematick*, Vol. 154, pp. 219–40

—— (1927a), "Mathematische Begrundung der Quantenmechanik", *Nachrichten von der Gesellschaft der Wissenschaften Zu Göttingen*, pp. 1–57. See also A. H. Taub (ed.) (1963), Vol. 1

_____ (1927b), "Wahrscheinlichkeitstheoretischer Aufbau der Quantenmechanik", *Nachrichten von der Gesellschaft der Wissenschaften Zu Göttingen*, pp. 245–72. See also A. H. Taub (ed.) (1963), Vol. 1

_____ (1927c), "Thermodynamik Quantenmechanischer Gesamtheiten", *Nachrichten von der Gesellschaft der Wissenschaften Zu Göttingen*, pp. 273–91. See also A. H. Taub (ed.) (1963), Vol. 1

_____ (1927d), "Über die Grundlagen der Quantenmechanik", with D. Hilbert and L. Nordheim, *Mathematische Annalen*, Vol. 98, pp. 1–30

_____ (1927e), "Zur Hilbertschen Beweistheorie", *Mathematische Zeitschrift*, 26: 1–46

_____ (1928a), "Calcul des Probabilités – Sur la Théorie des Jeux" presented by E. Borel, *Comptes Rendus Hebdomadaires des Séances de l'Académie des Sciences*, Tome 186, No. 25, Lundi 18 juin, pp. 1689–91

_____ (1928b), "Zur Theorie der Gesellschaftsspiele", *Mathematische Annalen*, Vol. 100, pp. 295–320. Translated by S. Bargmann as "On the Theory of Games of Strategy" in Tucker, A. W. and R. D. Luce (eds.) (1959), pp. 13–42

_____ (1928c), "Die Axiomatisierung der Mengenlehre", *Mathematische Zeitschrift*, Vol. 27, pp. 669–752. In Taub (ed.) (1963), Vol I, pp. 339–422

_____ (1928d), "Die Zerlegung eines Intervalles in abzahlbar viele kongruente Teilmengen", *Fundamenta Mathematicae* II, pp. 230–38. In Taub (ed.), Vol. I, pp. 302–11

_____ (1928e), "Über die Definition durch transfinite Induktion, und verwandte Fragen der allgemeinen Mengenlehre", *Mathematische Annalen* 99, pp. 373–91. In Taub (ed.), Vol. I, pp. 320–38

_____ (1929a), "Allgemeine Eigenwerttheorie Hermitescher Funktionaloperatoren", *Mathematisch Annalen*, Vol. 102, pp. 49–131

_____ (1929b), "Zur algebra der Funktionaloperatoren und Theorie der normalen Operatoren", *Mathematisch Annalen*, Vol. 102, pp. 370–427

_____ (1929c), "Zur Theorie der unbeschrankten Matrizen", *Journal für die Reine und Angewandte Mathematick*, Vol. 161, pp. 208–36

_____ (1937), "Über ein Ökonomisches Gleichungssystem und eine Verallgemeinerung des Brouwerschen Fixpunktsatzes", *Ergebnisse eines Mathematisches Kolloquium*, Vol. 8, pp. 73–83. Translated as "A model of general economic equilibrium" in *Review of Economic Studies*, Vol. 13, 1945, pp. 1–9

_____ (1947a), "The Mathematician", in R. B. Heywood. (ed.), *The Works of the Mind*, pp. 180–96, reprinted in Taub (ed.) (1963), Vol. I, pp. 1–9

_____ (1947b), "Discussion of a maximum problem" (unpublished working paper, Princeton, November)

_____ (1948), "A numerical Method for Determination of the Value and the Best Strategies of a Zero-sum Two-person Game with Large Numbers of Strategies", (mimeographed, Princeton, May)

_____ (1950), "Solutions of Games by Differential Equations" (with G. W. Brown), in *Contributions to the Theory of Games*, Vol. I (ed. by H. Kuhn and A. W. Tucker), *Annals of Mathematics Studies*, No. 24, Princeton.

_____ (1953), "A Certain Two-person Game Equivalent to the Optimal Assignment Problem", *Contributions to the Theory of Games*, Vol. II (edited by H. W. Kuhn and A. W. Tucker), *Annals of Mathematics Studies*, No. 28, Princeton, pp. 5–12

———— (1954a), "The Role of Mathematics in the Sciences and in Society", Address at 4th Conference of the Association of Princeton Graduate Alumni, pp. 16–29, reprinted in Taub (ed.) (1961), Vol. VI, pp. 477–90

———— (1954b), "A numerical method to determine optimum strategy", *Naval Research Logistics Quarterly*, Vol. 1, pp. 109–15

———— (1955a) [1932], *Mathematical Foundations of Quantum Mechanics*, translated By Robert Beyer, Princeton: Princeton University Press

———— (1955b), "The Impact of Recent Developments in Science on the Economy and on Economics", *Looking Ahead*, Vol. 4, No. 11, reproduced in Taub (ed.) (1961–63), Vol. VI, pp. 100–01

———— (1955c), "Can we survive technology?", in *Fortune*, June, reproduced in Taub (ed.) (1961–63), Vol. VI, pp. 504–19

———— (1984) [1931], "The formalist foundations of mathematics", in Benacerraf, P. and H. Putnam (eds.), *Philosophy of Mathematics, Selected Readings*, Cambridge: Cambridge University Press.

————, John von Neumann Papers, Institute for Advanced Study (VNIAS), Princeton

————, John von Neumann Papers, Library of Congress (VNLC), Washington D.C

———— and G. W. Brown (1950), "Solutions of games by differential equations" (with G.W. Brown), in *Contributions to the Theory of Games, Vol. I*, edited by H. Kuhn and A. W. Tucker, *Annals of Mathematics Studies*, No. 24, Princeton: Princeton University Press

———— and Oskar Morgenstern (1947) [1944], *The Theory of Games and Economic Behavior*, Princeton: Princeton University Press

von Neumann-Eckart, Klari, Klari von Neumann-Eckhart Papers. In possession of Marina von Neumann Whitman (KEMNW), Ann Arbor, Michigan

Neurath, Marie (1973), "26 September 1924 and After" (biographical note on Otto Neurath) in Neurath (1973), pp. 56–64

Neurath, Otto (1920), "Experiences of Socialization in Bavaria" (from a lecture given in 1920 to the Sociological Society of Vienna). Reprinted in Neurath (1973), pp. 18–28

———— (1925), "Gesellschafts- und Wirtschaftsmuseum in Wien", Österreichische Gemiende-Zeitung, Vol. 2, Jahrgang, No. 16, Wien. Translated as "The Social and Economic Museum in Vienna", in Neurath (1973), p. 214

———— (1928), *Lebensgestaltung und Klassenkampf*, p. 152 Berlin: E. Laub. Translated as "Personal Life and Class Struggle" in Neurath (1973), pp. 249–98

———— (1929), '*Wissenschaftliche Weltauffassung: Der Wiener Kreis*' (unsigned) (Preface signed by Hans Hahn, Otto Neurath, Rudolf Carnap), p. 64. Wien: Artur Wolf. Translated as "The Scientific Conception of the World: The Vienna Circle", in Neurath (1973), pp. 299–318

———— (1930), "*Wege der wissenschaftlichen Weltauffassung*", Erkenntnis, Heft 2–4, pp. 106–25. Translated as "Ways of the Scientific World-Conception" in Neurath (1983), pp. 32–47

———— (1931a), "*Bildhafte Pädagogik im Gesellschafts- und Wirtschaftsmuseum in Wien*", Museumkunde, Neue Folge, III, Heft 3, pp. 125–29, Berlin: Walter de Gruyter. Translated as "Visual Education and the Social and Economic Museum in Vienna", in Neurath (1973), pp. 215–17

———— (1931b), "Empirische Soziologie. Der wissenschaftliche Gehalt der Geschichte und Nationalökonomie", *Schriften zur wissenschaftlichen Weltauffassung*, her. von

Philipp Frank und Moritz Schlick, Bd. 5, p. 151, Wien: Julius Springer. Translated as "Empirical Sociology: The Scientific Content of History and Political Economy", in Neurath (1973), pp. 319–421

—— (1933), "Museums of the Future", *Survey Graphic*, Vol. 22, No. 9, pp. 458–63, reprinted in Neurath (1973), pp. 218–23

—— (1935), "Was bedeutet rationale Wirtschaftsbetrachtung?" *Einheitswissenschaft*, Heft 4, p. 46, Wien: Gerold & Co. Translated as "What is Meant by a Rational Economic Theory?" in McGuinness (ed.) (1987), pp. 67–109

—— (1937), "A New Language" from "Visual Education", *Survey Graphic*, Vol. 24, No. 1, pp. 25–28, reprinted in Neurath (1973), pp. 224–26

—— (1938), "Die neue Enzyklopaedie". *Zur Enzyklopaedie der Einheitswissenschaft*, Heft 6, pp. 6–16, Den Haag: Van Stockum & Zoon. Translated as "The New Encyclopedia" in McGuinness (ed.) (1987), pp. 132–41

—— (1939), *Modern Man in the Making*, London: Secker and Warburg

—— (1939/40), "The Social Sciences and Unified Science", *The Journal of Unified Science (Erkenntnis)*, Vol. 9, pp. 244–48. The Haag, Chicago, reprinted in Neurath (1983), pp. 209–12

—— (1945), "Visual Education: Humanisation versus Popularisation" (excerpt from unfinished 1945 manuscript *Empiricism and Sociology*, posthumously edited by M. Neurath and R. Cohen in Neurath [1973]), pp. 227–48

—— (1973), *Empiricism and Sociology*, edited by Marie Neurath and Robert S. Cohen, Vienna Circle Collection, Vol. 1, Dordrecht: Reidel

—— (1983), *Philosophical Papers, 1913–1946*, edited by Robert S. Cohen and Marie Neurath, Vienna Circle Collection, Vol. 16, Dordrecht: Reidel

Newell, Allen (1951), "An Example in the Theory of Organization", *RAND P-291*, Feb. 14.

—— and J.B. Kruskal Jr. (1951), "Formulating Precise Concepts in Organization Theory", RAND *RM-619-PR*

Nimzovich, Aron (1974), *My System*, edited by Fred Reinfeld, New York: David McKay, originally published in German in 1925–27 by Verlag B. Kagan, Berlin

Novick, David (ed.) (1965), *Program Budgeting*, Cambridge, Massachusetts: Harvard University

—— (1988), "Beginning of Military Cost Analysis, 1950–1961", *RAND P-7425*, March

O'Brien, D. P. (1988), *Lionel Robbins*, New York: St. Martin's Press

Owens, Larry (1989), "Mathematicians at War: Warren Weaver and the Applied Economics Panel 1942–1945", in David E. Rowe and John McCleary (eds.), *The History of Modern Mathematics*, Vol. II, Institutions and Applications: Academic Press, Harcourt, Brace, Jovanovitch, pp. 286–305

Oxaal, Ivar, Michael Pollak, and Gerhard Botz (eds.) (1987), *Jews, Antisemitism and Culture in Vienna*, London and New York: Routledge & Kegan Paul

Palmer, Greg (1978), *The McNamara Strategy and the Vietnam War*, Westport & London: Greenwood

Patai, Ralph (1996), *The Jews of Hungary: History, Culture, Psychology*, Detroit, Michigan: Wayne State University Press

Pauley, Bruce F. (1992), *From Prejudice to Perdition: A History of Austrian Anti-Semitism*, Chapel Hill & London: University of North Carolina Press

Paxson, E. W. (1948), "Games of Tactics", *RAND RM-45*, June 28

————— (1949), "Recent Developments in the Theory of Games", *Econometrica*, Vol. 17, pp. 72–73

Persons, Warren M. (1919a) "Indices of Business Conditions", *Review of Economic Statistics*, Vol. 1, pp. 5–110

————— (1919b) "An Index of General Business Conditions", *Review of Economic Statistics*, Vol. 1, pp. 111–205

Pfister, Oskar (1931), "Ein Hamlet am Schachbrett. Ein Beitrag zue Psychologie des Schachspiels", *Psychoanalytische Bewegung*, Vol. 3, May-June, pp. 217–22

Piaget, J. (1971), *Structuralism* (translated by Chaninah Maschler), London: Routledge and Kegan Paul

Pinch, T. (1977), "What Does a Proof do if it Does Not Prove?", in E. Mendelsohn, P. Weingart, and R. Whitley (eds.), *The Social Production of Scientific Knowledge*, Dordrecht: Kluwer, pp. 171–215

Pollack, B. (ed.) (1995), *The Experimental Psychology of Alfred Binet: Selected Papers*, New York: Springer Publishing Co.

Popper, Karl (1992), *Unended Quest: An Intellectual Autobiography*, London: Routledge

————— (1995), "Hans Hahn – Reminiscences of a Grateful Student", Introduction to *The Collected Works of Hans Hahn* (edited by L. Schmetterer and K. Sigmund), Vol. I, Vienna: Springer Verlag

Poundstone, William (1992), *Prisoner's Dilemma*, New York: Anchor

Powers, J. (1982), *Philosophy and the New Physics*, London: Methuen

Punzo, Lionello F. (1989), "Von Neumann and K. Menger's Mathematical Colloquium", in Dore, et al (eds.)

————— (1991), "The School of Mathematical Formalism and the Viennese Circle of Mathematical Economists", *Journal of the History of Economic Thought*, Vol. 13, Spring, pp. 1–18.

————— (1993), "On Robert Remak's Superponiertes Preissysteme", paper presented at the History of Economics Society, Philadelphia

————— (1994), "Karl Menger's Contribution to the Social Sciences", in Menger (1994), pp. xxi–xxiv

Quade, E. S. (ed.) (1964), *Analysis for Military Decisions*, Chicago, Illinois: Rand McNally

Rabinbach, Anson, (1983), *The Crisis of Austrian Socialism. From Red Vienna to Civil War 1927–1934*, Chicago: University of Chicago Press

————— (ed.) (1985), *The Austrian Socialist Experiment*, Boulder, Colorado: Westview Press

Rachman, Arnold W. (1997), *Sandor Ferenczi: The Psychotherapist of Tenderness and Passion*, New York: Jason Aronson

Radó, Tibor (1932), "On Mathematical Life in Hungary", *American Mathematical Monthly*, Vol. 37, pp. 85–90

RAND Corporation (March 1989), *A Bibliography of Selected RAND Publications*

Redéi, Miklós (ed.) (2005), *John von Neumann: Selected Letters*, Providence, Rhode Island: American Mathematical Society

Rees, Minah (1980), "The Mathematical Sciences and World War II", *American Mathematical Monthly*, Vol. 87, pp. 607–21

Regis, E. (1989), *Who Got Einstein's Office? Eccentricity and Genius at the Institute for Advanced Study*, London: Penguin

Reid, Constance (1970), *Hilbert*, New York: Springer-Verlag

Reisch, George (2005), *How the Cold War Transformed the Philosophy of Science*, Cambridge: Cambridge University Press

Rellstab, Urs (1991), "New Insights in the Collaboration between John von Neumann and Oskar Morgenstern on the Theory of Games and Economic Behavior" (mimeographed, Duke University, Department of Economics)

——— (1992a), "From German Romanticism to Game Theory: I. Oskar Morgenstern's Vienna in the 1920's" (mimeographed, Duke University, Department of Economics)

——— (1992b), *Ökonomie und Spiele*, Die Entstehungsgeschichte der Spieltheorie aus dem Blickwinkel des Ökonomen Oskar Morgenstern, Zurich: Verlag Rüegger, pp. x and 219

Remak, R (1929), "Kann die Volkswirtschaftslehre eine exakte Wissenschaft werden?" *Jahrbucher für Nationalökonomie und Statistik*, Vol. 131, pp. 703–35

——— (1933), "Können superponierte Preissysteme praktisch berechnet werden?" *Jahrbucher für Nationalökonomie und Statistik*, Vol. 138, pp. 839–42

Reti, Richard (1923a), "Do 'New Ideas' Stand Up in Practice?", translated from the Russian and reprinted in R. Tekel and M. Shibut in *Virginia Chess Newsletter*, Sept./Oct. issue, 1993, at www.vachess.org/content/Reti.pdf.

——— (1923b), *Modern Ideas in Chess*, translated by John Hart, London: G. Bell and Sons.

——— (1933), *Masters of the Chessboard*, translated by J. A. Schwendemann, London: Bell (orig. published 1932)

Rhodes, Richard (1986), *The Making of the Atomic Bomb*, New York: Touchstone

Rickert, H. (1902), *The Limits of Concept Formation in Natural Science*, New York: Cambridge University Press

Riesz, Frigyes, "Obituary", *Acta Scientiarum Mathematicarum Szeged*, 1956, Vol. 7, pp. 1–3

Rives, Norfleet W. Jr. (1975), "On the history of the mathematical theory of games", *History of Political Economy*, Vol. 7, No. 4, pp. 549–65

Robbins, L. (1935 [1932]), *An Essay on the Nature and Significance of Economic Science*, London: Macmillan

——— (1938), "The Methods of Economic Observation and the Problems of Prediction in Economics", in O'Brien (1988), pp. 170–78. O'Brien's translation of "Les méthodes d'observation économique et les problèmes de la prévision en matière économique", in *Cinq Conférences sur la Méthode dans les recherches économiques*, Paris: Recueil Sirey, with a preface by Charles Rist

Roos, Charles (1934), *Dynamic Economics*, Bloomington: Principia Press

Rosier, Michel (1987), "Otto Neurath, économiste et leader du Cercle de Vienne", *Oeconomica*, Vol. 7, pp. 113–45

Rosser, J. Barkley (1989), "Mathematics and Mathematicians in World War II", in Peter Duren, et al (eds.), *A Century of Mathematics in America*, Part III, Vol. 3, Providence: American Mathematical Society, pp. 303–09

Rostow, W. W. (1981), *Pre-Invasion Bombing Strategy*, Austin: University of Texas

Rota, Gian-Carlo (1997), *Indiscrete Thoughts*, Boston: Birkhäuser

Rothkirchen, Livia (1978), "Deep-Rooted Yet Alien: Some Aspects of the History of the Jews in Subcarpathian Ruthenia", *Yad Vashem Studies*, Vol. 12, pp. 147–91.

Rothschild, K. W. (1973), "Distributive Aspects of the Austrian Theory", in Hicks and Weber (eds.), pp. 207–25

Russell, Bertrand (1919), *Introduction to Mathematical Philosophy*, London: George Allen & Unwin

―――― and Alfred N. Whitehead (1910), *Principia Mathematica*, Cambridge: Cambridge University Press

Samuelson, Paul A. (1949), "Market Mechanisms and Maximization", *RAND P-69*, Mar. 28

―――― (1962), "Economists and the History of Ideas", *American Economic Review*, Vol. LII, No. 1, pp. 1–18

―――― (1991), Personal communication with R. Leonard, Nov. 19

Sapolsky, Harvey M. (1990), *Science and the Navy: The History of the Office of Naval Research*, Princeton: Princeton University Press

Saussure, Ferdinand de (1983) [1916], *Course in General Linguistics*, edited by Charles Bally and Albert Sechehaye. Translated by Roy Harris. La Salle, Illinois: Open Court

Scarf, Herbert E. (1955), "On Differential Games with Survival Payoffs", *RAND P-742*, September

Scarf, Herbert E. and L. S. Shapley (1956), "Games with Partial Information", *RAND P-797*, April

Schlesinger, Karl (1935), "Über die Produktionsgleichungen der ökonomischen Wertlehre", in Menger (ed.) (1935), *Ergebnisse eines mathematischen Kolloquiums*, 1933–34, Heft 6, pp. 10–11

Schlick, Moritz (1930), *Fragen der Ethik*, Vienna. Translated as *Problems of Ethics*, New York: Prentice Hall, 1939

Schorske, Carl (1981 [1961]), *Fin-de-siècle Vienna*, New York: Random House

Schotter, Andrew (ed.) (1976), *Selected Economic Writings of Oskar Morgenstern*, New York: NYU Press

Schumpeter, Joseph (1954), *History of Economic Analysis*, New York: Oxford University Press

Schwalbe, U. and P. Walker (2001), "Zermelo and the Early History of Game Theory", *Games and Economic Behavior*, Vol. 34, pp. 123–37

Science Letter News (Apr. 3, 1937), "Princeton Scientist Analyzes Gambling; You Can't Win" p. 216

Segal, Sanford L. (2003), *Mathematicians under the Nazis*, Princeton University Press

Selten, R. (1965), "Spieltheoretische Behandlung eines Oligopolmodells mit Nachfrageträgheit", *Zeitschrift für die gesamte Staatswissenschaft*, Vol. 121, pp. 301–24; 667–89

―――― (1975), "Reexamination of the Perfectness Concept for Equilibrium Points in Extensive Games", *International Journal of Game Theory*, Vol. 4, 1, pp. 25–55

Shapley, Lloyd (1949), "A Hidden-Target Model", *RAND RM-101*, Feb. 14

―――― (1949a), "The Silent Duel, One Bullet Versus Two, Equal Accuracy", *RAND RM-445-PR*

―――― (1949b), "Note on Duels with Continuous Firing", *RAND RM-118*, Mar. 11

―――― (1955), "Markets as Cooperative Games", *RAND P-629*, Mar. 7

―――― (1992), Interview with R. Leonard, April 21, University of California at Los Angeles

Shapley, L. and R. Snow (1950), "Basic solutions of discrete games" in Kuhn and Tucker (eds.) pp. 27–35

Shubik, M. (1952), "Information, Theories of Competition, and the Theory of Games", *Journal of Political Economy*, Vol. 60, pp. 145–50

———— (1959), *Strategy and Market Structure*, New York: Wiley

———— (1979), "Morgenstern, Oskar", *International Encyclopaedia of the Social Sciences*, Vol. 18, edited by David L. Sills, 1968, New York: Macmillan Co. and Free Press. (Vol. 18 a 1979 supplement), pp. 541–44

———— (1989), "Oskar Morgenstern", in Eatwell et al, (eds.), pp. 164–66

———— (ed.) (1964), *Game Theory and Related Approaches to Social Behavior*, New York: Wiley

———— (ed.) (1976), *Essays in Mathematical Economics in Honor of Oskar Morgenstern*, Princeton: Princeton University Press

———— (1982), *Game Theory in the Social Sciences*, Cambridge, Massachusetts: MIT Press

———— (1989), "Cournot, Antoine Augustin" in Eatwell, et al (eds.), pp. 117–28

———— (1991), Interview with R. Leonard, Dec. 6, New Haven, Connecticut

Sieg, Ulrich and Michael Dreyer (eds.) (2001), *Emanuel Lasker: Schach, Philosophie und Wissenschaft*, Berlin: Philo

Siegfried, Tom (2006), *A Beautiful Math: John Nash, Game Theory, and the Modern Quest for a Code of Nature*, Washington, D.C.: Joseph Henry Press

Siegmund-Schultze, Reinhard (1994), "Communications between German and American Mathematicians between World Wars", *Research Reports from the Rockefeller Archive Center*, Spring, pp. 14–16

Sigmund, Karl (1995a), "Hans Hahn and the Foundational Debate", in W. DePauli-Schimanovich, et al (eds.) (1995), *The Foundational Debate*, Amsterdam: Kluwer, pp. 235–45

———— (1995b), "A Philosopher's Mathematician: Hans Hahn and the Vienna Circle", *The Mathematical Intelligencer*, Vol. 17, No. 4, pp. 16–29

Silk, Leonard (1977), "The Game Theorist", *New York Times*, Sunday, Feb. 13.

Silverman, Paul (1984), "Law and Economics in Interwar Vienna" (unpublished PhD dissertation, University of Chicago)

Simon, H. (1945), Review of *Theory of Games and Economic Behavior*, *American Journal of Sociology*, Vol. 50, No. 6, pp. 558–60

———— (1991a), Interview with R. Leonard, Dec. 18, Carnegie-Mellon University

———— (1991b), Communication with R. Leonard, Dec. 20

Smith, Barry (1986), "Austrian Economics and Austrian Philosophy", in Wolfgang Grassl and Barry Smith (eds.) (1986), *Austrian Economics: Historical and Philosophical Background*, London and Sydney: Croom Helm

———— (1990), "On the Austrianness of Austrian Economics", *Critical Review*, Winter-Spring, pp. 212–38

———— (1995), "L'Autriche et la Naissance de la Philosophie Scientifique", *Actes de la Recherce en sciences sociales*, Vol. 109, Septembre, pp. 61–71

Smith, Bruce (1964), "Strategic Expertise and National Security Policy: a Case Study", *Public Policy*, Vol XIII, pp. 69–106

———— (1966), *The RAND Corporation*, Cambridge, Massachusetts: Harvard University Press

Smith, John Maynard (1988), *Games, Sex and Evolution*, New York and Toronto: Harvester-Wheatsheaf

Spann, Othmar (1924), *Kategorienlehre in Ergänzungsbände zur Sammlung Herdflamme*, Vol. I, Jena: Gustav Fischer

_____ (1931), *Der Wahre Staat, Vorlesungen über Abbruch und Neubau der Gesellschaft*, Jena: Fischer

_____ (1932), *Geschichtsphilosophie*, Jena: Fischer

_____ (1911), *Die Haupttheorien der Volkswirtschaftslehre*, Leipzig: Quelle & Meyer

Specht, Robert D. (1957), "War Games", *RAND P-1041*, Mar. 18.

_____ (1960), "RAND – A personal view of its history", *Operations Research*, Vol. 8, pp. 825–39

von Stackelberg, H. (1952), *The Theory of the Market Economy* (translated from German; original 1934), London: W. Hodge

Stadler, Friedrich (1991), "Otto Neurath: Encyclopedist, Adult Educationalist and School Reformer", in Uebel (ed.) (1991), pp. 255–64

_____ (1991), "Aspects of the Social Background and Position of the Vienna Circle at the University of Vienna", in Uebel (ed.), pp. 51–77.

Stigler, George J. (1965), "The Economist and the State", *American Economic Review*, Vol. LV, No. 1, pp. 1–18

Stockfisch, Jack (1987), "The Intellectual Foundations of Systems Analysis", *RAND P-7401*, December

Stocking, George W. (1959), "Institutional Factors in Economic Thinking", *American Economic Review*, Vol. XLIX, No. 1, pp. 1–21

Stone, Marshall (1983), Review: Steve J. Heims, *John von Neumann and Norbert Wiener, from mathematics to the technologies of life and death*, Bulletin of the American Mathematical Society (N.S.), Volume 8, Number 2, pp. 395–399

Stone, R. (1948), "The Theory of Games", *Economic Journal*, Vol. 58, pp. 185–201

Sturrock, J. (1993), *Structuralism*, London: Fontana

Swedberg, Richard (1991), *Schumpeter: A Biography*, Princeton: Princeton University Press

Tarrasch, Siegbert (1912), *Die moderne Schachpartie*, Nürnberg: Tarraschs Selbstrverlag

_____ (1940), *The Game of Chess: A Systematic Text-book for Beginners and More Experienced Players*, Philadelphia: D. McKay

Taub, Alfred H. (ed.) (1963), *John von Neumann, Collected Works*, Vols. I-VI, New York: Macmillan

Thrall, R. M., C. H. Coombs, and R. L. Davis (eds.) (1954), *Decision Processes*, New York: Wiley

Thucydides (1966), *The Peloponnesian War* (Books I, II [Chapters 1–7]) and V [Chapter 7]), Chicago: Great Books Foundation; based on the translation by Rex Warner (1954), Harmondsworth: Penguin Books

Tidman, Keith R. (1984), *The Operations Evaluation Group*, Annapolis, Maryland: Naval Institute Press

Tintner, Gerhard (1952), "Abraham Wald's Contributions to Econometrics", Annals of Mathematical Studies, Volume 23, Number 1, pp. 21–28

Tirole, Jean (1988), *The Theory of Industrial Organisation*, Cambridge, Massachusetts: MIT Press, pp. 479

Tucker, A. (1991), Interview with R. Leonard, Dec. 11, Princeton

Tucker, A. W. and R. D. Luce (eds.) (1959), *Contributions to the Theory of Games*, Vol. IV, Princeton: Princeton University Press

Turán, Paul (1949a), "Megemlékezés», *Mathematikai Lapok*", Vol. 1, pp. 3–16, translated as "Commemoration" in *Collected Papers*, Vol. 1, pp. 459–70.

_____ (1949b) "Fejér Lipót mathematikai munkásseaga", *Mathematikai Lapok*, Vol. I, pp. 160–70, translated as "Leopold Fejér's Mathematical Work", in *Collected Papers*, Vol. I, pp. 474–81

_____ (1960), "Fejér Lipót, 1880–1959", *Mathematikai Lapok*, Vol. 12, pp. 8–18, translated as "Leopold Fejér (1880–1959). His Life and Work", in *Collected Papers* Vol. 2, pp. 1204–12

_____ (1974), "Megemlékezés a fasizmus mathematikus áldozatairól", *Mat. Lapok*, Vol. 25, pp. 259–63, translated as "Commemoration of Mathematicians Who Were Victims of Fascism", in *Collected Papers*, Vol. 3, pp. 2622–26

_____ (1977), "A Note of Welcome", *Journal of Graph Theory*, Vol. 1, No. 1, pp. 7–9

Uebel, Thomas E. (ed.) (1991), *Rediscovering the Forgotten Vienna Circle*, Boston Studies in the Philosophy of Science, Vol. 133, Dordrecht and Boston: Kluwer

Ulam, Stan (1958), "John von Neumann", *Bulletin of the American Mathematical Society*, Vol. 64, No. 3, pp. 1–49

_____ (1976), *Adventures of a Mathematician*, New York: Scribners

_____ (1980), Unpublished Draft Biography of John von Neumann, located in Stanislaus Ulam Papers, American Philosophical Society Library, in Series IX, Manuscripts of Published Works, 1944–1984

Van Dalen, Dirk (1999), *Mystic, Geometer, and Intuitionist: The Life of L.E.J. Brouwer. Vol. 1: The Dawning Revolution*, Oxford: Oxford University Press

_____ (2005), *Mystic, Geometer, and Intuitionist: The Life of L.E.J. Brouwer. Vol. 2: Hope and Disillusion*, Oxford: Oxford University Press

Van Hove, Léon (1958), "Von Neumann's Contribution to Quantum Theory", *Bulletin of the American Mathematical Society*, Vol. 64, No. 3, Part 2, May, pp. 95–99, Special Issue on John von Neumann, 1903–1957

Van Stigt, W. P. (1981), "L.E.J. Brouwer, The Signific Interlude", in A. S. Troelstra and D. Van Dalen (eds.), *The L.E.J. Brouwer Centenary Symposium*, Amsterdam: North Holland

Vaughn, Karen (1994), *Austrian Economics in America: The Migration of a Tradition*. New York and Cambridge: Cambridge University Press

Veblen, Oswald, Oswald Veblen Papers. Library of Congress (VLC), Washington D.C.

Ville, Jean (1938), "Sur la Théorie Générale des Jeux de Hasard", in Borel (1938), *Traité du Calcul des Probabilités et de ses Applications*, Vol. IV, 2, Paris: Gauthier-Villars, pp. 105–13

Viteles, M.S. (1945), "The aircraft pilot: five years of research; a summary of outcomes", *Psychological Bulletin*, Vol. 42, pp. 489–526

Vonneuman, Nicholas A. (1987), *John von Neumann as Seen by His Brother*. Meadowbrook, Pennsylvania: N.A. Vonneumann.

Wald, Abraham (1931a) "Axiomatik des Zwischenbegriffes in metrischen Räumen", *Wiener Akademischer Anzeiger*, No. 16, pp. 1–3

_____ (1931b), "Axiomatik des Zwischenbegriffes in metrischen Räumen", *Mathematische Annalen*, Vol. 104, pp. 476–84

———— (1931c), "Axiomatik des metrischen Zwischenbegriffes", *Ergebnisse eines Mathematische Kolloquiums*, Vol. 2, p. 17

———— (1933), "Zur Axiomatik des Zwischenbegriffes", *Ergebnisse eines Mathematische Kolloquiums*, Vol. 4, pp. 23–24

———— (1934), "Über die eindeutige positive Lösbarkeit der neuen Produktionsgleichungen", in Karl Menger (ed.) (1935), *Ergebnisse eines Mathematischen Kolloquiums*, Vol. 6, 1933–34, Leipzig and Vienna: Franz Deuticke, pp. 12–18. Translated as "On the Unique Non-negative Solvability of the New Production Equations, Part I", in Baumol and Goldfeld (1968), pp. 281–88

———— (1935), "Über die Produktionsgleichungen der ökonomischen Wertlehre", in Karl Menger (ed.) (1936), *Ergebnisse eines Mathematischen Kolloquiums*, Vol. 7, 1934–35, Leipzig and Vienna: Franz Deuticke, pp. 1–6. Translated as "On the Production Equations of Economic Value Theory", in Baumol and Goldfeld (1968), pp. 289–93

———— (1936), *Berechnung und Ausschaltung von Saisonschwankungen.*, Vienna: Julius Springer

———— (1937), "Zur Theorie der Preisindexziffern", *Zeitschrift für Nationalökonomie*, Vol. 8, pp. 179–219

———— (1939), "A New Formula for the Index of Cost of Living", *Econometrica*, Vol. 7, pp. 280–306

———— (1940), "The Approximate Determination of Indifference Surfaces by Means of Engel Curves", *Econometrica*, Vol. 8, pp. 96–100

———— (1945a), "Statistical Functions which Minimize the Maximum Risk", *Annals of Mathematics*, Vol. 46, pp. 265–80

———— (1945b), "Generalization of a Theorem by v. Neumann Concerning Zero Sum Two Person Games", *Annals of Mathematics*, Vol. 46, No. 2, April, pp. 281–86

Wallis, W. Allen (1980), "The Statistical Research Group", *Journal of the American Statistical Association*, Vol. 75, No. 370, pp. 320–30 and "Rejoinder", pp. 334–35

Walras, L. (1863), "Compte rendu des 'Principes de la théorie des richesses'" in Lutfalla (ed.) (1938), pp. 224–32

———— (1900) [1874], Eléments d'économie politique pure, ou théorie de la richesse sociale. 4th ed. Lausanne: Rouge and Paris: Pichon. Translated by William Jaffé as *Elements of Pure Economics* London: Allen & Unwin, 1954.

War Ministry (1963), *Operational Research in the R.A.F.*, London: Her Majesty's Stationery Office

Weaver, Warren (1970), *Scene of Change: A Lifetime in American Science*, New York: Scribner's

Weber, M. (1921–22), *Economy and Society: an Outline of Intepretive Sociology*. 3-volume edition, New York: Bedminster Press (1967); 2-volume and 3-volume editions, Berkeley: University of California Press (1978)

———— (1949), *The Methodology of the Social Sciences*, edited by Edward Shils and Henry Finch, New York: Free Press

Weber, Wilhelm (1961), "Hans Mayer", *Handwörterbuch der Sozialwissenschaften*, Bd. 7, Stuttgart: Gustav Fischer, pp. 364–5

Weiner, Milton G. (1954), "Observations on the Growth of Information-Processing Centers", *RAND P-529*, May 21

Weininger, Otto (1903), *Geschlecht und Karakter: Eine prinzipielle Untersuchung*, Vienna and Leipzig: Wilhelm Braumüller. Translated as *Sex and Character*, from 6th German edition, London: Wm. Heinemann, and New York: G.P. Putnam's (1906).

Weintraub, E. Roy (1985), *General Equilibrium Analysis, Studies in Appraisal*, Cambridge: Cambridge University Press

―――― (1991), *Stabilizing Dynamics*, Cambridge: Cambridge University Press, 177pp.

―――― (1998), "From Rigor to Axiomatics: the Marginalization of Griffith C. Evans", in *History of Political Economy*, Special Issue, "From Interwar Pluralism to Postwar Neoclassicism", edited by Mary S. Morgan and Malcolm Rutherford, pp. 227–59

―――― (2002), *How Economics Became a Mathematical Science*, Duke University Press

Weintraub, E. Roy (ed.) (1992), "Towards a History of Game Theory", *History of Political Economy*, Special Issue, Vol. 24

Weisner, Louis (1945), "Review of *Theory of Games and Economic Behavior*", *Science and Society*, Vol. 9, pp. 366–69

Weisz, Jószef (1905), "Játékkülömbözetek Meghatározásáról", *KöMaL*, April, pp. 185–86

Werfel, Franz (1934, orig.1933), *The Forty Days of Musa-Dagh*. Translated from German by Geoffrey Dunlop. New York: Viking Press

Weyl, Hermann (1927), *Philosophie der Mathematick und Naturwissenschaften*, Munich: Leipzig Verlag. Translated as *Philosophy of Mathematics and Natural Science*, Princeton: Princeton University Press, 1949

―――― (1944), "David Hilbert, 1862–1943", Royal Society of London, *Obituary Notices of Fellows*, Vol. 4, No. 13, November, pp. 547 ff.

―――― (1949), "Ars Combinatoria", in *Philosophy of Mathematics and Natural Science*, Princeton: Princeton University Press, pp. 237–52

―――― (1950), "Elementary Proof of a minimax theorem Due to von Neumann", in Kuhn and Tucker (eds.) 1950, pp. 19–25

Whitehead, Alfred N. (1967 [1925]), *Science and the Modern World*, New York: Free Press

Who's Who in America: A Biographical Dictionary of Notable Living Men and Women, Vol. 34. 1966–67

von Wieser, F. (1914), *Theorie der gesellschaftlichen Wirtschaft*, translated as *Social Economics*, by A. Ford Hinrichs, 1927, with preface by Wesley Clair Mitchell, New York: Greenberg

von Wieser, Friedrich (1926), *Das Gesetz der Macht*, Vienna: J. Springer

Wiggershaus, Rolf (1995), *The Frankfurt School*, translated by Michael Robertson, Cambridge, Massachusetts: MIT Press.

Williams, John D. (1933), "The Use of Reticles for the Observation of Meteors", *Publications of the Astronomical Society of the Pacific*, Vol. XLV No. 266, August, pp. 175–79

―――― (1939), "Binocular Observation of 718 Meteors", *Proceedings of the American Philosophical Society*, Vol. 81, No. 4, pp. 505–20

―――― (1946), "Effect on Military Worth of Exchanging Bombing Accuracy for Bomber Safety by Increasing Range of Bomb", *RAND RA-15008*, Sept. 1

―――― (1949), "A Selection of Information on Coverage", *RAND RM-91*, Jan. 10

―――― (1952), "Conflicts with Imprecise Payoffs", *RAND P-354*, Dec. 15

―――― (1953), "Regarding Optimum Amount of Operational Training in GCI Centers", *RAND RM-1033*, Jan. 26

―――― (1954), *The Compleat Strategyst*, New York: McGraw-Hill

—— (1958), "Some Attributes of the Changing Society", *RAND RM-2285-RC*, Nov. 10. (later published as Part 1, The Place of Mathematics in a Changing Society, Report of Subcommittee 1, Secondary School Curriculum Committee, National Council of Teachers of Mathematics, October)

Wilson, F. (1981), *A Picture History of Chess*, New York: Dover

Wistrich, Robert S. (ed.) (1992), *Austrians and Jews in the Twentieth Century: From Franz Joseph to Waldheim*, New York: St. Martin's Press

Witte, Edwin (1957), "The Economist and Public Policy", *American Economic Review*, Vol. XLVII, No. 1, pp. 1–20

Wittgenstein, Ludwig (1921), *Tractatus Logico-Philosophicus*, translated by Pears, D. F. and B. F. McGuinness 1993, London: Routledge, pp. xxii and 89

Wittman, von W. (1967), "Die extremale Wirtschaft. Robert Remak – ein Vorläufer der Aktivitätsanalyse", *Jahrbucher für Nationalökonomie und Statistik*, Vol. 180, pp. 397–409

Wohlstetter, A. (1959), "The Delicate Balance of Terror", *Foreign Affairs*, January

—— (1964a), "Analysis and Design of Conflict Systems", in Hitch and McKean (1964), pp. 103–48

—— (1964b), "Sin and Games in America", in Shubik (ed.) (1964)

—— and H. Rowen (1951), "Economic and Strategic Considerations in Air Base Location: a Preliminary Review", *RAND D-1114*

Wolf, Theta Holmes (1973), *Alfred Binet*, Chicago and London: University of Chicago Press

Wolfe, P (ed.) (1955), "Report of an Informal Conference on Recent Developments in the Theory of Games, January 31 – February 1, 1955", Logistics Research Project, Dept. Mathematics, Princeton University

Zermelo, E. (1913), "Über eine Anwendung der Mengenlehre auf die Theorie des Schachspiels", *Proceedings of the 5th International Congress of Mathematicians*, Cambridge, Aug. 22–28, 1912, Vol. II, pp. 501–04, translated as "On an Application of Set Theory to the Theory of the Game of Chess" by Ulrich Schwalbe and Paul Walker in Schwalbe and Walker (2001), pp. 133–36

—— (1928), "Die Berechnung der Turnier – Ergebnisse als ein Maximumproblem der Wahrscheinlichkeitsrechnung", *Mathematische Zeitschrift*, Vol. 29, pp. 46–460

Zolo, Danilo (1990), "Reflexive Epistemology and Social Complexity: The Philosophical Legacy of Otto Neurath", *Philosophy of the Social Sciences*, Vol. 20, pp. 149–69

Zweig, Stefan (1981), *The Royal Game and Other Stories*, New York: Harmony Books (orig. *Schachnovelle*, written in late 1941, early 1942; translated as *The Royal Game*, New York: Viking Press, 1944)

Index

CPSIA information can be obtained at www.ICGtesting.com
Printed in the USA
LVOW11s0346021113

359552LV00003B/20/P